STUDY GUIDE
to Accompany Ouellette's

INTRODUCTION TO

General, Organic, AND Biological CHEMISTRY

STUDY GUIDE
to Accompany Ouellette's

INTRODUCTION TO General, Organic, AND Biological CHEMISTRY

Wendy Walton Sonntag
Claire Rothenbuhler

The Ohio State University

Macmillan Publishing Company
NEW YORK

Collier Macmillan Publishers
LONDON

Printed in the United States of America

Earlier edition entitled Learning Chemistry by Problem Solving: Supplementary Questions copyright © 1981 by Macmillan Publishing Company.

Macmillan Publishing Company
866 Third Avenue, New York, New York 10022

Collier Macmillan Canada, Inc.

ISBN 0-02-413780-4

Printing: 1 2 3 4 5 6 7 8 Year: 4 5 6 7 8 9 0 1

ISBN 0-02-413780-4

Preface

This Study Guide has been developed from sets of problems written for our students over a period of years. Our present students are primarily nonscience majors who have little, if any, background in science.

Ideally, students in a basic chemistry course should read their text, study their lecture notes, and memorize basic definitions and equations pertaining to each concept presented in class. They will then be ready to work problems in the appropriate section of this workbook. As they proceed through a section, they will find increasingly challenging problems. To answer these will require applying or extrapolating the basic knowledge acquired in earlier problems. The "Solution" provides immediate feedback on each question and thoroughly explains each concept. Some specific suggestions are made concerning student study methods at the beginning of Part Two.

Although written to accompany Introduction to General, Organic, and Biological Chemistry by Dr. Robert J. Ouellette, this book would be beneficial to students in many general, organic, and biochemistry courses.

We gratefully thank

- Our patient, supportive husbands who encouraged us in the writing of this book.
- Our colleagues and teaching associates for their assistance and encouragement.
- Our students whose enthusiastic response to our earlier "Supplementary Questions" stimulated us to develop this book.

W. W. S.
C. R.

Contents

Introduction

To the Student

Learning chemistry effectively requires a systematic approach to problem solving. The format of this book is designed to help you develop such an approach.

This book presumes that you have utilized your text and lecture notes to acquire a basic understanding of each topic. You should also memorize the definitions that your lecturer presents in class. You will then be ready to proceed through the set of problems in this workbook dealing with the topic. For each problem, keep the "Solution" covered until you have worked, really worked, on "Your Answer." To maximize the use of your book, we suggest that you answer the question only in the space for "Your Answer". Leaving the questions and chemical structures entirely unmarked will permit you to return to a clean question to test your skills again. When you have finished working on "Your Answer," compare what you have written to the "Solution." Read the "Solution" thoroughly, because it will give you a detailed explanation of the problem in addition to the answer.

The Authors

Part One

General and Inorganic Chemistry

1 Significant Figures, Scientific Notation, and Unit Conversions

I. Use of the Proper Number of Significant Figures

Express each of the following numbers in scientific notation and indicate the number of significant figures in each answer.

1. 0.000007062

Your answer:

Solution: 7.062×10^{-6}; four significant figures
Move the decimal point of 0.000007062 to the right until you have one digit to the left of the decimal point (7.062). Express the move of six positions to the right as 10^{-6}. The zero between the 7 and the 6 is significant.
All zeroes between nonzero digits are significant.

---

2. 900.0

Your answer:

Solution: 9.000×10^{2}; four significant figures
There are four significant figures in this number. A final zero to the right of the decimal point is significant.

---

3. 0.08020

Your answer:

Solution: 8.020×10^{-2}; four significant figures
There are four significant figures in 0.08020. Zeroes to the left of nonzero digits are not significant; they are merely "place holders."

---

4. 21.070

Your answer:

Solution: 2.1070×10^{1}; five significant figures

---

5. 3500

Your answer:

Solution: 3500 is ambiguous; it has at least two significant figures, but it might have three or four.
3.5×10^3 indicates 3500 with two significant figures; 3.50×10^3 indicates 3500 with three significant figures; and 3.500×10^3 or 3500. indicates 3500 with four significant figures.

Solve each of the following and express each answer with the correct number of significant figures.

1. 4.72 + 201.4 + 75 =

Your answer:

Solution: 281

4.72 201.4 75. 281.12	4.72 is uncertain in the hundredths place; 201.4 is uncertain in the tenths place; and 75 is uncertain in the units place. The sum of these numbers would be uncertain in the units, tenths, and hundredths place. However, since any number can have only one uncertain digit, you should round the answer to the units place and record it as 281.

---

2. 7.1462 + 3.75 + 5.210 =

Your answer:

Solution: 16.11

7.1462 3.75 5.210 16.106	3.75 is the least precise measurement and is uncertain in the hundredths place. Therefore you should record the answer to the hundredths place.

---

3. 32.7945 + 121.5 - 326.37 =

Your answer:

Solution: -172.1

32.7945 +121.5 154.2945 -326.37 -172.0755	121.5 is the least precise measurement, so the final answer is expressed to the nearest tenths place.

---

4. $(2.00 \times 10^2)^3 =$

Your answer:

Solution: 8.00×10^6

$2.00 \times 2.00 \times 2.00 \times 10^2 \times 10^2 \times 10^2 = 8.00 \times 10^6$

When an exponential expression (10^2) is raised to a power (3), the exponent is multiplied by the power. The answer is recorded to three significant figures because the original number (2.00) is limited to three significant figures.

---

5. $\frac{55.2 \times 0.90}{0.4007} =$

Your answer:

Solution: 1.2×10^2

The answer read from the calculator is 123.983. However, whenever the operations of multiplication and division are performed, you can have only as many significant figures in the answer as in the least precise factor. The two significant figures of 0.90 limit the answer to two significant figures, so the answer can only be expressed using scientific notation.

---

6. $\frac{2.75 \times -0.0126}{3.14} =$

Your answer:

Solution: -0.0110

Each number in the problem has three significant figures, so the answer read from the calculator (-0.01103503) must be rounded to three significant figures.

---

7. $\frac{105.2 \times 3.01}{3.1 \times 10^3} =$

Your answer:

Solution: 1.0×10^{-1} or 0.10

The expression can be simplified by multiplying the numbers in the numerator and dividing by 3.1 to give $\frac{102.14}{10^3}$. The numerator can be expressed as 1.0214×10^2. When numbers in exponential form are divided, the exponent of the denominator (3) is subtracted from the exponent of the numerator (2), and the answer, 1.0214×10^{-1}, is rounded to two significant figures, 1.0×10^{-1}.

II. Unit Conversions Using the Factor-Label Method

To indicate subdivisions and multiples of the basic units of measurement in the metric system (meter, gram, liter), you will use the following Latin and Greek prefixes.

deci- means one tenth or 10^{-1}
centi- means one hundredth or 10^{-2}
milli- means one thousandth or 10^{-3}
micro- means one millionth or 10^{-6}
nano- means one billionth or 10^{-9}
kilo- means one thousand or 10^{3}
mega- means one million or 10^{6}

These prefixes are applied to the meter as shown. To solve the problems in this section, you should learn these metric relationships.

1 meter (m) = 10 decimeters (dm)
1 meter = 100 centimeters (cm)
1 meter = 1000 millimeters (mm)
1 meter = 1,000,000 micrometers (μm)
1 meter = 1,000,000,000 nanometers (nm)
1000 meters = 1 kilometer (km)
1,000,000 meters = 1 megameter (Mm)

Similar sets of relationships exist for mass (1 gram = 10 decigrams, etc.) and for volume (1 liter = 10 deciliters, etc.)

You should also learn the following English/metric relationships.

1 inch (in.) = 2.54 centimeters
1 angstrom (Å) = 10^{-8} centimeter
1 pound (lb) = 454 grams
1 quart (qt) = 946 milliliters

"Factor-label" method refers to the use of factors derived from the equality between two kinds of units. For example, 2.54 cm = 1 in. can be converted to either $\frac{2.54\text{ cm}}{1\text{ in.}} = 1$ or $\frac{1\text{ in.}}{2.54\text{ cm}} = 1$. The appropriate factor can then be used to convert the quantity in question to the desired units. Often it is necessary to use several factors to achieve the desired units.

A. UNIT CONVERSIONS

Use the factor-label method to perform the following conversions. Record all answers to the correct number of significant figures.

1. 2 m to centimeters

Your answer:

Solution:

$$?\text{ cm} = 2\text{ m} \times \frac{100\text{ cm}}{1\text{ m}} = 2 \times 10^{2}\text{ cm}$$

The answer is limited to one significant figure by 2 m. Note that the metric conversions are exact (defined) relationships and will never control the number of significant figures in the answer.

2. 55 m to kilometers

Your answer:

Solution:

$$? \text{ km} = 55 \text{ m} \times \frac{1 \text{ km}}{1000 \text{ m}} = 0.055 \text{ km} \Rightarrow 5.5 \times 10^{-2} \text{ km}$$

---

3. 5 in. to centimeters

Your answer:

Solution:

$$? \text{ cm} = 5 \text{ in.} \times \frac{2.54 \text{ cm}}{1 \text{ in.}} = 12.7 \text{ cm} \Rightarrow 1 \times 10^{1} \text{ cm}$$

The answer is limited to one significant figure by the measurement of 5 in.

---

4. 5 mi to kilometers

Your answer:

Solution:

$$? \text{ km} = 5 \text{ mi} \times \frac{5280 \text{ ft}}{1 \text{ mi}} \times \frac{12 \text{ in.}}{1 \text{ ft}} \times \frac{2.54 \text{ cm}}{1 \text{ in.}} \times \frac{1 \text{ m}}{100 \text{ cm}} \times \frac{1 \text{ km}}{1000 \text{ m}}$$

$$= 8.04672 \text{ km} \Rightarrow 8 \text{ km}$$

---

5. 625 m^2 to square centimeters

Your answer:

Solution:

$$? \text{ cm}^2 = 625 \text{ m}^2 \times \frac{100 \text{ cm}}{1 \text{ m}} \times \frac{100 \text{ cm}}{1 \text{ m}}$$

$$= 625 \times 10^4 \text{ cm}^2 \Rightarrow 6.25 \times 10^6 \text{ cm}^2$$

---

6. 10. cm^2 to square millimeters

Your answer:

Solution:

$$? \text{ mm}^2 = 10. \text{ cm}^2 \times \frac{1 \text{ m}}{100 \text{ cm}} \times \frac{1 \text{ m}}{100 \text{ cm}} \times \frac{1000 \text{ mm}}{1 \text{ m}} \times \frac{1000 \text{ mm}}{1 \text{ m}}$$

$$= \frac{10. \times 10^6}{10^4} \text{ mm}^2 \Rightarrow 1.0 \times 10^3 \text{ mm}^2$$

Note that the decimal point in 10. indicates that the measurement is precise to the units position.

---

7. 5736 ft^2 to square centimeters

Your answer:

Solution:

$$? \text{ cm}^2 = 5736 \text{ ft}^2 \times \frac{12 \text{ in.}}{1 \text{ ft}} \times \frac{12 \text{ in.}}{1 \text{ ft}} \times \frac{2.54 \text{ cm}}{1 \text{ in.}} \times \frac{2.54 \text{ cm}}{1 \text{ in.}}$$

$$= 5328918.4 \text{ cm}^2 \Rightarrow 5.33 \times 10^6 \text{ cm}^2$$

The answer is limited to three significant figures because there are three significant figures in the metric/English conversion. The English/English conversion factor is a defined relationship and will consequently not limit significant figures in the answer.

---

8. 62 lb to grams

Your answer:

Solution:

$$? \text{ g} = 62 \text{ lb} \times \frac{454 \text{ g}}{1 \text{ lb}} = 28148 \text{ g} \Rightarrow 2.8 \times 10^4 \text{ g}$$

---

9. 3657 lb/in.^{2} to grams per square centimeter

Your answer:

Solution:

$$? \frac{\text{g}}{\text{cm}^2} = \frac{3657 \text{ lb}}{\text{in.}^2} \times \frac{454 \text{ g}}{1 \text{ lb}} \times \frac{1 \text{ in.}}{2.54 \text{ cm}} \times \frac{1 \text{ in.}}{2.54 \text{ cm}}$$

$$= 257343.6 \text{ g/cm}^2 \Rightarrow 2.57 \times 10^5 \text{ g/cm}^2$$

The significant figures in the answer are limited to three places by the English/metric conversion factor.

---

10. 22 mi/gal to kilometers per liter

Your answer:

Solution:

$$\frac{? \text{ km}}{\text{L}} = \frac{22 \text{ mi}}{\text{gal}} \times \frac{5280 \text{ ft}}{1 \text{ mi}} \times \frac{12 \text{ in.}}{1 \text{ ft}} \times \frac{2.54 \text{ cm}}{1 \text{ in.}} \times \frac{1 \text{ m}}{100 \text{ cm}} \times \frac{1 \text{ km}}{1000 \text{ m}} \times \frac{1 \text{ gal}}{4 \text{ qt}} \times \frac{1 \text{ qt}}{946 \text{ mL}} \times \frac{1000 \text{ mL}}{1 \text{ L}}$$

$$= 9.4 \text{ km/L}$$

The answer is recorded to two significant figures. Notice that 4 qt/gal is a defined (exact) relationship and does not limit the number of significant figures in the answer.

B. PROBLEMS INVOLVING CONVERSION OF UNITS

1. Add 0.65 m, 350. cm, 800. mm, 0.02 dm, 2.5 cm, and 35 Å

Your answer:

Solution: 4.98 m or 498 cm

Before adding the measurements, you should choose a common unit, (for example, meters or centimeters) into which you will convert all the measurements.

0.65 m	0.65 m	65 cm
350. cm	3.50 m	350. cm
800. mm	0.800 m	80.0 cm
0.02 dm	0.002 m	0.2 cm
2.5 cm	0.025 m	2.5 cm
35 Å	0.0000000035 m	0.00000035 cm
	4.977 m	497.7 cm

Two measurements in the meter column are given only to the hundredths place, so the sum is rounded to the nearest hundredths place.

---

2. $(4.802 \times 10^{-2}\ \text{m}) + (2.05 \times 10^{-3}\ \text{m}) + 5.522\ \text{m} =$

Your answer:

Solution: 5.572 m

Before adding the numbers together, you should rewrite the numbers so that they all have the same exponent. Alternatively, the numbers can be converted to decimal form.

4.802×10^{-2} m	0.04802 m
0.205×10^{-2} m	0.00205 m
552.2×10^{-2} m	5.522 m
557.2×10^{-2} m	5.572 m

---

3. If 6.02×10^{23} atoms of hydrogen have a mass of 1.008 g, what is the mass (in grams) of one hydrogen atom?

Your answer:

Solution: 1.67×10^{-24} g

$$? \text{ g} = 1 \text{ H atom} \times \frac{1.008 \text{ g H}}{6.02 \times 10^{23} \text{ H atoms}} = 0.167 \times 10^{-23} \text{ g}$$

$$= 1.67 \times 10^{-24} \text{ g}$$

---

4. What is the mass of a sample containing 6.02×10^{23} atoms of sodium (Na)? Each sodium atom has a mass of 3.817×10^{-23} g.

Your answer:

Solution:

$$? \text{ g} = 6.02 \times 10^{23} \text{ Na atoms} \times \frac{3.817 \times 10^{-23} \text{ g}}{1 \text{ Na atom}} = 23.0 \text{ g}$$

---

5. During an experiment a student weighed a crucible on the triple-beam balance and recorded 12.6 g in a notebook. After adding tin to the crucible, the student used an analytical balance and determined the combined mass, which he recorded as 13.8642 g. What is the mass of the tin sample according to the data obtained?

Your answer:

Solution: 1.3 g tin

13.8642 g -12.6 g 1.2642 g	The precision of the mass of the tin is limited to the tenths place, because the mass of the crucible is uncertain in the tenths place.

---

6. If 1.00 cc (cm^3) of oil forms a film over an area of 1000 m^2, what is its thickness in angstroms (1 Å = 10^{-8} cm)?

Your answer:

Solution: 10.0 Å

Visualize a very thin rectangular, three-dimensional object that has a length l, a width w, and a thickness th.

Volume (V) of oil = l × w × th = area × th

If V = area × th, then th = V/area and th = 1.00 cm^3/1000 m^2.

$$? \text{ Å} = \frac{1.00 \text{ cm}^3}{1000 \text{ m}^2} \times \frac{1 \text{ m}}{100 \text{ cm}} \times \frac{1 \text{ m}}{100 \text{ cm}} \times \frac{1.00 \text{ Å}}{10^{-8} \text{ cm}}$$

$$= \frac{1.00 \text{ Å}}{10^7 \times 10^{-8}} = 10.0 \text{ Å}$$

---

7. How much does a 5.0 L sample of a liquid weigh, if its density is 0.972 g/mL?

Your answer:

Solution: 4.9×10^3 g

Density = mass/volume. This can be rearranged to give mass = density × volume.

$$? \text{ g} = \frac{0.972 \text{ g}}{\text{mL}} \times 5.0 \text{ L} \times \frac{1000 \text{ mL}}{1 \text{ L}} = 4.9 \times 10^3 \text{ g}$$

---

8. If a medication is to be given at a dosage of 15.00 mg per kilogram of body weight, how many ounces of this medication should be prescribed for a 155 lb patient?

Your answer:

Solution: 0.0372 oz or 3.72×10^{-2} oz

$$? \text{ oz} = \frac{15.00 \text{ mg}}{\text{kg body wt}} \times \frac{1 \text{ g}}{1000 \text{ mg}} \times \frac{1 \text{ lb}}{454 \text{ g}} \times \frac{16 \text{ oz}}{1 \text{ lb}} \times \frac{1 \text{ kg}}{1000 \text{ g}} \times \frac{454 \text{ g}}{\text{lb}} \times 155 \text{ lb}$$

$$= 0.0372 \text{ oz} = 3.72 \times 10^{-2} \text{ oz}$$

The first three factors are needed to convert milligrams to ounces of medication. The next two factors convert kilograms to pounds of body weight. Then, multiply by the weight (155 lb) of this patient.

---

9. What is the volume (in cubic centimeters) of a piece of metal that has a specific gravity of 6.98 and a mass of 15.54 oz?

Your answer:

Solution: 63.2 cc or 63.2 cm^3

Because specific gravity is defined as the ratio of the density of a substance to the density of water (1 g/cc), the density of this metal is 6.98 g/cc.

density = mass/volume; therefore, volume = mass/density

$$\text{volume} = \frac{15.54 \text{ oz}}{6.98 \text{ g/cc}} \times \frac{1 \text{ lb}}{16 \text{ oz}} \times \frac{454 \text{ g}}{1 \text{ lb}} = 63.2 \text{ cc } (63.2 \text{ cm}^3)$$

---

10. The diameter of an atom is approximately 10^{-8} cm (1 Å). How many atoms placed side by side would total 1 in. in length?

Your answer:

Solution: 2.54×10^8 atoms

$$? \text{ atoms} = 1 \text{ in.} \times \frac{1 \text{ atom}}{10^{-8} \text{ cm}} \times \frac{2.54 \text{ cm}}{1 \text{ in.}} = 2.54 \times 10^8 \text{ atoms}$$

---

11. How many helium atoms (each with mass 6.646×10^{-24} g) would occupy a balloon that contains 12 lb of helium gas?

Your answer:

Solution: 8.2×10^{26} atoms

$$? \text{ helium atoms} = 12 \text{ lb} \times \frac{454 \text{ g}}{1 \text{ lb}} \times \frac{1 \text{ atom}}{6.646 \times 10^{-24} \text{ g}} = 8.2 \times 10^{26} \text{ atoms}$$

---

2 Temperature Scales and Calories

I. Temperature Scales

To convert between the Celsius and Fahrenheit scales, use one of these equations.

$$^{o}C = \frac{5}{9}\ (^{o}F - 32^{o}) \qquad \text{or} \qquad ^{o}F = \frac{9}{5}\ (^{o}C) + 32^{o}$$

In the equation $^{o}C = \frac{5}{9}\ (^{o}F - 32^{o})$, 32° is first subtracted from the given Fahrenheit temperature to adjust for the fact that the freezing point of water on the Fahrenheit scale is $32^{o}F$ above zero, but is zero on the Celsius scale. The result (after subtracting) is then multiplied by $\frac{5}{9}$. The factor $\frac{5}{9}$ is determined by the difference between the boiling and freezing points of water on the two scales.

$$\frac{100^{o}C - 0^{o}C}{212^{o}F - 32^{o}F} = \frac{100 \text{ degrees Celsius}}{180 \text{ degrees Fahrenheit}} = \frac{5 \text{ degrees Celsius}}{9 \text{ degrees Fahrenheit}}$$

To use the equation $^{o}F = \frac{9}{5}\ (^{o}C) + 32^{o}$, first convert the degrees Celsius to degrees Fahrenheit by multiplying by the inverse of the conversion factor described in the previous sentence. Then add $32^{o}F$ to adjust for the difference between the freezing points on the two scales. To convert between the Celsius and Kelvin scales, use the equation:

$$K = {^{o}C} + 273^{o}$$

Notice that the degree symbol (^{o}) is not used for Kelvin temperatures.

Convert the given temperatures to the indicated temperature scale.

1. $0^{o}F$ to ___^{o}C

Your answer:

Solution:

$$?\ ^{o}C = \frac{5}{9}\ (^{o}F - 32^{o}) = \frac{5}{9}\ (0^{o} - 32^{o}) = -17.78^{o}C = -18^{o}C$$

---

2. $25^{o}C$ to ___^{o}F

Your answer:

Solution:

$$?\ ^{o}F = \frac{9}{5}\ (^{o}C) + 32^{o} = \frac{9}{5}\ (25^{o}) + 32^{o} = 77^{o}F$$

---

3. $25^\circ C$ to ___K

Your answer:

Solution:

$? \ K = {}^\circ C + 273^\circ = 25^\circ + 273^\circ = 298 \ K$

---

4. 546 K to ___$^\circ C$

Your answer:

Solution:

$? \ {}^\circ C = K - 273^\circ = 546 - 273^\circ = 273^\circ C$

---

5. $100^\circ C$ to ___$^\circ F$

Your answer:

Solution:

$? \ {}^\circ F = \frac{9}{5} ({}^\circ C) + 32^\circ = \frac{9}{5} (100^\circ) + 32^\circ = 212^\circ F$

---

6. 263 K to ___$^\circ C$

Your answer:

Solution:

$? \ {}^\circ C = K - 273^\circ = 263 - 273^\circ = -10^\circ C$

---

7. 263 K to ___$^\circ F$

Your answer:

Solution: Before you can convert 263 K to the Fahrenheit scale, you need to convert it to the Celsius scale (as you did in Exercise 6). The equivalent Celsius temperature can then be converted to the Fahrenheit scale.

$? \ {}^\circ C = 263 \ K - 273 = -10^\circ C$

$? \ {}^\circ F = \frac{9}{5} ({}^\circ C) + 32 = \frac{9}{5} (-10) + 32 = 14^\circ F$

---

8. $-16.5^{\circ}F$ to ___$^{\circ}C$

Your answer:

Solution:

$$? ^{\circ}C = \frac{5}{9} (^{\circ}F - 32^{\circ}) = \frac{5}{9} (-16.5 - 32) = -26.9^{\circ}C$$

---

9. Which is the preferable temperature for the task or occasion indicated?
   (a) Boiling potatoes: $100^{\circ}C$ or $100^{\circ}F$?

Your answer:

Solution: $100^{\circ}C$

To compare two temperatures, both temperatures must be in the same temperature scale. However, for this problem, we need not actually make any temperature conversions; rather, we can logically deduce the answer from the given information. We realize that to boil potatoes, the water we use must be at its boiling point ($100^{\circ}C$ or $212^{\circ}F$). $100^{\circ}F$ is far below the boiling point, so it could not possibly be used.

(b) Ice skating on a pond: $20^{\circ}C$ or $20^{\circ}F$?

Your answer:

Solution: $20^{\circ}F$

The pond could not be frozen at $20^{\circ}C$ because the freezing point of water is $0^{\circ}C$ ($32^{\circ}F$).

(c) A summer day: $27^{\circ}C$ or $80^{\circ}F$?

Your answer:

Solution: $27^{\circ}C$ and $80^{\circ}F$

$$? ^{\circ}F = \frac{9}{5} (27^{\circ}) + 32^{\circ} = 81^{\circ}F$$

Both $81^{\circ}F$ ($27^{\circ}C$) and $80^{\circ}F$ are warm temperatures and would typify a summer day.

(d) Absolute zero: $0^{\circ}F$, $0^{\circ}C$, or 0 K?

Your answer:

Solution: 0 K

(e) Lab room temperature: $20^{\circ}C$ or $58^{\circ}F$?

Your answer:

Solution: $20^{\circ}C$

$$? ^{\circ}F = \frac{9}{5} (20^{\circ}) + 32^{\circ} = 68^{\circ}F$$

$20^{\circ}C$ ($68^{\circ}F$) would be much more comfortable than $58^{\circ}F$.

II. Calories

1. How many calories of heat must be added to 1.5 L of water to raise the temperature of the water from 22°C to 62°C?

Your answer:

Solution: 6.0×10^4 cal

$$? \text{ cal} = \frac{1.00 \text{ cal}}{\text{g} \times \text{deg}} \times \text{g} \times \text{deg} \qquad \left(\frac{1.00 \text{ cal}}{\text{g} \times \text{deg}} \text{ is the specific heat of water}\right)$$

1.5 L of water must be expressed in grams. To do this, use the density of water.

$$1.5 \text{ L} \times \frac{1000 \text{ mL}}{1 \text{ L}} \times \frac{1.0 \text{ g}}{\text{mL}} = 1.5 \times 10^3 \text{ g}$$

Degrees are calculated by subtracting the final temperature from the initial temperature: $62°\text{C} - 22°\text{C} = 40°$.

$$? \text{ cal} = \frac{1.00 \text{ cal}}{\text{g} \times \text{deg}} \times 1.5 \times 10^3 \text{ g} \times 40.° = 6.0 \times 10^4 \text{ cal}$$

---

2. Should a person who is on a 1200 Cal diet eat enough food to furnish 1200 cal or 1200 kcal of energy?

Your answer:

Solution: 1200 kcal

1200 Cal means 1200 × 1000 cal or 1200 kcal. This energy could heat 1 g of water 1.2×10^6 degrees (theoretically) or 1.2×10^6 g of water 1 degree.

---

3. How many degrees Celsius will the temperature rise if 5.0 kcal of heat is transferred to 2.0 kg of water?

Your answer:

Solution: 2.5°C

This problem can be solved using the factor-label method; to calculate degrees Celsius you should use the specific heat of water, $\frac{1.00 \text{ cal}}{\text{g} \times \text{deg}}$, in its inverted form.

$$? °\text{C} = \frac{\text{g} \times \text{deg}}{1.00 \text{ cal}} \times 5.0 \text{ kcal} \times \frac{1000 \text{ cal}}{1 \text{ kcal}} \times \frac{1}{2.0 \text{ kg}} \times \frac{1 \text{ kg}}{1000 \text{ g}}$$

$$= 2.5°\text{C}$$

---

4. When 1.0 g of coal is burned, 7600 cal of heat is evolved. How many kilograms of coal would be necessary to supply 975 kcal of heat?

Your answer:

Solution: 0.13 kg

You can solve this problem with the factor-label method.

$$? \text{ kg} = \frac{1 \text{ g coal}}{7600 \text{ cal}} \times 975 \text{ kcal} \times \frac{10^3 \text{ cal}}{1 \text{ kcal}} \times \frac{1 \text{ kg}}{10^3 \text{ g}} = 0.13 \text{ kg}$$

0.13 kg is the preferred answer rather than 0.128 kg or 0.1283 kg. The precision and number of significant figures in 7600 are ambiguous: without additional information, we can be sure of only two significant figures and would therefore express the answer as 0.13 kg. If this measurement were precise to three significant figures, it would have been recorded as 7.60×10^3, and the answer could be expressed as 0.128 kg; if the measurement were precise to four significant figures, it would have been recorded as either 7600. or 7.600×10^3, and the answer could be expressed as 0.1283 kg.

---

5. If 0.12 cal is required to heat 1.0 g of iron by 1°C, how many calories are needed to raise the temperature of 625 g of iron from 25°C to 50°C?

Your answer:

Solution: 1.9×10^3 cal

$\frac{0.12 \text{ cal}}{\text{g} \times \text{deg}}$ is the specific heat of iron, which you can use to initiate the factor-label method of problem solving.

$$? \text{ cal} = \frac{0.12 \text{ cal}}{\text{g} \times \text{deg}} \times 625 \text{ g} \times (50^\circ - 25^\circ) = 0.12 \times 625 \times 25 \text{ cal} = 1.9 \times 10^3 \text{ cal}$$

---

3 Isotopes, Subatomic Particles, and Ions

For some of these questions you will need to use the periodic table inside the front cover of this book.

I. Subatomic Particles in Isotopes and Ions

A. SUBATOMIC PARTICLES IN ATOMS

Use these symbols for the questions of this section.

(a) $^{16}_{8}O$ (b) $^{18}_{8}O$ (c) $^{58}_{26}Fe$ (d) $^{54}_{26}Fe$ (e) $^{48}_{20}X$

1. What term relates the atoms represented in (a) and (b)?

Your answer:

Solution: isotope
(a) and (b) represent two forms of the same element, oxygen. The atoms differ only in the mass number, which means that they differ in the number of neutrons contained in the nucleus.

---

2. What is the number 8 in (a) and (b) called? This number indicates the number of which type of subatomic particles in the oxygen atom? Where are these particles located in the atom?

Your answer:

Solution: atomic number (symbol: Z); protons; in the nucleus
In all oxygen atoms there are eight protons in the nucleus and eight electrons in the region outside the nucleus.

---

3. What is the relationship between (c) and (d)?

Your answer:

Solution: isotopes of iron

---

4. What does the number 54 in (d) represent? For which subatomic particles does this number account? How many of each kind of subatomic particles are in an atom represented by (d)? Where is each of these particles located?

Your answer:

Solution: 54 is the mass number (symbol: A). The mass number is the sum of the number of protons and neutrons in the nucleus of an atom. The mass number minus the atomic number equals the number of neutrons in the nucleus. In (d), an isotope of iron, there are 26 protons and 28 neutrons in the nucleus and 26 electrons in the region outside the nucleus.

---

5. Locate the element with atomic number 8 in the periodic table. What term describes the number 15.999? What does this number represent? Express this number to four significant figures.

Your answer:

Solution: 15.999 is the atomic weight. This number represents the weighted average of all naturally occurring isotopes of oxygen. The three naturally occurring isotopes of oxygen have mass numbers of 16, 17, and 18, and because the weighted average of the masses of these three isotopes most closely approximates 16, we can conclude that $^{16}_{8}O$ is the most abundant isotope of the three. 15.999 expressed to four significant figures is 16.00.

---

6. What element does (e) represent? If an isotope with a mass number of 40 is the lightest-weight naturally occurring isotope of this element, what would you surmise about the relative abundance of the isotope represented by (e)?

Your answer:

Solution: (e) is $^{48}_{20}Ca$. An element is identified by its atomic number (Z); the atomic number indicated in (e) is 20, which is the atomic number corresponding to calcium. Because the average atomic weight of calcium is 40.08, we can hypothesize that the percent abundance of ^{48}Ca is very small; in fact, the natural abundance of this isotope is 0.18%.

B. SUBATOMIC PARTICLES IN IONS

In each of the following questions, indicate what the given symbol represents. Then state the number of protons, electrons, and neutrons contained in an ion formed from the isotope having a mass number closest to the atomic weight.

1. H^+

Your answer:

Solution: H^+ is a hydrogen ion. A H^+ ion forms when an H atom loses one electron to another atom. $^{1}H^+$ has one proton, zero electrons, and zero neutrons.

---

2. Ca^{2+}

Your answer:

Solution: Ca^{2+} is a calcium ion. A Ca^{2+} ion forms when a Ca atom loses two electrons to another atom or atoms. $^{40}Ca^{2+}$ has twenty protons, eighteen electrons, and twenty neutrons.

---

3. O^{2-}

Your answer:

Solution: O^{2-} is an oxide ion. An oxygen atom gains two electrons from another atom or atoms to form an O^{2-} ion. $^{16}O^{2-}$ consists of eight protons, ten electrons, and eight neutrons.

---

4. F^-

Your answer:

Solution: F^- is a fluoride ion. A F^- ion forms when an F atom gains one electron from another atom. $^{19}F^-$ consists of nine protons and ten neutrons in the nucleus and ten electrons in the region outside the nucleus.

C. SUBATOMIC PARTICLES IN ATOMS AND IONS: MISCELLANEOUS QUESTIONS

1. What is the composition of the nuclei of ^{1}H, ^{2}H, and ^{3}H? What is the relationship among these three species?

Your answer:

Solution: ^{1}H (often referred to as protium) contains one proton and no neutrons in the nucleus; ^{2}H (referred to as deuterium) has one proton and one neutron in its nucleus; and ^{3}H (tritium) contains one proton and two neutrons in its nucleus. ^{1}H, ^{2}H, and ^{3}H are three isotopes of hydrogen.

---

2. How does the oxide ion differ from the neon atom? How are they similar?

Your answer:

Solution: $^{16}_{8}O^{2-}$ contains 8 protons and 8 neutrons. $^{20}_{10}Ne$ contains 10 protons and 10 neutrons. However, both the oxide ion and the neon atom contain 10 electrons (and are isoelectronic).

---

3. Compare the nuclei of $^{36}_{16}S$ and $^{36}_{18}Ar$.

Your answer:

Solution: Both atoms have the same number of nucleons (particles in the nucleus); however, the sulfur isotope indicated has 16 protons and 20 neutrons in its nucleus, and the argon isotope indicated has 18 protons and 18 neutrons in its nucleus.

---

4. What similarities exist between the $^{39}_{19}K^{+}$ ion and the $^{37}_{17}Cl^{-}$ ion?

Your answer:

Solution: Both ions contain 20 neutrons in the nucleus. They also have the same number of electrons (are isoelectronic).

II. Isotope Abundance Calculations

1. An element occurs as two isotopes, 19.78% of the atoms weighing 10.0129 amu and 80.22% of the atoms weighing 11.00931 amu. Calculate the "atomic weight," that is, the weighted average of the naturally occurring isotopes. Identify the element from the periodic table and write symbols for the two isotopic forms.

Your answer:

Solution: The atomic weight of the element is 10.813 amu.

To calculate the atomic weight; let the isotopes be represented by ^{10}E and ^{11}E. 19.78% = 0.1978 in decimal form; 80.22% = 0.8022 in decimal form.

relative mass contribution by ^{10}E atoms = 0.1978 × 10.0129 amu
= 1.981 amu

relative mass contribution by ^{11}E atoms = 0.8022 × 11.00931 amu
= 8.832 amu

atomic weight = 1.981 + 8.832 = 10.813 amu

This element is boron; it is recognized by the atomic weight. The isotopes are $^{10}_{5}B$ and $^{11}_{5}B$.

---

2. An element has three isotopic forms with the following masses and abundances.

Mass (amu)	Abundance (%)
23.98504	78.70
24.98584	10.13
25.98259	11.17

What is the atomic weight? What is the element? Give the symbol for each isotope.

Your answer:

Solution: The atomic weight is 24.31 amu.

mass contribution by ^{24}E = 0.7870 × 23.98504 amu = 18.88 amu
mass contribution by ^{25}E = 0.1013 × 24.98584 amu = 2.531 amu
mass contribution by ^{26}E = 0.1117 × 25.98259 amu = 2.902 amu
atomic weight = 24.31 amu

The element is magnesium and can be recognized by its atomic weight. The three isotopes can be symbolized by $^{24}_{12}Mg$, $^{25}_{12}Mg$, and $^{26}_{12}Mg$.

---

3. If the only naturally occurring isotopes of silver have masses of 106.90509 and 108.9047, what is the relative abundance of each? (Hint: Use the atomic weight of silver given in the periodic table.) Write the symbol of each isotope.

Your answer:

Solution: The relative abundance of ^{107}Ag is 51.745%; the relative abundance of ^{109}Ag is 48.255%. These two isotopes can be symbolized by writing ^{107}Ag and ^{109}Ag. Notice that the mass numbers have been obtained by rounding the atomic masses to the nearest integer. Let x = the decimal fraction of the atoms that are ^{107}Ag and 1 - x = the decimal fraction of the atoms that are ^{109}Ag.

mass contribution of ^{107}Ag = (106.90509)(x)
mass contribution of ^{109}Ag = (108.9047)(1 - x)

The sum of these mass contributions = atomic weight of Ag = 107.87

$$106.90509x + 108.9047(1 - x) = 107.87$$
$$106.90509x - 108.9047x = 107.87 - 108.9047$$
$$-1.9996x = -1.03$$
$$x = 0.515, \text{ therefore } 51.5\% \text{ of } ^{107}Ag$$
$$1 - x = 0.485, \text{ therefore } 48.5\% \text{ of } ^{109}Ag$$

III. Nomenclature and Formula Writing of Inorganic Compounds

1. Memorize the formulas and names of the ions indicated.

(1) The list of simple anions and cations given in your text.
(2) The cation NH_4^+, ammonium ion.
(3) The following polyatomic anions.

OH^-	hydroxide	CrO_4^{2-}	chromate
$C_2H_3O_2^-$	acetate	PO_4^{3-}	phosphate
CN^-	cyanide	CO_3^{2-}	carbonate
HCO_3^-	bicarbonate (or hydrogen carbonate)		

(4) The following oxyanions. Notice that the prefix and/or suffix indicates the relative number of oxygen atoms in the ion (i.e., the suffix -ite implies that the ion has one less oxygen than does an ion with suffix -ate).

ClO_4^-	perchlorate		
ClO_3^-	chlorate	NO_3^-	nitrate
ClO_2^-	chlorite	NO_2^-	nitrite
ClO^-	hypochlorite		
SO_4^{2-}	sulfate	HSO_4^-	bisulfate (or hydrogen sulfate)
SO_3^{2-}	sulfite	HSO_3^-	bisulfite (or hydrogen sulfite)

If you have difficulty memorizing and using the ion formulas and names you might try this: on one side of an index card write the formula of an ion (include the correct charge) and on the back side, write the name. Put the anions in one "hat" and the cations in another.

(1) Look at a name, and then write the formula (check your answer with the formula on the reverse side of the card). After practicing all the ions this way, try the reverse procedure: from the formula, write the name of the ion.
(2) Combine any anion and cation and write the name or formula of the ionic compound that might be formed from this combination.

2. Write the chemical formulas for the following ionic compounds by combining the oppositely charged ions in the simplest ratios to give neutral formulas. After you have completed this exercise, try Exercise 3. The answers for Exercise 2 exactly correspond to the questions in Exercise 2. The reverse is also true; the answers for Exercise 3 exactly correspond to the questions in Exercise 2.

Your answers:

(a) calcium chloride ______

(b) silver(I) nitrate ______

(c) barium bromide ______

(d) potassium hydroxide ______

(e) sodium sulfate ______

(f) ammonium chloride ______

(g) iron(III) oxide ______

(h) sodium acetate ______

(i) calcium phosphate ______

(j) magnesium sulfate ______

(k) barium chromate ______

(l) copper(II) hydroxide ______

(m) ammonium acetate ______

(n) lithium carbonate ______

(o) sodium bicarbonate (sodium hydrogen carbonate) ______

(p) potassium cyanide ______

(q) magnesium hypochlorite ______

(r) iron(III) nitrate ______

3. Write the name corresponding to each structure.

Your answers:

(a) $CaCl_2$ ______

(b) $AgNO_3$ ______

(c) $BaBr_2$ ______

(d) KOH ______

(e) Na_2SO_4 ______

(f) NH_4Cl ______

(g) Fe_2O_3 ______

(h) $NaC_2H_3O_2$ ______

(i) $Ca_3(PO_4)_2$ ______

(j) $MgSO_4$ ______

(k) $BaCrO_4$ ______

(l) $Cu(OH)_2$ ______

(m) $NH_4C_2H_3O_2$ ______

(n) Li_2CO_3 ______

(o) $NaHCO_3$ ______

(p) KCN ______

(q) $Mg(ClO)_2$ ______

(r) $Fe(NO_3)_3$ ______

IV. Nomenclature of Nonionic Compounds (Compounds Between Two Nonmetals)

1. Write the chemical formulas for the compounds named. Notice that prefixes are used to designate the relative proportion of elements.
   After you have completed this exercise, do Exercise 2. The answers for Exercise 1 are the questions for Exercise 2.

Your answers:

(a) carbon dioxide ______

(b) carbon monoxide ______

(c) boron tribromide ______

(d) ammonia ______

(e) silicon dioxide ______

(f) carbon tetrachloride ______

(g) sulfur dioxide ______

(h) sulfur trioxide ______

(i) phosphorus trichloride ______

(j) nitrogen dioxide ______

(k) dinitrogen pentoxide ______

2. Write the name corresponding to the structure shown.

Your answers:

(a) CO_2 ______

(b) CO ______

(c) BBr_3 ______

(d) NH_3 ______

(e) SiO_2 ______

(f) CCl_4 ______

(g) SO_2 ______

(h) SO_3 ______

(i) PCl_3 ______

(j) NO_2 ______

(k) N_2O_5 ______

4 Atomic and Molecular Weights, Moles, Empirical and Molecular Formulas, Chemical Equations, and Stoichiometry

For all calculations, use atomic weights from the periodic table rounded to four significant figures.

I. Atomic, Molecular, and Formula Weights

1. What are the atomic weights of aluminum, oxygen, fluorine, and argon?

Your answer:

Solution: Al = 26.98 amu; O = 16.00 amu; F = 19.00 amu; Ar = 39.95 amu
The atomic weight of an element given in the periodic table is the weighted average of the masses of the naturally occurring isotopes of the element. (Problems relating isotope abundance and atomic weights are found in Chapter 3.)

---

2. Calculate the molecular weights of water, carbon monoxide, and carbon tetrachloride.

Your answer:

Solution: One molecule of water, H_2O, has a mass of 18.02 amu.

$(2 \times 1.008) + 16.00 = 18.02$

One molecule of carbon monoxide, CO, has a mass of 28.01 amu.

$12.01 + 16.00 = 28.01$

One molecule of carbon tetrachloride, CCl_4, has a mass of 153.81 amu.

$12.01 + (4 \times 35.45) = 153.81$

---

3. Calculate the formula weights of these ionic compounds: sodium hydroxide, NaOH, and barium nitrate, $Ba(NO_3)_2$.

Your answer:

Solution: One "formula unit" of NaOH (one Na^+ ion and one OH^- ion of the entire crystal) has a mass of 40.00 amu.

$22.99 + 16.00 + 1.008 = 40.00$

One "formula unit" of $Ba(NO_3)_2$ has a mass of 261.5 amu.

$137.3 + (2 \times 14.10) + (6 \times 16.00) = 261.5$

II. Atoms, Molecules, and Formula Units (in amu) Contrasted to *Moles of* Atoms, Molecules, and Formula Units (in grams); Relationship of Avogadro's Number (6.02×10^{23}) to the Mole Concept.

1. Contrast an atom of hydrogen to a mole of hydrogen atoms.

Your answer:

Solution: A hydrogen atom is the smallest particle of hydrogen that can exist. It is submicroscopic in size and has a mass of 1.008 amu. A mole of hydrogen atoms has a mass of 1.008 g and contains 6.02×10^{23} hydrogen atoms. A mole of any substance is the standard conveniently sized package for laboratory work and calculations. A mole of a given substance contains 6.02×10^{23} of its smallest kind of particle and has a mass in grams numerically equal to the sum of the atomic weights represented in its symbol or formula.

---

2. Contrast a molecule of water and a mole of water molecules.

Your answer:

Solution: A water molecule is the smallest particle of water that can exist. It is far too small to be measured in the ordinary chemistry laboratory. One water molecule has a mass of 18.02 amu. One mole of water molecules (6.02×10^{23} H_2O molecules) has a mass of 18.02 g, a conveniently sized sample. Notice again that the mass in grams of a mole of any substance is numerically equal to the mass of its individual particles expressed in atomic mass units.

---

3. Contrast a formula unit of sodium hydroxide to a mole of sodium hydroxide.

Your answer:

Solution: In the crystal lattice of alternating Na^+ and OH^- ions (which are attracted to each other), one sodium ion and one hydroxide ion constitute a "formula unit," which has a mass of 40.00 amu. 6.02×10^{23} Na^+ ions and 6.02×10^{23} OH^- ions (or 1 mole of Na^+ ions and 1 mole of OH^- ions) constitute 1 mole of NaOH, which has a mass of 40.00 g.

III. Problems Relating Moles, Number of Particles, and Mass

In the following problems write all answers to the correct number of significant figures. Use the atomic weights in the periodic table to four significant figures.

1. One mole of lithium has a mass of 6.939 g and contains 6.02×10^{23} atoms. Write four unit factors relating these numbers, for example,

$$\frac{6.939 \text{ g Li}}{1 \text{ mole Li}}$$

Your answer:

Solution:

$$\frac{6.939 \text{ g Li}}{1 \text{ mole Li}} \; ; \; \frac{1 \text{ mole Li}}{6.939 \text{ g Li}} \; ; \; \frac{6.02 \times 10^{23} \text{ atoms Li}}{1 \text{ mole Li}} \; ; \; \frac{1 \text{ mole Li}}{6.02 \times 10^{23} \text{ atoms Li}}$$

---

2. How many moles of lithium are in a 25.92 g sample of lithium?

Your answer:

Solution: 3.735 moles Li

$$? \text{ moles Li} = 25.92 \text{ g Li} \times \frac{1 \text{ mole Li}}{6.939 \text{ g Li}} = 3.735 \text{ moles Li}$$

---

3. What is the mass of 3.50 moles of lithium?

Your answer:

Solution: 24.3 g Li

$$? \text{ g Li} = 3.50 \text{ moles Li} \times \frac{6.939 \text{ g Li}}{1 \text{ mole Li}} = 24.3 \text{ g Li}$$

---

4. How many atoms are contained in 10.4 g of lithium?

Your answer:

Solution: 9.02×10^{23} atoms Li

$$? \text{ atoms Li} = 10.4 \text{ g Li} \times \frac{1 \text{ mole Li}}{6.939 \text{ g Li}} \times \frac{6.02 \times 10^{23} \text{ atoms Li}}{1 \text{ mole Li}}$$

$$= 9.02 \times 10^{23} \text{ atoms Li}$$

---

5. What is the mass of an "average" lithium atom, expressed in atomic mass units?

Your answer:

Solution: 6.939 amu (from the periodic table)

---

6. What is the mass in grams of an "average" lithium atom?

Your answer:

Solution: 1.15×10^{-23} g Li

$$? \text{ g Li} = 1 \text{ atom Li} \times \frac{1 \text{ mole Li}}{6.02 \times 10^{23} \text{ atoms Li}} \times \frac{6.939 \text{ g Li}}{1 \text{ mole Li}}$$

$$= 1.15 \times 10^{-23} \text{ g Li}$$

---

7. How many moles of carbon atoms are in 1 mole of C_2H_6O (ethanol)?

Your answer:

Solution: The formula C_2H_6O means that in one molecule there are two carbon atoms, six hydrogen atoms, and one oxygen atom. Therefore, in 1 mole of C_2H_6O molecules, there are 2 moles of carbon atoms, 6 moles of hydrogen atoms, and 1 mole of oxygen atoms.

---

8. How many moles of hydrogen atoms are in 4.5 moles of ethanol, C_2H_6O?

Your answer:

Solution: 27 moles H atoms

$$? \text{ moles H atoms} = 4.5 \text{ moles } C_2H_6O \times \frac{6 \text{ moles H atoms}}{1 \text{ mole } C_2H_6O} = 27 \text{ moles H atoms}$$

The answer is written to two significant figures because the 6 (from the chemical formula) is an exact number and so is not used to determine significant figures needed in the answer.

---

9. How many oxygen atoms are in 1.56 moles of C_2H_6O?

Your answer:

Solution: 9.39×10^{23} oxygen atoms

$$? \text{ atoms O} = 1.56 \text{ moles } C_2H_6O \times \frac{6.02 \times 10^{23} \text{ molecules } C_2H_6O}{1 \text{ mole } C_2H_6O} \times \frac{1 \text{ atom O}}{1 \text{ molecule } C_2H_6O}$$

$$= 9.39 \times 10^{23} \text{ atoms O}$$

---

10. What is the mass in grams of 1 mole of C_2H_6O?

Your answer:

Solution: 46.07 g/mole

2 carbons	$= 2 \times 12.01 =$	24.02
6 hydrogens	$= 6 \times 1.008 =$	6.048
1 oxygen	$= 1 \times 16.00 =$	16.00
		46.068 (round to 46.07 g/mole)

---

11. How many moles of C_2H_6O are in a 9.123 g sample of the compound?

Your answer:

Solution: 0.1980 mole C_2H_6O

$$? \text{ moles } C_2H_6O = 9.123 \text{ g } C_2H_6O \times \frac{1 \text{ mole } C_2H_6O}{46.07 \text{ g } C_2H_6O} = 0.1980 \text{ mole } C_2H_6O$$

---

12. How many molecules are present in the 9.123 g sample of C_2H_6O?

Your answer:

Solution: 1.19×10^{23} molecules C_2H_6O

$$? \text{ molecules } C_2H_6O = 9.123 \text{ g } C_2H_6O \times \frac{1 \text{ mole } C_2H_6O}{46.07 \text{ g } C_2H_6O} \times \frac{6.02 \times 10^{23} \text{ molecules } C_2H_6O}{1 \text{ mole } C_2H_6O}$$

$$= 1.19 \times 10^{23} \text{ molecules } C_2H_6O$$

---

13. What is the mass in grams of one molecule of C_2H_6O?

Your answer:

Solution: 7.65×10^{-23} g C_2H_6O

$$? \text{ g } C_2H_6O = 1 \text{ molecule } C_2H_6O \times \frac{1 \text{ mole } C_2H_6O}{6.02 \times 10^{23} \text{ molecules } C_2H_6O} \times \frac{46.07 \text{ g } C_2H_6O}{1 \text{ mole } C_2H_6O}$$

$$= 7.65 \times 10^{-23} \text{ g } C_2H_6O$$

---

14. Which of the following contains the most magnesium atoms?

(i) 1.216 g Mg　　(ii) 3.013×10^{22} Mg atoms

(iii) 13.15 g $Mg_3(PO_4)_2$　　(iv) 0.05002 mole Mg

Your answer:

Solution: (iii) has the greatest number of magnesium atoms.

To compare the number of atoms in each number, you should convert each to atoms or to moles (because a mole contains 6.02×10^{23} atoms).

(i) $? \text{ atoms Mg} = 1.216 \text{ g Mg} \times \frac{1 \text{ mole Mg}}{24.31 \text{ g Mg}} \times \frac{6.02 \times 10^{23} \text{ atoms Mg}}{1 \text{ mole Mg}}$

$= 3.01 \times 10^{22}$ atoms Mg

(ii) 3.013×10^{22} atoms Mg

(iii) $? \text{ atoms Mg} = 13.15 \text{ g } Mg_3(PO_4)_2 \times \frac{1 \text{ mole } Mg_3(PO_4)_2}{262.87 \text{ g } Mg_3(PO_4)_2} \times \frac{3 \text{ moles Mg}}{1 \text{ mole } Mg_3(PO_4)_2}$

$\times \frac{6.02 \times 10^{23} \text{ atoms Mg}}{1 \text{ mole Mg}}$

$= 9.03 \times 10^{22}$ atoms Mg

(iv) $? \text{ atoms Mg} = 0.05002 \text{ mole Mg} \times \frac{6.02 \times 10^{23} \text{ atoms Mg}}{1 \text{ mole Mg}} = 3.01 \times 10^{22} \text{ atoms Mg}$

IV. Percent Composition (by Mass)

Calculate the percent by mass of each element in the formulas indicated.

1. H_3PO_4

Your answer:

Solution: 3.086% H; 31.61% P; 65.31% O

The percent composition by mass of a compound is calculated from

$$\frac{\text{mass of the element in a molecule or formula unit of the compound}}{\text{molecular weight or formula weight of the compound}} \times 100$$

The molecular weight of H_3PO_4 is calculated as follows.

$(3 \times 1.008) + 30.97 + (4 \times 16.00) = 97.99$ amu

$\% \text{ H} = \frac{3.024}{97.99} \times 100 = 3.086\%$

$$\% \text{ P} = \frac{30.97}{97.99} \times 100 = 31.61\%$$

$$\% \text{ O} = \frac{64.00}{97.99} \times 100 = 65.31\%$$

To calculate percent of an element from molar quantities,

$$\frac{\text{mass in grams of the element in 1 mole of the compound}}{\text{mass in grams of 1 mole of the compound}} \times 100$$

---

2. $Ba(NO_3)_2$

Your answer:

Solution: 52.54% Ba; 10.72% N; 36.74% O

$$\% \text{ Ba} = \frac{137.3}{261.3} \times 100 = 52.54\%$$

$$\% \text{ N} = \frac{28.02}{261.3} \times 100 = 10.72\%$$

$$\% \text{ O} = \frac{96.00}{261.3} \times 100 = 36.74\%$$

---

3. $HC_2H_3O_2$

Your answer:

Solution: 6.714% H; 40.00% C; 53.29% O

$$\% \text{ H} = \frac{4.032}{60.05} \times 100 = 6.714\%$$

$$\% \text{ C} = \frac{24.02}{60.05} \times 100 = 40.00\%$$

$$\% \text{ O} = \frac{32.00}{60.05} \times 100 = 53.29\%$$

---

4. $CuSO_4 \cdot 5H_2O$

Your answer:

Solution: 25.45% Cu; 12.84% S; 57.67% O; 4.037% H

The formula shows that there are five water molecules associated with each $CuSO_4$ formula unit; therefore you should add five times the molecular weight of water to the formula weight of $CuSO_4$. A compound such as $CuSO_4 \cdot 5H_2O$ is called a hydrate; $CuSO_4$ alone is said to be anhydrous.

The molecular weight of $CuSO_4 \cdot 5H_2O$ is

$$63.54 + 32.06 + (9 \times 16.00) + (10 \times 1.008) = 249.68 \text{ amu}$$

and its percent composition is

$$\% \text{ Cu} = \frac{63.54}{249.68} \times 100 = 25.45\%$$

$$\% \text{ S} = \frac{32.06}{249.68} \times 100 = 12.84\%$$

$$\% \text{ O} = \frac{144.00}{249.68} \times 100 = 57.67\%$$

$$\% \text{ H} = \frac{10.08}{249.68} \times 100 = 4.037\%$$

You have seen that chemical formulas of compounds can be used to give information such as percent composition or the relative number of moles or grams of any element in a mole of compound. In the next section, percent composition and relative weights of elements are used to determine the empirical formulas of compounds.

V. Empirical and Molecular Formula Calculations

The following procedure is suggested for working an empirical formula problem.

(1) Calculate the number of moles of atoms of each element from the mass. If the percent of each element in the compound is given, you can assume that in a mass of exactly 100.00 g of the compound, the percentage of each element represents the element's mass in grams.

(2) Divide the number of moles of each element by the smallest number of moles to get the mole ratios among the various elements.

(3) If the calculated ratios are not whole numbers (to the nearest tenth place), convert them to whole numbers by multiplying each ratio by 2; if the ratios are not yet whole numbers to the nearest tenth, multiply the original ratios by 3; continue trying each consecutive integer until the ratios are whole numbers to the nearest tenth place.

(4) The integer for each element that you obtain in step 3 becomes the subscript for that element in the empirical formula.

Determine the empirical formula of each compound from the percent composition.

1. 74.20% Na; 25.80% O

Your answer:

Solution: Na_2O

In a 100.00 g sample of the compound you can assume that there are 74.20 g of Na and 25.80 g O.

$$74.20 \text{ g Na} \times \frac{1 \text{ mole Na}}{22.99 \text{ g Na}} = 3.227 \text{ moles Na}; \quad \div 1.612 = 2.002$$

$$25.80 \text{ g O} \times \frac{1 \text{ mole O}}{16.00 \text{ g O}} = 1.612 \text{ moles O}; \quad \div 1.612 = 1.000$$

2.0 moles of Na and 1.0 mole of O are expressed in a formula as Na_2O.

---

2. 26.5% K; 35.4% Cr; 38.1% O

Your answer:

Solution: $K_2Cr_2O_7$

$$26.5 \text{ g K} \times \frac{1 \text{ mole K}}{39.10 \text{ g K}} = 0.678 \text{ mole K}; \quad \div 0.678 = 1.00; \quad \times 2 = 2.00$$

$$35.4 \text{ g Cr} \times \frac{1 \text{ mole Cr}}{52.00 \text{ g Cr}} = 0.681 \text{ mole Cr}; \quad \div 0.678 = 1.00; \quad \times 2 = 2.00$$

$$38.1 \text{ g O} \times \frac{1 \text{ mole O}}{16.00 \text{ g O}} = 2.38 \text{ mole O}; \quad \div 0.678 = 3.51; \quad \times 2 = 7.02$$

---

3. 53.73% Fe; 46.27% S

Your answer:

Solution: Fe_2S_3

$$53.73 \text{ g Fe} \times \frac{1 \text{ mole Fe}}{55.85 \text{ g Fe}} = 0.9620 \text{ mole Fe}; \quad \div 0.9620 = 1.000; \quad \times 2 = 2.0$$

$$46.27 \text{ g S} \times \frac{1 \text{ mole S}}{32.06 \text{ g S}} = 1.443 \text{ moles S}; \quad \div 0.9620 = 1.500; \quad \times 2 = 3.0$$

---

4. 26.52% Cr; 24.52% S; 48.96% O

Your answer:

Solution: $Cr_2S_3O_{12}$

$$26.52\text{ g Cr} \times \frac{1\text{ mole Cr}}{52.00\text{ g Cr}} = 0.5100\text{ mole Cr}; \quad \div 0.5100 = 1.000; \quad \times 2 = 2.0$$

$$24.52\text{ g S} \times \frac{1\text{ mole S}}{32.06\text{ g S}} = 0.7648\text{ mole S}; \quad \div 0.5100 = 1.500; \quad \times 2 = 3.0$$

$$48.96\text{ g O} \times \frac{1\text{ mole O}}{16.00\text{ g O}} = 3.060\text{ mole O}; \quad \div 0.5100 = 6.000; \quad \times 2 = 12.0$$

Determine the empirical formula of each compound from the information provided in the problem.

1. When 5.49 g of manganese reacts with oxygen, 7.62 g of oxide forms. What is the empirical formula of the product?

Your answer:

Solution: Mn_3O_4

Calculate the grams of oxygen in the oxide by subtraction.

O + Mn: 7.62 g
Mn: 5.49 g
2.13 g O

$$5.49\text{ g Mn} \times \frac{1\text{ mole Mn}}{54.94\text{ g Mn}} = 0.0999\text{ mole Mn}; \quad \div 0.0999 = 1.00; \quad \times 3 = 3.0$$

$$2.13\text{ g O} \times \frac{1\text{ mole O}}{16.00\text{ g O}} = 0.133\text{ mole O}; \quad \div 0.0999 = 1.33; \quad \times 3 = 4.0$$

---

2. A sample of a compound contains 1.89 g sodium, 0.0820 mole aluminum atoms, and 2.62 g oxygen. What is the empirical formula?

Your answer:

Solution: $NaAlO_2$

$$1.89 \text{ g Na} \times \frac{1 \text{ mole Na}}{22.99 \text{ g Na}} = 0.0822 \text{ mole Na}; \quad \div 0.0820 = 1.00$$

$$0.0820 \text{ mole Al}; \quad \div 0.0820 = 1.00$$

$$2.62 \text{ g O} \times \frac{1 \text{ mole O}}{16.00 \text{ g O}} = 0.164 \text{ mole O}; \quad \div 0.0820 = 2.00$$

---

3. A compound contains 1.035×10^{24} atoms of element X, 0.4300 mole of element Y, and 1.150 moles of element Z. What is the empirical formula of the compound?

Your answer:

Solution: $X_{12}Y_3Z_8$

$$1.035 \times 10^{24} \text{ atoms X} \times \frac{1 \text{ mole X}}{6.02 \times 10^{23} \text{ atoms X}} = 1.72 \text{ moles X}$$

$$1.72 \text{ moles X} \div 0.4300 = 4.00; \quad \times 3 = 12.0$$

$$0.4300 \text{ mole Y} \div 0.4300 = 1.000; \quad \times 3 = 3.000$$

$$1.150 \text{ moles Z} \div 0.4300 = 2.674; \quad \times 3 = 8.022$$

Determine the empirical and the molecular formulas from the information provided in each problem.

1. 0.108 g of boron, 0.0100 mole of nitrogen atoms, and 1.205×10^{22} atoms of hydrogen are present in a sample of a compound. The molecular weight of the compound is approximately 81. What are the empirical and molecular formulas for the compound?

Your answer:

Solution: BH_2N; $B_3H_6N_3$

$$0.108 \text{ g B} \times \frac{1 \text{ mole B}}{10.81 \text{ g B}} = 0.00999 \text{ mole B}; \quad \div 0.00999 = 1.00$$

0.0100 mole N; ÷ 0.00999 = 1.00

$$1.205 \times 10^{22} \text{ atoms H} \times \frac{1 \text{ mole H}}{6.02 \times 10^{23} \text{ atoms H}} = 0.0200 \text{ mole H; } \div 0.00999 = 2.00$$

The empirical formula for the compound is BH_2N.
How are the empirical formula weight and the molecular weight of the same compound related? If they are not the same, then the molecular weight is a multiple of the empirical formula weight. Divide the molecular weight by the empirical formula weight to get this multiple. The empirical formula weight is the sum of the atomic weights in the empirical formula.

10.81 + 14.01 + 2.016 = 26.84 amu; 81 ÷ 26.84 = 3.0

Each subscript of the empirical formula is multiplied by 3 to give the molecular formula, $B_3H_6N_3$; each molecule of this compound contains three atoms B, six atoms H, and three atoms N.

---

2. A compound that is 85.6% carbon and 14.4% hydrogen has an approximate molecular weight of 56. What are the empirical and molecular formulas for the compound?

Your answer:

Solution: CH_2; C_4H_8

$$85.6 \text{ g C} \times \frac{1 \text{ mole C}}{12.01 \text{ g C}} = 7.13 \text{ moles C;} \quad \div 7.13 = 1.00$$

$$14.4 \text{ g H} \times \frac{1 \text{ mole H}}{1.008 \text{ g H}} = 14.3 \text{ moles H;} \quad \div 7.13 = 2.01$$

The empirical formula of the compound is CH_2 and the empirical formula weight corresponding to this formula is 14.03. Therefore, 56 ÷ 14.03 = 4.0, and the molecular formula of the compound is C_4H_8.

VI. Completing and Balancing Chemical Equations

Balance the following chemical equations.

1. $H_2 + Cl_2 \rightarrow HCl$

Your answer:

Solution: $H_2 + Cl_2 \rightarrow 2\ HCl$
According to the law of conservation of mass, the number of hydrogen atoms on the reactant side must be equal to the number of hydrogen atoms on the product side, and the number of chlorine atoms on the reactant side must be equal to the number of chlorine atoms on the product side. To satisfy this requirement, a "2" is placed in front of the HCl; whenever

you balance a chemical equation, you may adjust only the coefficients of the compounds.

---

2. $SO_2 + O_2 \rightarrow SO_3$

Your answer:

Solution: $2\ SO_2 + O_2 \rightarrow 2\ SO_3$

---

3. $Fe + O_2 \rightarrow Fe_2O_3$

Your answer:

Solution: $4\ Fe + 3O_2 \rightarrow 2\ Fe_2O_3$

---

4. $KOH + H_2SO_4 \rightarrow K_2SO_4 + H_2O$

Your answer:

Solution: $2\ KOH + H_2SO_4 \rightarrow K_2SO_4 + 2\ H_2O$

---

5. $C_3H_8 + O_2 \rightarrow CO_2 + H_2O$

Your answer:

Solution: $C_3H_8 + 5O_2 \rightarrow 3\ CO_2 + 4\ H_2O$

It is easiest if you balance the atoms in the following order: carbon first, hydrogen next, and oxygen last.

---

6. $C_7H_{16} + O_2 \rightarrow CO_2 + H_2O$

Your answer:

Solution: $C_7H_{16} + 11\ O_2 \rightarrow 7\ CO_2 + 8\ H_2O$

---

7. $C_7H_{14} + O_2 \rightarrow CO_2 + H_2O$

Your answer:

Solution: $2\ C_7H_{14} + 21\ O_2 \rightarrow 14\ CO_2 + 14\ H_2O$

When you balance in the order carbon, hydrogen, and oxygen, you obtain the following results initially.

$$C_7H_{14} + \frac{21}{2}\ O_2 \rightarrow 7\ CO_2 + 7\ H_2O$$

Multiply all coefficients by the number in the denominator to remove the fraction because a correctly balanced equation has whole-number coefficients.

---

8. $C_6H_{14} + O_2 \rightarrow CO_2 + H_2O$

Your answer:

Solution: $2\ C_6H_{14} + 19\ O_2 \rightarrow 12\ CO_2 + 14\ H_2O$

---

9. $K + H_2O \rightarrow KOH + H_2$

Your answer:

Solution: $2\ K + 2\ H_2O \rightarrow 2\ KOH + H_2$

---

10. $NH_3 + O_2 \rightarrow NO + H_2O$

Your answer:

Solution: $4\ NH_3 + 5\ O_2 \rightarrow 4\ NO + 6\ H_2O$
The NH_3 appears to be the most complicated formula in the equation, so would be a suggested starting place for balancing.

---

11. $Al_2S_3 + H_2O \rightarrow Al(OH)_3 + H_2S$

Your answer:

Solution: $Al_2S_3 + 6\ H_2O \rightarrow 2\ Al(OH)_3 + 3\ H_2S$

---

12. $Na_2CO_3 + HCl \rightarrow NaCl + CO_2 + H_2O$

Your answer:

Solution: $Na_2CO_3 + 2\ HCl \rightarrow 2\ NaCl + CO_2 + H_2O$

Complete and balance the following equations, which represent chemical reactions that occur in laboratory tests or illustrate complete combustion of hydrocarbons (compounds of hydrogen and carbon). In the reactions of two ionic substances, you may assume that the ions "trade partners" because a precipitate or a gaseous product forms.

1. $C_2H_6 + O_2 \rightarrow$

Your answer:

Solution: $2\ C_2H_6 + 7\ O_2 \rightarrow 4\ CO_2 + 6\ H_2O$
If you assume that the combustion of a hydrocarbon occurs with 100% efficiency, then the products for this type of reaction will always be carbon dioxide and water.

---

2. $KOH + NiCl_2 \rightarrow$

Your answer:

Solution: $2\ KOH + NiCl_2 \rightarrow 2\ KCl + Ni(OH)_2$

To predict the products in this type of ionic reaction, combine the cation of the first reactant with the anion of the second reactant (to form a neutral compound), and then combine the cation of the second reactant with the anion of the first reactant (to form a second neutral compound).

---

3. $Ba(NO_3)_2 + Li_2SO_4 \rightarrow$

Your answer:

Solution: $Ba(NO_3)_2 + Li_2SO_4 \rightarrow BaSO_4 + 2\ LiNO_3$

---

4. $C_6H_{14} + O_2 \rightarrow$

Your answer:

Solution: $2\ C_6H_{14} + 19\ O_2 \rightarrow 12\ CO_2 + 14\ H_2O$

---

5. $CaS + FeBr_3 \rightarrow$

Your answer:

Solution: $3\ CaS + 2\ FeBr_3 \rightarrow Fe_2S_3 + 3\ CaBr_2$

---

6. $MgCl_2 + K_2CO_3 \rightarrow$

Your answer:

Solution: $MgCl_2 + K_2CO_3 \rightarrow 2\ KCl + MgCO_3$

---

7. $Li_3PO_4 + Cu(NO_3)_2 \rightarrow$

Your answer:

Solution: $2\ Li_3PO_4 + 3\ Cu(NO_3)_2 \rightarrow 6\ LiNO_3 + Cu_3(PO_4)_2$

---

8. $C_{11}H_{24} + O_2 \rightarrow$

Your answer:

Solution: $C_{11}H_{24} + 17\ O_2 \rightarrow 11\ CO_2 + 12\ H_2O$

---

9. $Mg(C_2H_3O_2)_2 + Na_2CO_3 \rightarrow$

Your answer:

Solution: $Mg(C_2H_3O_2)_2 + Na_2CO_3 \rightarrow MgCO_3 + 2\ NaC_2H_3O_2$

---

10. MgS + HCl →

Your answer:

Solution: $MgS + 2HCl \rightarrow MgCl_2 + H_2S$
Hydrogen sulfide, H_2S is a poisonous gas that smells like rotten eggs.

---

11. $CaCO_3$ + HCl →

Your answer:

Solution: $CaCO_3 + 2\ HCl \rightarrow CaCl_2 + H_2CO_3$
$$H_2CO_3 \rightarrow H_2O + CO_2$$

Carbonic acid, H_2CO_3, is unstable and readily decomposes to water and carbon dioxide gas. Consequently, whenever H_2CO_3 occurs as a product, you should write H_2O and CO_2 instead. In a test tube reaction, you will observe the bubbling of the gaseous CO_2 from the reaction mixture.

VII. Stoichiometry

A. PROBLEMS INVOLVING CHEMICAL REACTIONS

For all stoichiometric problems, use the atomic weights from the periodic table in this book rounded to four significant figures.

1. Phosphorus, P_4, reacts with chlorine, Cl_2, to form PCl_5.

(a) Write the balanced equation for the reaction.

Your answer:

Solution: $P_4 + 10\ Cl_2 \rightarrow 4\ PCl_5$

(b) What do the coefficients in the balanced equation represent (in addition to being the numbers required to balance the equation)?

Your answer:

Solution: Because P_4 can represent either 1 molecule of P_4 or 1 mole of P_4, the coefficients in the balanced equation may be thought of either on the submicroscopic level as 1 molecule P_4 + 10 molecules Cl_2 will yield 4 molecules PCl_5 or on the macroscopic level as 1 mole P_4 + 10 moles Cl_2 will yield 4 moles PCl_5. Once a chemical equation is correctly written and balanced, the coefficients represent the mole relationships between the reactants and the products.

(c) Calculate the number of grams of chlorine that are required for complete conversion of

5.000 g of phosphorus to PCl_5. Use dimensional analysis to solve the problem and show all units.

Your answer:

Solution: 28.62 g Cl_2

$$? \text{ g } Cl_2 = \underset{(1)}{5.000 \text{ g } P_4} \times \underset{(2)}{\frac{1 \text{ mole } P_4}{123.88 \text{ g } P_4}} \times \underset{(3)}{\frac{10 \text{ moles } Cl_2}{1 \text{ mole } P_4}} \times \underset{(4)}{\frac{70.90 \text{ g } Cl_2}{1 \text{ mole } Cl_2}}$$

$$= 28.62 \text{ g } Cl_2$$

The steps utilized to solve stoichiometry problems are

(1) Start with the grams of the given substance.

(2) Convert the grams of the given substance to moles using the corresponding ratio of grams per mole (the numerical value of the atomic or molecular weight expressed in grams).

(3) Convert moles of the given substance to moles of the substance requested with the mole ratio (ratio of the coefficients in the balanced equation).

(4) Convert the moles of the requested substance to grams of the requested substance using the ratio of grams per mole for that requested substance.

---

2. Consider the reaction

$$8\ KClO_3 + C_{12}H_{22}O_{11} \rightarrow 8\ KCl + 12\ CO_2 + 11\ H_2O$$

(a) How many moles of $KClO_3$ react with 0.5000 mole $C_{12}H_{22}O_{11}$?

Your answer:

Solution: 4.000 moles $KClO_3$

$$? \text{ moles } KClO_3 = 0.5000 \text{ mole } C_{12}H_{22}O_{11} \times \frac{8 \text{ moles } KClO_3}{1 \text{ mole } C_{12}H_{22}O_{11}} = 4.000 \text{ moles } KClO_3$$

(b) How many grams of $KClO_3$ react with 0.5000 mole $C_{12}H_{22}O_{11}$?

Your answer:

Solution: 490.2 g $KClO_3$

$$? \text{ g } KClO_3 = 0.5000 \text{ mole } C_{12}H_{22}O_{11} \times \frac{8 \text{ moles } KClO_3}{1 \text{ mole } C_{12}H_{22}O_{11}} \times \frac{122.55 \text{ g } KClO_3}{1 \text{ mole } KClO_3}$$

$$= 490.2 \text{ g } KClO_3$$

(c) How many moles of CO_2 can be produced from 2.00 moles $KClO_3$?

Your answer:

Solution: 3.00 moles CO_2

$$? \text{ moles } CO_2 = 2.00 \text{ moles } KClO_3 \times \frac{12 \text{ moles } CO_2}{8 \text{ moles } KClO_3} = 3.00 \text{ moles } CO_2$$

(d) How many moles of water can be produced from 2.00 moles of $KClO_3$?

Your answer:

Solution: 2.75 moles H_2O

$$? \text{ moles } H_2O = 2.00 \text{ moles } KClO_3 \times \frac{11 \text{ moles } H_2O}{8 \text{ moles } KClO_3} = 2.75 \text{ moles } H_2O$$

(e) 2.00 moles $KClO_3$ will produce ___ g of CO_2.

Your answer:

Solution: 132 g CO_2

$$? \text{ g } CO_2 = 2.00 \text{ moles } KClO_3 \times \frac{12 \text{ moles } CO_2}{8 \text{ moles } KClO_3} \times \frac{44.01 \text{ g } CO_2}{1 \text{ mole } CO_2}$$

$$= 132 \text{ g } CO_2$$

132 g can be termed the "theoretical yield," the number of grams that can be obtained if the reaction is 100% efficient.

(f) If a 0.2450 g sample of potassium chlorate reacts with sucrose, what is the theoretically expected yield of water?

Your answer:

Solution: 0.04953 g H_2O

$$? \text{ g } H_2O = 0.2450 \text{ g } KClO_3 \times \frac{1 \text{ mole } KClO_3}{122.55 \text{ g } KClO_3} \times \frac{11 \text{ moles } H_2O}{8 \text{ moles } KClO_3} \times \frac{18.02 \text{ g } H_2O}{1 \text{ mole } H_2O}$$

$$= 0.04953 \text{ g } H_2O$$

(g) If 3.332 g of water forms in the reaction of potassium chlorate with sucrose, how many grams of $KClO_3$ reacted?

Your answer:

Solution: 16.48 g $KClO_3$

$$? \text{ g } KClO_3 = 3.332 \text{ g } H_2O \times \frac{1 \text{ mole } H_2O}{18.02 \text{ g } H_2O} \times \frac{8 \text{ moles } KClO_3}{11 \text{ moles } H_2O} \times \frac{122.55 \text{ g } KClO_3}{1 \text{ mole } KClO_3}$$

$$= 16.48 \text{ g } KClO_3$$

(h) How many grams of carbon dioxide will form when 23.45 g of water form?

Your answer:

Solution: 62.48 g CO_2

$$? \text{ g } CO_2 = 23.45 \text{ g } H_2O \times \frac{1 \text{ mole } H_2O}{18.02 \text{ g } H_2O} \times \frac{12 \text{ moles } CO_2}{11 \text{ moles } H_2O} \times \frac{44.01 \text{ g } CO_2}{1 \text{ mole } CO_2}$$

$$= 62.48 \text{ g } CO_2$$

---

3. How many moles of NO will form when 72.06 g of water form in the following reaction?

$$NH_3 + O_2 \longrightarrow NO + H_2O$$

Your answer:

Solution: 2.666 moles NO

The equation must first be balanced.

$$4\ NH_3 + 5\ O_2 \rightarrow 4\ NO + 6\ H_2O$$

$$?\ \text{moles NO} = 72.06\ \text{g}\ H_2O \times \frac{1\ \text{mole}\ H_2O}{18.02\ \text{g}\ H_2O} \times \frac{4\ \text{moles NO}}{6\ \text{moles}\ H_2O} = 2.666\ \text{moles NO}$$

---

4. Aluminum sulfide can be hydrolyzed according to the equation

$$Al_2S_3 + H_2O \rightarrow Al(OH)_3 + H_2S$$

How many grams of aluminum sulfide will exactly react with 72.06 g of water?

Your answer:

Solution: 100.1 g Al_2S_3

The balanced equation is

$$Al_2S_3 + 6\ H_2O \rightarrow 2\ Al(OH)_3 + 3\ H_2S$$

$$?\ \text{g}\ Al_2S_3 = 72.06\ \text{g}\ H_2O \times \frac{1\ \text{mole}\ H_2O}{18.02\ \text{g}\ H_2O} \times \frac{1\ \text{mole}\ Al_2S_3}{6\ \text{moles}\ H_2O} \times \frac{150.14\ \text{g}\ Al_2S_3}{1\ \text{mole}\ Al_2S_3}$$

$$= 100.1\ \text{g}\ Al_2S_3$$

---

5. How many grams of heptene, C_7H_{14}, yield 98.23 g of carbon dioxide when burned?

Your answer:

Solution: 31.31 g C_7H_{14}

The equation must first be written and balanced.

$$2\ C_7H_{14} + 21\ O_2 \longrightarrow 14\ CO_2 + 14\ H_2O$$

$$?\ g\ C_7H_{14} = 98.23\ g\ CO_2 \times \frac{1\ mole\ CO_2}{44.01\ g\ CO_2} \times \frac{2\ moles\ C_7H_{14}}{14\ moles\ CO_2} \times \frac{98.18\ g\ C_7H_{14}}{1\ mole\ C_7H_{14}}$$

$$= 31.31\ g\ C_7H_{14}$$

B. PROBLEMS RELATING ACTUAL YIELD AND THEORETICAL YIELD; DETERMINATION OF LIMITING REAGENT, THEORETICAL YIELD, AND PERCENT YIELD

For all problems in previous sections, it has been assumed that

(1) For a specified sample of one reactant the available quantity of any other reactant is sufficient for complete conversion to products.

(2) For the given sample of reactant the conversion to product is 100% efficient.

In these problems, then, the answer has always indicated the "theoretical yield" of a product or the amount of a reactant needed "theoretically."

1. 2.0 moles of magnesium and 2.5 moles of chlorine gas are mixed together.
   (a) Write the balanced equation for the reaction.

Your answer:

Solution: $Mg + Cl_2 \longrightarrow MgCl_2$

(b) Which reactant is limiting and which is in excess?

Your answer:

Solution: The balanced equation shows that magnesium and chlorine react in a one-to-one molar ratio: 2.0 moles Mg and 2.0 moles Cl_2 react to form 2.0 moles $MgCl_2$. Therefore chlorine is the "reagent in excess" (2.5 moles are provided), and magnesium is the "limiting reagent."

(c) What is the "theoretical yield" of the product in this reaction?

Your answer:

Solution: 190 g $MgCl_2$

The "theoretical yield" of $MgCl_2$ is 2.0 moles (determined in (b) from the molar ratio).

$$2.0 \text{ moles } MgCl_2 \times \frac{95.21 \text{ g } MgCl_2}{1 \text{ mole } MgCl_2} = 190 \text{ g } MgCl_2 \text{ (or } 1.9 \times 10^2 \text{ g } MgCl_2\text{)}$$

(d) When a student reacts 2.0 moles Mg with 2.0 moles Cl_2, she actually obtains 75 g $MgCl_2$. Calculate her percent yield.

Your answer:

Solution: 39%

$$\% \text{ yield} = \frac{\text{actual yield}}{\text{theoretical yield}} \times 100$$

$$= \frac{75 \text{ g}}{190 \text{ g}} \times 100 = 39\%$$

---

2. 5.0 moles of hydrogen gas and 2.0 moles of nitrogen gas are mixed under reaction conditions to form ammonia, NH_3.

   (a) Write a balanced equation for the reaction.

Your answer:

Solution: $3 \; H_2 + N_2 \rightarrow 2 \; NH_3$

(b) Which reactant is limiting and which is present in excess?

Your answer:

Solution: From the given amount of either reactant, calculate the exact amount of the other reactant needed for reaction.

$$? \text{ moles } N_2 = 5.0 \text{ moles } H_2 \times \frac{1 \text{ mole } N_2}{3 \text{ moles } H_2} = 1.7 \text{ moles } N_2$$

The nitrogen is present in excess because 2.0 moles has been provided for the reaction (more than the 1.7 moles required.) Therefore hydrogen is the "limiting reagent" and is used to calculate the "theoretical yield" of the product.

(c) Calculate the theoretical yield of ammonia in grams.

Your answer:

Solution: 57 g NH_3

$$\text{theoretical yield} = 5.0 \text{ moles } H_2 \times \frac{2 \text{ moles } NH_3}{3 \text{ moles } H_2} \times \frac{17.03 \text{ g } NH_3}{1 \text{ mole}}$$

$$= 57 \text{ g } NH_3$$

(d) What is the percent yield of NH_3 if 37 g of NH_3 is produced when this reaction is performed?

Your answer:

Solution: 65%

$$\% \text{ yield} = \frac{37 \text{ g}}{57 \text{ g}} \times 100 = 65\%$$

---

3. An aluminum wire with a mass of 0.1079 g reacts with a solution containing 0.8236 g of $CuSO_4$ according to the equation

$$Al + CuSO_4 \longrightarrow Cu + Al_2(SO_4)_3$$

(a) How much copper metal can be theoretically produced?

Your answer:

Solution: 0.3279 g Cu

First balance the equation.

$$2\ Al + 3\ CuSO_4 \longrightarrow 3\ Cu + Al_2(SO_4)_3$$

Determine the limiting reagent.

$$? \text{ g } CuSO_4 = 0.1079 \text{ g Al} \times \frac{1 \text{ mole Al}}{26.98 \text{ g}} \times \frac{3 \text{ moles } CuSO_4}{2 \text{ moles Al}} \times \frac{159.60 \text{ g } CuSO_4}{1 \text{ mole } CuSO_4}$$

$$= 0.9574 \text{ g } CuSO_4 \text{ needed for complete reaction of } 0.1079 \text{ g Al}$$

Only 0.8236 g $CuSO_4$ is available for reaction; $CuSO_4$ is the limiting reagent (and Al is present in excess).

$$\text{theoretical yield Cu} = 0.8236 \text{ g } CuSO_4 \times \frac{1 \text{ mole } CuSO_4}{159.60 \text{ g } CuSO_4} \times \frac{3 \text{ moles Cu}}{3 \text{ moles } CuSO_4} \times \frac{63.54 \text{ g Cu}}{1 \text{ mole Cu}}$$

$$= 0.3279 \text{ g Cu}$$

(b) Calculate the percent yield of Cu if 0.1967 g Cu is actually obtained when the reaction is run.

Your answer:

Solution: 59.99%

$$\% \text{ yield} = \frac{0.1967 \text{ g}}{0.3279 \text{ g}} \times 100 = 59.99\%$$

---

5 Atomic Structure and Periodicity

I. Electron Configurations and Orbital Diagrams

1. Use the periodic table in this book to write the electron configurations for each of the first 36 elements. Show the orbital diagram of the valence (outer) electrons for each of these elements.

Your answer:

Solution:

Element	Electron configuration	Valence electron(s) orbital diagram
$_{1}H$	$1s^1$	1s: ↑
$_{2}He$	$1s^2$	1s: ↑↓
$_{3}Li$	$1s^2 2s^1$	2s: ↑
$_{4}Be$	$1s^2 2s^2$	2s: ↑↓
$_{5}B$	$1s^2 2s^2 2p^1$	2s: ↑↓; 2p: ↑ __ __
$_{6}C$	$1s^2 2s^2 2p^2$	2s: ↑↓; 2p: ↑ ↑ __
$_{7}N$	$1s^2 2s^2 2p^3$	2s: ↑↓; 2p: ↑ ↑ ↑
$_{8}O$	$1s^2 2s^2 2p^4$	2s: ↑↓; 2p: ↑↓ ↑ ↑
$_{9}F$	$1s^2 2s^2 2p^5$	2s: ↑↓; 2p: ↑↓ ↑↓ ↑
$_{10}Ne$	$1s^2 2s^2 2p^6$	2s: ↑↓; 2p: ↑↓ ↑↓ ↑↓
$_{11}Na$	$1s^2 2s^2 2p^6 3s^1$	3s: ↑
$_{12}Mg$	$1s^2 2s^2 2p^6 3s^2$	3s: ↑↓
$_{13}Al$	$1s^2 2s^2 2p^6 3s^2 3p^1$	3s: ↑↓; 3p: ↑ __ __
$_{14}Si$	$1s^2 2s^2 2p^6 3s^2 3p^2$	3s: ↑↓; 3p: ↑ ↑ __
$_{15}P$	$1s^2 2s^2 2p^6 3s^2 3p^3$	3s: ↑↓; 3p: ↑ ↑ ↑
$_{16}S$	$1s^2 2s^2 2p^6 3s^2 3p^4$	3s: ↑↓; 3p: ↑↓ ↑ ↑
$_{17}Cl$	$1s^2 2s^2 2p^6 3s^2 3p^5$	3s: ↑↓; 3p: ↑↓ ↑↓ ↑
$_{18}Ar$	$1s^2 2s^2 2p^6 3s^2 3p^6$	3s: ↑↓; 3p: ↑↓ ↑↓ ↑↓
$_{19}K$	$1s^2 2s^2 2p^6 3s^2 3p^6 4s^1$	4s: ↑
$_{20}Ca$	$1s^2 2s^2 2p^6 3s^2 3p^6 4s^2$	4s: ↑↓

Valence electrons are considered to be only the electrons in the s and p subshells; however, the d subshell electrons will be shown here for the transition elements to illustrate Hund's rule.

Element	Electron configuration	Valence electron(s) orbital diagram
$_{21}Sc$	$1s^2 2s^2 2p^6 3s^2 3p^6 4s^2 3d^1$	4s: ↑↓; 3d: ↑ __ __ __ __

Element	Electron configuration	Valence electron(s) orbital diagram
$_{22}Ti$	$1s^2 2s^2 2p^6 3s^2 3p^6 4s^2 3d^2$	4s: ↑↓ 3d: ↑ ↑ — — —
$_{23}V$	$1s^2 2s^2 2p^6 3s^2 3p^6 4s^2 3d^3$	4s: ↑↓ 3d: ↑ ↑ ↑ — —
$_{24}Cr$	The configuration predicted from the periodic table is	
	$1s^2 2s^2 2p^6 3s^2 3p^6 4s^2 3d^4$	4s: ↑↓ 3d: ↑ ↑ ↑ ↑ —
	The actual configuration (shown below) has one 4s electron and five 3d electrons. This half-filled 3d subshell is more stable.	
	$1s^2 2s^2 2p^6 3s^2 3p^6 4s^1 3d^5$	4s: ↑ 3d: ↑ ↑ ↑ ↑ ↑
$_{25}Mn$	$1s^2 2s^2 2p^6 3s^2 3p^6 4s^2 3d^5$	4s: ↑↓ 3d: ↑ ↑ ↑ ↑ ↑
$_{26}Fe$	$1s^2 2s^2 2p^6 3s^2 3p^6 4s^2 3d^6$	4s: ↑↓ 3d: ↑↓ ↑ ↑ ↑ ↑
$_{27}Co$	$1s^2 2s^2 2p^6 3s^2 3p^6 4s^2 3d^7$	4s: ↑↓ 3d: ↑↓ ↑↓ ↑ ↑ ↑
$_{28}Ni$	$1s^2 2s^2 2p^6 3s^2 3p^6 4s^2 3d^8$	4s: ↑↓ 3d: ↑↓ ↑↓ ↑↓ ↑ ↑
$_{29}Cu$	The configuration predicted from the periodic table is	
	$1s^2 2s^2 2p^6 3s^2 3p^6 4s^2 3d^9$	4s: ↑↓ 3d: ↑↓ ↑↓ ↑↓ ↑↓ ↑
	The actual configuration (shown below) has a filled 3d subshell, which is more stable.	
	$1s^2 2s^2 2p^6 3s^2 3p^6 4s^1 3d^{10}$	4s: ↑ 3d: ↑↓ ↑↓ ↑↓ ↑↓ ↑↓
$_{30}Zn$	$1s^2 2s^2 2p^6 3s^2 3p^6 4s^2 3d^{10}$	4s: ↑↓ 3d: ↑↓ ↑↓ ↑↓ ↑↓ ↑↓
$_{31}Ga$	$1s^2 2s^2 2p^6 3s^2 3p^6 4s^2 3d^{10} 4p^1$	4s: ↑↓ 4p: ↑ — —
$_{32}Ge$	$1s^2 2s^2 2p^6 3s^2 3p^6 4s^2 3d^{10} 4p^2$	4s: ↑↓ 4p: ↑ ↑ —
$_{33}As$	$1s^2 2s^2 2p^6 3s^2 3p^6 4s^2 3d^{10} 4p^3$	4s: ↑↓ 4p: ↑ ↑ ↑
$_{34}Se$	$1s^2 2s^2 2p^6 3s^2 3p^6 4s^2 3d^{10} 4p^4$	4s: ↑↓ 4p: ↑↓ ↑ ↑
$_{35}Br$	$1s^2 2s^2 2p^6 3s^2 3p^6 4s^2 3d^{10} 4p^5$	4s: ↑↓ 4p: ↑↓ ↑↓ ↑
$_{36}Kr$	$1s^2 2s^2 2p^6 3s^2 3p^6 4s^2 3d^{10} 4p^6$	4s: ↑↓ 4p: ↑↓ ↑↓ ↑↓

II. Periodic Table—Quiz on Terminology

Use the periodic table for these questions.

1. What element has an outer-shell configuration of $4s^2$?

Your answer:

Solution: calcium

---

2. What element has $2s^2 2p^6$ as the configuration of the outer shell?

Your answer:

Solution: neon

---

3. What group of elements has seven electrons in the valence shell?

Your answer:

Solution: Group VIIA, which is known as the halogen family

---

4. What element has only two electrons in the 3d subshell?

Your answer:

Solution: titanium (element 22)
The 3d subshell is filled in the series of transition elements, 21-30, in period 4. Titanium is the second element in this series and therefore is the element with only two electrons in the 3d subshell.

---

5. What group of elements has one electron in a p subshell?

Your answer:

Solution: elements in Family IIIA

---

6. How many elements are in the fourth period?

Your answer:

Solution: 18
The fourth period contains elements 19-36.

---

7. How many electrons are in the fourth shell (energy level) when it is full?

Your answer:

Solution: The fourth energy level has a capacity of $2(4^2) = 32$ electrons.
Notice that the sum of the capacities of the four possible subshells (s, p, d, and f) of the fourth energy level is 32 electrons.

---

8. How many p orbitals are in a p subshell?

Your answer:

Solution: three: p_x, p_y, and p_z

---

9. How many elements within a given period are involved with filling the d subshell?

Your answer:

Solution: ten (two electrons in each of the five d orbitals)

---

10. How many orbitals are in an f subshell?

Your answer:

Solution: seven

Notice the pattern for the number of orbitals per subshell: one for s, three for p, five for d, and seven for f.

---

11. How many unpaired electrons are in the outer shell of sodium?

Your answer:

Solution: The outer configuration of the sodium atom is $3s^1$; there is one electron in the outer shell.

---

12. How many electrons are in the valence shell of each of the following: K, Br, P, Al, Bi, Ar, Na^+ (count only s and p electrons)?

Your answer:

Solution: The group number (of the "A" elements) indicates the number of valence electrons; notice how this corresponds to the electrons in the outer-shell configuration.

Element	Outer shell configuration	Group number	Valence electrons
$_{19}K$	$4s^1$	IA	1
$_{35}Br$	$4s^2 4p^5$	VIIA	7
$_{15}P$	$3s^2 3p^3$	VA	5
$_{13}Al$	$3s^2 3p^1$	IIIA	3
$_{83}Bi$	$6s^2 6p^3$	VA	5
$_{18}Ar$	$3s^2 3p^6$	0 (sometimes VIIIA)	8
$_{11}Na^+$	$3s^0$	IA	0

The sodium atom has lost its one valence electron to form the sodium ion.

---

13. How many unpaired electrons are in a vanadium (element 23) atom?

Your answer:

Solution: three (refer to the orbital diagram for vanadium in Section I).

---

14. Which element has the larger radius, sodium or rubidium (element 37)?

Your answer:

Solution: rubidium

Rubidium has one valence electron in the fifth shell; the one valence electron of sodium is in the third shell. Because rubidium has more shells than does sodium, rubidium has the larger radius. The four filled shells of electrons beneath the valence shell (in the rubidium atom) shield the outer electron from the larger nuclear charge.

---

15. Which element has the higher ionization energy, sodium or rubidium?

Your answer:

Solution: sodium

Ionization energy is the energy required to remove an electron from a gaseous state atom. Because rubidium has a larger radius than does sodium, the electron of rubidium can be more easily removed from the atom; therefore, more energy is required to remove the electron of the sodium atom, and consequently, the sodium atom has the higher ionization energy.

---

16. Which has the larger radius, lithium or fluorine?

Your answer:

Solution: lithium

Lithium and fluorine are both in the second period of the periodic table and therefore have their valence electrons in the second shell. The greater nuclear charge of the fluorine atom pulls the outer electrons closer to the nucleus, decreasing the radius; lithium has the larger radius.

---

17. Which atom has the higher ionization energy, lithium or fluorine?

Your answer:

Solution: fluorine

The atom with higher nuclear charge requires more energy for the removal of an electron from the corresponding ($n = 2$) valence shell. Thus fluorine has a higher ionization energy than lithium.

---

18. Rank the atoms in each group from greatest electronegativity to least electronegativity.
    (a) Na, O, S (b) F, N, O, Cl (c) Ge, Si, P (d) C, Al, N

Your answer:

Solution: Electronegativity is the relative attraction for electrons (electron "greediness"). The trend for electronegativity is similar to the trend for ionization energy; an atom requiring much energy for the removal of its electron has a high attraction for that electron.

(a) O > S > Na (b) F > O > N ≈ Cl (c) P > Si > Ge (d) N > C > Al

All of these answers are predictable by trend (electronegativity increases toward the right in a period and toward the top in a group).

---

19. Which atoms would be the most likely to lose electrons to form positive ions: P, Rb, Co, or Si? Which is the most metallic?

Your answer:

Solution: Positive ions are formed by elements in families IA, IIA, and (sometimes) IIIA and by the transition elements. Rb (IA) and Co (transition element) will form positive ions. Rb is the most metallic.

---

20. Which elements have a single unpaired electron in a p subshell?

Your answer:

Solution: One unpaired electron in a p subshell implies the configuration ns^2np^1, which is the configuration of IIIA elements, and ns^2np^5, which is the configuration of VIIA elements.

---

21. Which elements in period 2 are diamagnetic? Which elements in this period are paramagnetic?

Your answer:

Solution: Diamagnetic elements are those in which all the electrons are paired (the element has no unpaired electrons), and therefore the element is not attracted to a magnetic field. The diamagnetic elements in period 2 are beryllium and neon. Paramagnetic elements have at least one unpaired electron and are attracted to a magnetic field. Paramagnetic elements in period 2 are lithium, boron, carbon, nitrogen, oxygen, and fluorine.

---

22. What do helium and neon have in common?

Your answer:

Solution: Helium and neon atoms have filled valence shells and are therefore chemically unreactive.

---

23. Consider the elements Mg, Mn, Zn, and Cl: Which are representative elements? What name is given to the others? Which are metals? Which are nonmetals?

Your answer:

Solution: Mg and Cl are representative elements (they are found in "A" groups or families). Mn and Zn are transition elements. Mg, Mn, and Zn are metals. Cl is a nonmetal.

---

24. Give the electron dot formulas for each of the following atoms: nitrogen, sodium, aluminum, phosphorus, potassium, and arsenic. Arrange together the elements that have the same number of "dots." What is the special relationship among these elements?

Your answer:

Solution:

Group IA	Group IIIA	Group VA
Na·	·Ȧl·	·N̈·
K·		:Ṗ·
		·Ȧs:

Each symbol represents the kernel of the atom (the nucleus and the filled shells of electrons). Each dot represents a valence electron. Elements with the same number of valence electrons belong to the same family (or group) of elements in the periodic table. Members of a family have similar chemical properties; their chemical and physical properties exhibit trends within the given group.

---

6 Bonding and Lewis Dot Formulas; Oxidation Numbers and Oxidation - Reduction Reactions

Use the periodic table for each question.

I. Bonding and Lewis Formulas

1. For each formula shown, answer the questions and draw the Lewis dot formula. The questions represent the sequential steps needed for describing the bonding and for giving the Lewis dot formula. From the periodic table relationships and trends, predict the type of bonding that occurs: pure covalent, polar covalent, or ionic. If desired, check your conclusion with the differences in electronegativity values.

(a) H_2

Your answers:

(i) What are the group number and period number of each kind of element in the molecule?

(ii) What is the electron configuration of each atom?

(iii) What is the orbital diagram for the valence electrons?

(iv) Predict the type of bond formed between the two atoms.

(v) Draw the Lewis dot formula of H_2.

Solution:

(i) Group IA, period 1

(ii) $1s^1$

(iii) $\underset{1s}{\underline{\uparrow}}$

(iv) Because the molecule consists of two hydrogen atoms, there are identical attractions by the atoms for the pair of bonding electrons. This type of bond is called a pure covalent (sometimes, a nonpolar covalent) bond.

(v) H : H

(b) HBr

Your answers:

(i) What are the group number and period number of each kind of element in the molecule?

(ii) What is the electron configuration of each atom?

(iii) What is the orbital diagram for the valence electrons?

(iv) Predict the type of bond formed between the two atoms.

(v) Draw the Lewis dot formula of HBr.

Solution:

(i) H: Group IA, period 1; Br: Group VIIA, period 4

(ii) H: $1s^1$

Br: $1s^2 2s^2 2p^6 3s^2 3p^6 4s^2 3d^{10} 4p^5$

(iii) H: $\underset{1s}{\uparrow}$

Br: $\underset{4s}{\uparrow\downarrow}$ $\quad \underbrace{\uparrow\downarrow \quad \uparrow\downarrow \quad \uparrow}_{4p}$

(iv) Hydrogen is not a typical member of Group IA. The hydrogen atom is a very small atom with its one valence electron close to its positive nucleus and, consequently, does not easily lose an electron to form a cation, even to the most electronegative atom, fluorine. With a valence shell of seven electrons, bromine can either share electrons or form an anion. Because hydrogen and bromine do not differ sufficiently in attraction for electrons (electronegativity), they share (unequally) a pair of electrons; the bond formed by this pair of electrons is called a polar covalent bond. A polar covalent bond may be shown $^{\delta+}H{-}Br^{\delta-}$ or $\overset{\longmapsto}{H{-}Br}$ indicating that a partial positive charge exists on the hydrogen end of the bond and a partial negative charge exists on the bromine end of the bond. This molecule is said to be a dipole (or to have a dipole moment). (From the table of electronegativity values, ΔE.N. = 2.8 - 2.1 = 0.7. A difference in electronegativity values between zero and 1.7 indicates that the bond between the two atoms is a polar covalent bond.)

(v) $H:\underset{\cdot\cdot}{\overset{\cdot\cdot}{Br}}:$

(c) NaCl

Your answers:

(i) What are the group number and period number of each kind of element in the compound?

(ii) What is the electron configuration of each atom?

(iii) What is the orbital diagram for the valence electrons?

(iv) Predict the type of bond formed between the two atoms.

(v) Draw the Lewis dot formula of NaCl.

Solution:

(i) Na: Group IA, period 3; Cl: Group VIIA, period 3

(ii) Na: $1s^2 2s^2 2p^6 3s^1$

Cl: $1s^2 2s^2 2p^6 3s^2 3p^5$

(iii) Na: $\underset{3s}{\uparrow}$

Cl: $\underset{3s}{\uparrow\downarrow}$ $\quad \underbrace{\uparrow\downarrow \quad \uparrow\downarrow \quad \uparrow}_{3p}$

(iv) A sodium atom has one valence electron that it loses to form a +1 ion (cation). However, this can occur only if another atom or group of atoms is available that has space for that electron and that has sufficient attraction for electrons (electronegativity). Since chlorine is a member of Group VIIA, you can predict that it would attract the electron to form Cl^-. The attraction between any Na^+ ion and any Cl^- ion is called an ionic bond. (From a table of electronegativity values, ΔE.N. = 3.0 - 0.9 = 2.1. Because 2.1 is greater than 1.7, the bond is predicted to be ionic.)

(v) Na^+ $:\underset{\cdot\cdot}{\overset{\cdot\cdot}{Cl}}:^-$

(d) BrCl

Your answers:

(i) What are the group number and period number of each kind of element in the molecule?

(ii) What is the electron configuration of each atom?

(iii) What is the orbital diagram for the valence electrons?

(iv) Predict the type of bond formed between the two atoms.

(v) Draw the Lewis dot formula of BrCl.

Solution:

(i) Br: Group VIIA, period 4; Cl: Group VIIA, period 3

(ii) Br: $1s^22s^22p^63s^23p^64s^23d^{10}4p^5$

Cl: $1s^22s^22p^63s^23p^5$

(iii) Br: ↑↓ (4s) ↑↓ ↑↓ ↑ (4p)

Cl: ↑↓ (3s) ↑↓ ↑↓ ↑ (3p)

(iv) What kind of bond can be formed between two atoms with identical valence shells and with just one filled shell of electrons difference?

The difference in electronegativity will be slight, so the bond between them will be a polar covalent bond. Chlorine, with the fewer shells, will have a greater electronegativity value, so the dipole can be represented as $^{\delta+}Br{-}Cl^{\delta-}$. Typically, two different nonmetals will form a polar covalent bond. (For the Br-Cl molecule, ΔE.N. = 3.0 - 2.8 = 0.2. This difference is greater than zero and less than 1.7, so the bond is polar covalent.)

(v) $:\underset{\cdot\cdot}{\overset{\cdot\cdot}{Br}}:\underset{\cdot\cdot}{\overset{\cdot\cdot}{Cl}}:$

(e) $SrCl_2$

Your answers:

(i) What are the group number and period number of each kind of element in the compound?

(ii) What is the electron configuration of each atom?

(iii) What is the orbital diagram for the valence electrons?

(iv) Predict the type of bond formed between the atoms.

(v) Draw the Lewis dot formula of $SrCl_2$.

Solution:

(i) Sr: Group IIA, period 5; Cl: Group VIIA, period 3

(ii) Sr: $1s^22s^22p^63s^23p^64s^23d^{10}4p^65s^2$

Cl: $1s^22s^22p^63s^23p^5$

(iii) Sr: ↑↓ (5s)

Cl: ↑↓ (3s) ↑↓ ↑↓ ↑ (3p)

(iv) Sr, similar to other Group IIA members, forms its ion by losing two valence electrons to become a Sr^{+2} cation. A chlorine atom (Group VIIA) is predicted to have sufficient electronegativity to pull a valence electron from a strontium atom. However, the strontium atom must lose both electrons or neither, so two chlorine atoms are required to each accept one of the electrons and consequently form two chloride anions. (ΔE.N. = 3.0 - 1.0 = 2.0. Because the ΔE.N. is greater than 1.7, each bond is ionic.)

(v) $:\ddot{\underset{..}{Cl}}:^-$ Sr^{2+} $:\ddot{\underset{..}{Cl}}:^-$

(f) $KClO_4$

Your answers:

(i) What are the group number and period number of each kind of element in the compound?

(ii) What is the electron configuration of each atom?

(iii) What is the orbital diagram for the valence electrons?

(iv) Predict the type of bond formed between the atoms.

(v) Draw the Lewis dot formula of $KClO_4$. (Hint: Consider K^+ and ClO_4^-.)

Solution:

(i) K: Group IA, period 4; Cl: Group VIIA, period 3; O: Group VIA, period 2

(ii) K: $1s^22s^22p^63s^23p^64s^1$

Cl: $1s^22s^22p^63s^23p^5$

O: $1s^22s^22p^4$

(iii) K: ↑ (4s)

Cl: ↑↓ (3s) ↑↓ ↑↓ ↑ (3p)

O: ↑↓ (2s) ↑↓ ↑ ↑ (2p)

(iv) Potassium is a typical member of Group IA and therefore is predicted to lose its one valence electron to form the K^+ cation. Chlorine and oxygen, two nonmetals from groups VIIA and VIA, respectively, are expected to form polar covalent bonds between each other but to form ionic bonds with an active metal such as potassium. (For Cl and O, ΔE.N. = 3.5 - 3.0 = 0.5, polar covalent bond.)

(v) To show the Lewis dot formula for ClO_4^- (perchlorate ion), add the number of valence electrons for the chlorine (7), each oxygen (4 × 6 = 24), and the one extra electron (from the potassium atom), which gives the perchlorate ion its negative charge. Arrange the oxygens around the chlorine and draw the 32 electrons so that every atom has an octet. Be sure to show the charges on the ions.

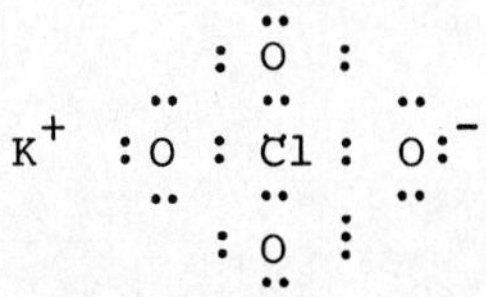

(g) Rb_2O

Your answers:

(i) What are the group number and period number of each kind of element in the compound?

(ii) What is the electron configuration of each atom?

(iii) What is the orbital diagram for the valence electrons?

(iv) Predict the type of bond formed between the atoms.

(v) Draw the Lewis dot formula of Rb_2O.

Solution:

(i) Rb: Group IA, period 5; O: Group VIA, period 2

(ii) Rb: $1s^2 2s^2 2p^6 3s^2 3p^6 4s^2 3d^{10} 4p^6 5s^1$

O: $1s^2 2s^2 2p^4$

(iii) Rb: 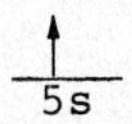

O: 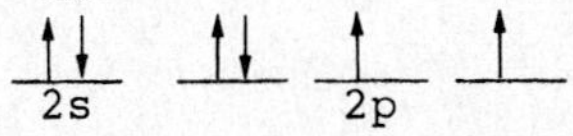

(iv) Atomic size increases down the column in Group IA; the one valence electron is farther from the positive nucleus in a larger atom and therefore is more easily lost. Rubidium forms its cation more readily than do the atoms above it (in the column). Oxygen, the smallest atom in Group VIA, has a high attraction for electrons. (In fact, it has the second highest electronegativity of all the atoms.) Therefore rubidium and oxygen are expected to form an ionic bond. Because oxygen lacks two electron and rubidium has only one to lose, there are two Rb^+ cations for every one O^{2-} anion. (ΔE.N. = 3.5 - 0.8 = 2.7. Because the ΔE.N. is greater than 1.7, the bond is ionic.)

(v) Rb^+ $:\ddot{\underset{\cdot\cdot}{O}}:^{2-}$ Rb^+

(h) CO_2

Your answers:

(i) What are the group number and period number of each kind of element in the molecule?

(ii) What is the electron configuration of each atom?

(iii) What is the orbital diagram for the valence electrons?

(iv) Predict the type of bond formed between the two atoms.

(v) Draw the Lewis dot formula of CO_2.

Solution:

(i) C: Group IVA, period 2; O: Group VIA, period 2

(ii) C: $1s^2 2s^2 2p^2$

O: $1s^2 2s^2 2p^4$

(iii) C: ↑↓ (2s) ↑ ↑ ___ (2p)

This is the orbital diagram of the orbital electrons of the carbon atom before it bonds with two oxygen atoms to form CO_2.

O: ↑↓ (2s) ↑↓ ↑ ↑ (2p)

(iv) Between the nonmetals carbon and oxygen, there is not enough difference in relative attraction for electrons to enable the atoms to form ions by transfer of electrons. Instead, the bonding electrons are shared in a polar covalent bond. (ΔE.N. = 3.5 - 2.5 = 1.0.)

(v) When the sixteen valence electrons (four from each carbon, six from each oxygen) are arranged around the atoms, forming single (sigma) bonds between the atoms, it is impossible for the carbon atom to have an octet: :Ö : C : Ö: To obtain this required octet, a pair of electrons from each oxygen is placed between each oxygen and the carbon. The four electrons between each oxygen and carbon constitute a double bond.

:O :: C :: O:

(i) MgO

Your answers:

(i) What are the group number and period number of each kind of element in the compound?

(ii) What is the electron configuration of each atom?

(iii) What is the orbital diagram for the valence electrons?

(iv) Predict the type of bond formed between the two atoms.

(v) Draw the Lewis dot formula of MgO.

Solution:

(i) Mg: Group IIA, period 3; O: Group VIA, period 2

(ii) Mg: $1s^2 2s^2 2p^6 3s^2$
O: $1s^2 2s^2 2p^4$

(iii) Mg: ↑↓ (3s)

O: ↑↓ (2s) ↑↓ ↑ ↑ (2p)

(iv) A magnesium atom has two valence electrons and would be expected to form a +2 cation if combined with a nonmetal of great enough electronegativity to attract the electrons from magnesium. Oxygen is second only to fluorine in electronegativity and therefore does form an oxide ion when reacted with magnesium. There is an ionic bond between Mg^{2+} and O^{2-}. (ΔE.N. = 3.5 - 1.2 = 2.3; ΔE.N.>1.7, so the bond is ionic.)

(v) Mg^{2+} $:\ddot{\underset{..}{O}}:^{2-}$

(j) $MgSO_4$

Your answers:

(i) What are the group number and period number of each kind of element in the compound?

(ii) What is the electron configuration of each atom?

(iii) What is the orbital diagram for the valence electrons?

(iv) Predict the type of bond formed between the atoms.

(v) Draw the Lewis dot formula of $MgSO_4$.

Solution:

(i) Mg: Group IIA, period 3; S: Group VIA, period 3; O: Group VIA, period 2

(ii) Mg: $1s^2 2s^2 2p^6 3s^2$
S: $1s^2 2s^2 2p^6 3s^2 3p^4$
O: $1s^2 2s^2 2p^4$

(iii) Mg: ↑↓ (3s)

S: ↑↓ (3s) ↑↓ ↑ ↑ (3p)

O: ↑↓ (2s) ↑↓ ↑ ↑ (2p)

(iv) As in the previous example, magnesium will lose its two valence electrons to oxygen and form an ionic bond between Mg^{2+} and SO_4^{2-}. Sulfur is predicted to have a lower electronegativity than oxygen, so the sulfur-oxygen bond will be polar covalent.

(v) Mg^{2+} 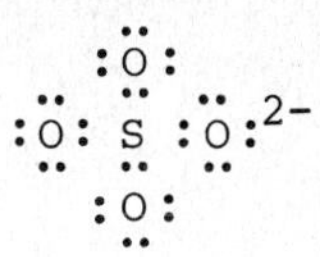

(k) CaF_2

Your answers:

(i) What are the group number and period number of each kind of element in the compound?

(ii) What is the electron configuration of each atom?

(iii) What is the orbital diagram for the valence electrons?

(iv) Predict the type of bond formed between the atoms.

(v) Draw the Lewis dot formula of CaF_2.

Solution:

(i) Ca: Group IIA, period 4; F: Group VIIA, period 2

(ii) Ca: $1s^2 2s^2 2p^6 3s^2 3p^6 4s^2$

F: $1s^2 2s^2 2p^5$

(iii) Ca: $\underline{\uparrow\downarrow}$ (4s)

F: $\underline{\uparrow\downarrow}$ (2s) $\underline{\uparrow\downarrow}$ $\underline{\uparrow\downarrow}$ $\underline{\uparrow\ }$ (2p)

(iv) An ionic bond is formed between calcium (an active metal of Group IIA) and fluorine (the smallest atom of VIIA, and thus the most electronegative atom). Because the calcium atom has two valence electrons to lose, two fluorine atoms are required to each accept one of the electrons and become fluoride ions. (ΔE.N. = 4.0 - 1.0 = 3.0.)

(v) $:\ddot{\underset{..}{F}}:^-$ Ca^{2+} $:\ddot{\underset{..}{F}}:^-$

(l) H_2O

Your answers:

(i) What are the group number and period number of each kind of element in the molecule?

(ii) What is the electron configuration of each atom?

(iii) What is the orbital diagram for the valence electrons?

(iv) Predict the type of bond formed between the atoms.

(v) Draw the Lewis dot formula of H_2O.

Solution:

(i) H: Group IA, period 1; O: Group VIA, period 2

(ii) H: $1s^1$

O: $1s^2 2s^2 2p^4$

(iii) H: $\underline{\uparrow\ }$ (1s)

O: $\underline{\uparrow\downarrow}$ (2s) $\underline{\uparrow\downarrow}$ $\underline{\uparrow\ }$ $\underline{\uparrow\ }$ (2p)

(iv) Although oxygen has a great attraction for electrons, it will not form an ionic bond with hydrogen because hydrogen has a stronger hold on its valence electron than do the other elements in IA. (Hydrogen is not a typical member of IA; its actual electronegativity value is between the values of boron and carbon.) Each hydrogen-oxygen bond in H_2O is polar covalent. (ΔE.N. = 3.5 - 2.1 = 1.4.)

(v) 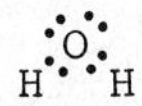

Notice that of the four pairs of electrons in the valence shell of oxygen in this molecule, two of the pairs are shared with hydrogen and are therefore called "bonding pairs." The other two pairs are called "lone pairs" or "nonbonding pairs." These "lone pairs" of electrons play a major role in determining the chemical and physical properties of water.

(m) NH_3

Your answers:

(i) What are the group number and period number of each kind of element in the molecule?

(ii) What is the electron configuration of each atom?

(iii) What is the orbital diagram for the valence electrons?

(iv) Predict the type of bond formed between the atoms.

(v) Draw the Lewis dot formula of NH_3.

Solution:

(i) N: Group VA, period 2; H: Group IA, period 1

(ii) N: $1s^2 2s^2 2p^3$

H: $1s^1$

(iii) N: 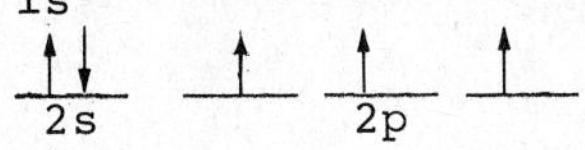

H: 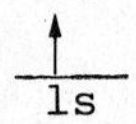

(iv) The nitrogen atom is a highly electronegative atom, but it forms a polar covalent (not ionic) bond with a hydrogen atom because both nitrogen and hydrogen are non-metals. (ΔE.N. = 3.0 - 2.1 = 0.9.)

(v) 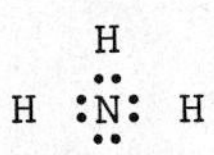

The nitrogen atom has one "lone pair" and three "bonding pairs" of electrons. The "lone pair" helps to determine the chemical and physical properties of ammonia.

(n) C_2H_6

Your answers:

(i) What are the group number and period number of each kind of element in the molecule?

(ii) What is the electron configuration of each atom?

(iii) What is the orbital diagram for the valence electrons?

(iv) Predict the type of bond formed between the two carbons and between a carbon and a hydrogen.

(v) Draw the Lewis dot formula of C_2H_6.

Solution:

(i) C: Group IVA, period 2; H: Group IA, period 1

(ii) C: $1s^22s^22p^2$

H: $1s^1$

(iii) C: ↑↓ (2s) ↑ ↑ __ (2p)

H: ↑ (1s)

(iv) The bond between the two carbon atoms is a pure covalent bond. The difference in electronegativity between C and H is 0.4; the bond between the atoms is (barely) polar covalent.

(v)
```
    H  H
    ..  ..
H : C : C : H
    ..  ..
    H  H
```

---

2. For each of the molecules indicated, answer the series of questions.

(a) CH_4

Your answers:

(i) How many valence electrons does each atom in the molecule have?

(ii) Draw the Lewis dot structure of the molecule.

(iii) Sum the number of atoms bonded to the central atom and the number of lone pairs (nonbonding pairs) of electrons around the central atom.

(iv) How many orbitals are required to accommodate the electron pairs around the central atom?

(v) What type of hybridization does the central atom require?

(vi) What is the orbital geometry (spatial arrangement) of this type of hybridization?

(vii) What are the bond angles predicted by this type of hybridization?

(viii) What shape of the molecule is specified by the arrangement of the atoms?

(ix) Are the bonds polar? Is the molecule polar?

Refer to your text for an orbital diagram of this molecule. Notice that the elemental symbol is used to represent the nucleus and the nonvalence electrons and that the orbitals containing any lone pair (nonbonding) electrons are shown for the central atom, but orbitals containing the lone pair electrons on any other atom (chlorine, oxygen, etc.) are not shown.

Solution:

(i) C has four valence electrons; H has one valence electron.

(ii)
```
   H
   ··
H : C : H
   ··
   H
```

(iii) Four atoms bonded to central atom + 0 lone pair of electrons = 4.

(iv) Four orbitals are needed because there are four (atoms plus lone pairs of electrons) around the central atom.

(v) sp^3. Consider the valence shell electrons of carbon (the central atom)

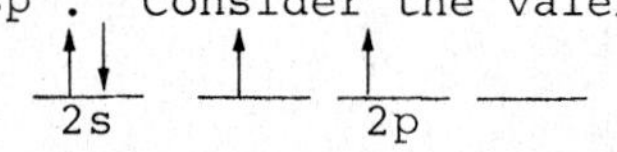

Carbon appears to permit the formation of only two bonds because there are only two half-filled orbitals available for bonding. However, in the previous two questions, you calculated a requirement for four orbitals; to provide these four orbitals, you can hypothesize hybrid orbitals. A hybrid orbital is a blend of the available atomic orbitals; by hypothetically "blending" the one s and three p orbitals, you can postulate four new hybrid orbitals. These hypothetical orbitals are called sp^3 hybrid orbitals, to indicate that each has 1/4 s character (in shape and energy) and 3/4 p character (in shape and energy).

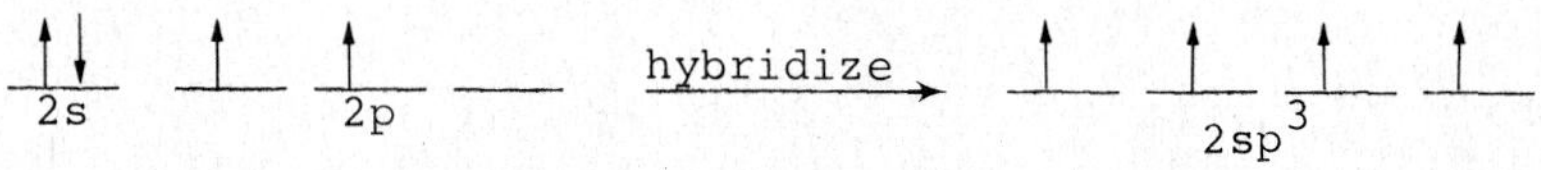

Observe that by hybridizing the atomic orbitals into hybrid orbitals, you provide four bonding orbitals (and can therefore fulfill the requirement for four orbitals).

(vi) The four electron pairs in the four hybrid orbitals repel each other and, consequently, orient themselves in three-dimensional space as far from each other as possible, generating a tetrahedral geometry.

(vii) 109.5° ($109^\circ 28'$).

(viii) Because all four orbitals are involved in bonds, the shape of the molecule is tetrahedral (the same as the orbital geometry).

(ix) The difference in electronegativity between the C and H is 0.4; the bond between these atoms is (barely) polar covalent. However, the molecule is not polar because the symmetry of the molecule permits the cancellation of the dipoles in the bonds.

(b) $CHCl_3$ (chloroform)

Your answers:

(i) How many valence electrons does each atom in the molecule have?

(ii) Draw the Lewis dot structure of the molecule.

(iii) Sum the number of atoms bonded to the central atom and the number of lone pairs of electrons around the central atom.

(iv) How many orbitals are required to accommodate the electron pairs around the central atom?

(v) What type of hybridization does the central atom require?

(vi) What is the orbital geometry (spatial arrangement) of this type of hybridization?

(vii) What are the bond angles predicted by this type of hybridization?

(viii) What shape of the molecule is specified by the arrangement of the atoms?

(ix) Are the bonds polar? Is the molecule polar?

Refer to your text for an orbital diagram of a similar molecule.

Solution:

(i) C has four valence electrons; H has one valence electron; Cl has seven valence electrons.

(ii)
```
     ..
    :Cl:
     ..  ..
  H :C : Cl:
     ..  ..
    :Cl:
     ..
```

(iii) Four atoms bonded to central atom + 0 lone pair of electrons = 4.

(iv) Four orbitals are needed because there are four (atoms plus lone pairs of electrons) around the central atom.

(v) sp^3.

(vi) The four electron pairs in the four hybrid orbitals repel each other and, consequently, orient themselves in three-dimensional space as far from each other as possible, generating a tetrahedral geometry.

(vii) 109.5°.

(viii) Because all four orbitals are involved in bonds, the shape of the molecule is tetrahedral (the same as the orbital geometry).

(ix) The difference in electronegativity between C and H is 0.4; the bond between the C and the H is (barely) polar covalent. The difference in electronegativity between C and Cl is 0.9; the bond between these atoms is polar covalent. The molecule is polar because the dipoles of the bonds do not cancel (Cl is much more electronegative than is H).

(c) NH_3

Your answers:

(i) How many valence electrons does each atom in the molecule have?

(ii) Draw the Lewis dot structure of the molecule.

(iii) Sum the number of atoms bonded to the central atom and the number of lone pairs of electrons around the central atom.

(iv) How many orbitals are required to accommodate the electron pairs around the central atom?

(v) What type of hybridization does the central atom require?

(vi) What is the orbital geometry (spatial arrangement) of this type of hybridization?

(vii) What are the bond angles predicted by this type of hybridization?

(viii) What shape of the molecule is specified by the arrangement of the atoms?

(ix) Are the bonds polar? Is the molecule polar?

Refer to your text for an orbital diagram of a similar molecule.

<u>Solution</u>:

(i) N has five valence electrons; H has one valence electron.

(ii)
```
   ..
H :N: H
   ..
   H
```

(iii) Three atoms bonded to central atom + 1 lone pair of electrons = 4.

(iv) Four orbitals are needed because there are four (atoms plus lone pairs of electrons) around the central atom.

(v) sp^3.

(vi) Tetrahedral.

(vii) The bond angle is slightly less than 109.5° (107°) due to lone-pair electron repulsion of the bonding pairs.

(viii) Trigonal pyramidal.

(ix) The bonds and the molecule are polar.

(d) NF_3

Your answers:

(i) How many valence electrons does each atom in the molecule have?

(ii) Draw the Lewis dot structure of the molecule.

(iii) Sum the number of atoms bonded to the central atom and the number of lone pairs of electrons around the central atom.

(iv) How many orbitals are required to accommodate the electron pairs around the central atom?

(v) What type of hybridization does the central atom require?

(vi) What is the orbital geometry (spatial arrangement) of this type of hybridization?

(vii) What are the bond angles predicted by this type of hybridization?

(viii) What shape of the molecule is specified by the arrangement of the atoms?

(ix) Are the bonds polar? Is the molecule polar?

Refer to your text for an orbital diagram of a similar molecule.

<u>Solution</u>:

(i) N has five valence electrons; F has seven valence electrons.

(ii)
```
   ..
F :N: F
   ..
   F
```

(iii) Three atoms bonded to central atom + 1 lone pair of electrons = 4.
(iv) Four orbitals.
(v) sp^3.
(vi) Tetrahedral.
(vii) The bond angle is somewhat less than 109.5° due to lone-pair electron repulsion.
(viii) Trigonal pyramidal.
(ix) The bonds and the molecule are polar.

(e) H_2O

Your answers:

(i) How many valence electrons does each atom in the molecule have?

(ii) Draw the Lewis dot structure of the molecule.

(iii) Sum the number of atoms bonded to the central atom and the number of lone pairs of electrons around the central atom.

(iv) How many orbitals are required to accommodate the electron pairs around the central atom?

(v) What type of hybridization does the central atom require?

(vi) What is the orbital geometry (spatial arrangement) of this type of hybridization?

(vii) What are the bond angles predicted by this type of hybridization?

(viii) What shape of the molecule is specified by the arrangement of the atoms?

(ix) Are the bonds polar? Is the molecule polar?

Refer to your text for an orbital diagram of a similar molecule.

Solution:

(i) H has one valence electron; O has six valence electrons.
(ii)
```
      ..
H  :O:
      ..
      H
```
(iii) Two atoms bonded to central atom + 2 lone pairs of electrons = 4.
(iv) Four orbitals.
(v) sp^3.
(vi) Tetrahedral.
(vii) The bond angle is slightly less than 109.5° (104.5°) due to lone-pair electron repulsion of the bonding pairs.
(viii) Bent or angular.
(ix) The bonds and the molecule are polar.

(f) CH_2O

Your answers:

(i) How many valence electrons does each atom in the molecule have?

(ii) Draw the Lewis dot structure of the molecule.

(iii) Sum the number of atoms bonded to the central atom and the number of lone pairs of electrons around the central atom.

(iv) How many orbitals are required to accommodate the electron pairs around the central atom?

(v) What type of hybridization does the central atom require?

(vi) What is the orbital geometry (spatial arrangement) of this type of hybridization?

(vii) What are the bond angles predicted by this type of hybridization.

(viii) What shape of the molecule is specified by the arrangement of the atoms?

(ix) Are the bonds polar? Is the molecule polar?

Refer to your text for an orbital diagram of a similar molecule.

Solution:

(i) H has one valence electron; C has four valence electrons; O has six valence electrons.

(ii) 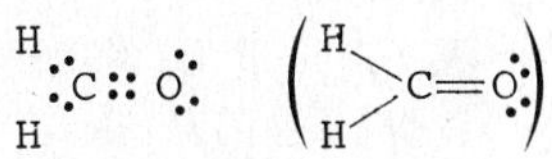

(iii) Three atoms bonded to central atom + 0 lone pair of electrons = 3.

(iv) Three orbitals.

(v) sp^2. To provide three orbitals, we hypothetically "blend" the one s and two of the three p orbitals from the valence shell of carbon. These hypothetical orbitals are called sp^2 hybrid orbitals, to indicate that each has 1/3 s character (in shape and energy) and 2/3 p character (in shape and energy). The leftover p orbital overlaps (sideways) with a p orbital from the oxygen atom to form a pi bond.

(vi) The orbital geometry predicted for the atoms involved in the sigma bonds is trigonal planar; however, there is electron density from the overlapping p orbitals (forming the pi bond) above and below the plane formed by the trigonal planar structure.

(vii) 120°.

(viii) Trigonal planar.

(ix) The bonds are polar and the molecule is polar.

(g) C_2H_2

Your answers:

(i) How many valence electrons does each atom in the molecule have?

(ii) Draw the Lewis dot structure of the molecule.

(iii) Sum the number of atoms bonded to the central atom and the number of lone pairs of electrons around the central atom.

(iv) How many orbitals are required to accommodate the electron pairs around the central atom?

(v) What type of hybridization does the central atom require?

(vi) What is the orbital geometry (spatial arrangement) of this type of hybridization?

(vii) What are the bond angles predicted by this type of hybridization?

(viii) What shape of the molecule is specified by the arrangement of the atoms?

(ix) Are the bonds polar? Is the molecule polar?

Refer to your text for an orbital diagram of a similar molecule.

Solution:

(i) C has four valence electrons; H has one valence electron.

(ii) H : C ::: C : H (H—C≡C—H)

(iii) Two atoms bonded to central atom (consider just one of the carbon atoms) + 0 lone pair electrons = 2.

(iv) Two orbitals.

(v) sp. To provide two orbitals, we hypothetically "blend" the one s and one of the three p orbitals from the valence shell of carbon. These hypothetical orbitals are called sp hybrid orbitals, to indicate that each has 1/2 s character (in shape and energy) and 1/2 p character (in shape and energy). Each of the leftover p orbitals overlaps (sideways) with the leftover p orbitals from the adjacent carbon atom to form two pi bonds.

(vi) Linear.

(vii) 180°.

(viii) Linear.

(ix) The bond between each carbon and hydrogen is (slightly) polar, but because the molecule is linear, the dipoles from these bonds cancel and the molecule is nonpolar.

(h) CO_2

Your answers:

(i) How many valence electrons does each atom in the molecule have?

(ii) Draw the Lewis dot structure of the molecule.

(iii) Sum the number of atoms bonded to the central atom and the number of lone pairs of electrons around the central atom.

(iv) How many orbitals are required to accommodate the electron pairs around the central atom?

(v) What type of hybridization does the central atom require?

(vi) What is the orbital geometry (spatial arrangement) of this type of hybridization?

(vii) What are the bond angles predicted by this type of hybridization?

(viii) What shape of the molecule is specified by the arrangement of the atoms?

(ix) Are the bonds polar? Is the molecule polar?

Solution:

(i) C has four valence electrons; O has six valence electrons.

(ii) :O::C::O:

(iii) Two atoms bonded to the central atom + 0 lone pair of electrons = 2.

(iv) Two orbitals.

(v) sp.

(vi) Linear.

(vii) 180°.

(viii) Linear.

(ix) The bond between each carbon and hydrogen is polar, but because the molecule is linear, the dipoles from these bonds cancel and the molecule is nonpolar.

II. Oxidation Numbers (Oxidation States)

1. Write the oxidation number of each kind of atom above its symbol, and state the rule that applies to the given type of formula.

(a) Na, Fe, P_4, O_2

Your answer:

Solution: $\overset{0}{Na}$, $\overset{0}{Fe}$, $\overset{0}{P}_4$, $\overset{0}{O}_2$

The oxidation number of each element in these formulas is zero. The oxidation number (also called the oxidation state) of any atom in its elemental form is zero.

(b) Cu^{2+}, Cl^-, H^+

Your answer:

Solution: $\overset{+2}{Cu}{}^{2+}$, $\overset{-1}{Cl}{}^-$, $\overset{+1}{H}{}^+$

The oxidation number of the element in a monoatomic ion is the same as the charge on the ion.

(c) NaI, FeS, N_2O_5, NH_3

Your answer:

Solution:

+1 −1	+2 −2	+5 −2	−3 +1
Na I,	Fe S,	$N_2 O_5$,	$N H_3$
		$2x - 10 = 0$	$-3 + 3x = 0$
		$2x = 10$	$3x = 3$
		$x = +5$	$x = +1$

The sum of the oxidation numbers in a neutral compound equals zero. Assume that the most electronegative atom fills its valence shell; then calculate the remaining "apparent charge" on the more electropositive kind of atom or atoms. Metals of families IA and IIA always have the oxidation number corresponding to the loss of all valence electrons. Hydrogen has a +1 oxidation state unless it is combined with the active metals of IA or IIA (i.e., in NaH, the oxidation number of sodium is +1, and the oxidation number of hydrogen is −1). Some transition metals and some nonmetals show different positive states in their various compounds.

(d) SO_3^{2-}, $S_2O_3^{2-}$, NH_4^+, ClO_4^-

Your answer:

Solution:

+4 −2	+2 −2	−3 +1	+7 −2
$S O_3^{2-}$	$S_2 O_3^{2-}$	$N H_4^+$	$Cl O_4^-$
$x - 6 = -2$	$2x - 6 = -2$	$-3 + 4x = +1$	$x - 8 = -1$
$x = +4$	$2x = 4$	$4x = 4$	$x = +7$
	$x = +2$	$x = +1$	

The sum of all oxidation numbers in complex ions equals the charge on the ion. In each ion the more electronegative atom is assumed to have a filled valence shell.

(e) $K_2Cr_2O_7$, $RbMnO_4$, OF_2

Your answer:

Solution:

+1 +6 −2	+1 +7 −2	+2 −1
$K_2 Cr_2 O_7$	Rb Mn O_4	O F_2
$2 + 2x - 14 = 0$	$1 + x - 8 = 0$	
$2x = 12$	$x = +7$	
$x = +6$		

Write the oxidation number for the most electronegative element and the most electropositive element. Then calculate the value for any remaining elements. OF_2 is an unusual example in which oxygen is combined with a more electronegative element than itself.

III. Definition and Identification of Oxidation and Reduction

1. An element of Group IIA typically reacts as shown for magnesium.

   $Mg \longrightarrow Mg^{2+} + 2\ e^-$.

   Does this represent an oxidation or a reduction?

Your answer:

Solution: A magnesium atom loses two electrons; consequently its oxidation number (oxidation state) increases. This is known as oxidation.

---

2. Write a balanced equation that shows a bromine molecule gaining electrons to form bromide ions. Is this an oxidation or a reduction?

Your answer:

Solution: $Br_2 + 2\ e^- \longrightarrow 2\ Br^-$

Reduction is the gain of electrons (and the consequent decrease in oxidation number).

---

3. Oxidation and reduction occur simultaneously. The number of electrons lost must equal the number of electrons gained in the balanced chemical equation. Write a balanced equation for the reaction of magnesium and bromine.

Your answer:

Solution: $Mg + Br_2 \longrightarrow Mg^{2+} + 2\ Br^-$

---

4. The following equations are oxidation-reduction reactions (also called redox reactions). Write the oxidation number above the element in each formula. Indicate which elements are oxidized and which are reduced.

   (a) $2\ Al + 6\ HCl \longrightarrow 2\ AlCl_3 + 3\ H_2$

Your answer:

Solution: $2\ \overset{0}{Al} + 6\overset{+1}{H}\ \overset{-1}{Cl} \longrightarrow 2\ \overset{+3}{Al}\ \overset{-1}{Cl_3} + 3\ \overset{0}{H_2}$

Al is oxidized because each atom of Al loses three electrons. H^+ is reduced because each ion of H^+ gains one electron.

(b) $2\ C_5H_{10} + 15\ O_2 \longrightarrow 10\ CO_2 + 10\ H_2O$

Your answer:

Solution: $2\ \overset{-2}{C}_5\overset{+1}{H}_{10} + 15\ \overset{0}{O}_2 \longrightarrow 10\ \overset{+4}{C}\ \overset{-2}{O}_2 + 10\ \overset{+1}{H}_2\overset{-2}{O}$

Carbon (in C_5H_{10}) is oxidized; each atom loses six electrons. Oxygen is reduced; each atom gains two electrons.

(c) $10\ HClO + P_4 + 6\ H_2O \longrightarrow 4\ H_3PO_4 + 10\ HCl$

Your answer:

Solution: $10\ \overset{+1}{H}\ \overset{+1}{Cl}\ \overset{-2}{O} + \overset{0}{P}_4 + 6\ \overset{+1}{H}_2\overset{-2}{O} \longrightarrow 4\ \overset{+1}{H}_3\overset{+5}{P}\ \overset{-2}{O}_4 + 10\ \overset{+1}{H}\ \overset{-1}{Cl}$

Phosphorus is oxidized; each atom loses five electrons and the oxidation state of each P atom increases from 0 to +5. Chlorine (in HClO) is reduced; each atom gains two electrons, and the oxidation state decreases from +1 to -1.

IV. Recognition of Oxidizing and Reducing Agents

1. What is an oxidizing agent? What is a reducing agent? Use the reactants of the equation shown to illustrate your answers.

$2\ Al + 3\ Br_2 \longrightarrow 2\ AlBr_3$

Your answer:

Solution: Indicate the oxidation numbers above each element in the formulas. Next decide which element is oxidized and which element is reduced. (It is possible, although not common, to have more than one element oxidized or more than one element reduced in a reaction.)

$2\ \overset{0}{Al} + 3\ \overset{0}{Br}_2 \longrightarrow 2\ \overset{+3}{Al}\overset{-1}{Br}_3$

Al is oxidized. What caused the oxidation? Br_2 caused the oxidation to occur and is therefore called the oxidizing agent; Br_2 is itself reduced (it gains electrons). Al causes the reduction of Br_2; so Al is called the reducing agent and is itself oxidized (it loses electrons).

---

2. Identify the oxidizing and reducing agents in the reactions shown in question 4 of III.

Your answer:

(a)

(b)

(c)

Solution:
(a) Al is the reducing agent, HCl is the oxidizing agent.
(b) C_5H_{10} is the reducing agent, O_2 is the oxidizing agent.
(c) HClO is the oxidizing agent, P_4 is the reducing agent.

---

3. Indicate the oxidation numbers of each element in the following equation. Also indicate the oxidizing and reducing agents.

$$3\ N_2O_4 + 2\ H_2O \longrightarrow 4\ HNO_3 + 2\ NO$$

Your answer:

Solution: $3\ \overset{+4}{N_2}\overset{-2}{O_4} + 2\ \overset{+1}{H_2}\overset{-2}{O} \longrightarrow 4\ \overset{+1}{H}\ \overset{+5}{N}\ \overset{-2}{O_3} + 2\ \overset{+2}{N}\ \overset{-2}{O}$

Some of the nitrogen atoms in N_2O_4 molecules are oxidized to the +5 oxidation state in HNO_3, and some are reduced to the +2 oxidation state in NO; therefore N_2O_4 is both the reducing agent and the oxidizing agent.

V. Oxidation and Reduction of Organic Compounds

In organic chemistry (the study of carbon compounds) oxidation is seen as the increase (by carbon) of oxygen and/or the decrease (by carbon) of hydrogen, and reduction is seen as the decrease of oxygen and/or the increase of hydrogen. Does this agree with the definitions of oxidation and reduction you have been using? Consider these two equations in your answer.

$$2\ CH_3OH + 2\ O_2 \longrightarrow 2\ H_2CO + 2\ H_2O$$

$$C_2H_2 + 2\ H_2 \longrightarrow C_2H_6$$

Your answer:

$$\overset{-2}{\mathrm{C}}\,\overset{+1}{\mathrm{H_3}}\,\overset{-2}{\mathrm{O}}\,\overset{+1}{\mathrm{H}}$$

Solution: $2\,\overset{-2}{C}\overset{+1}{H_3}\overset{-2}{O}\overset{+1}{H} + 2\,\overset{0}{O_2} \longrightarrow 2\,\overset{+1}{H_2}\overset{0}{C}\overset{-2}{O} + 2\,\overset{+1}{H_2}\overset{-2}{O}$

The carbon in CH_3OH increases in oxidation number from -2 to 0 (in H_2CO). Notice that the CH_3OH loses hydrogens when it becomes the product H_2CO. Thus the carbon in CH_3OH has been oxidized.

$\overset{-1}{C_2}\overset{+1}{H_2} + 2\,\overset{0}{H_2} \longrightarrow \overset{-3}{C_2}\overset{+1}{H_6}$

The carbon in C_2H_2 decreases in oxidation number when it becomes C_2H_6. The carbon also gains hydrogens, and therefore is reduced.

7 The Gas Laws

I. The Combined Gas Laws

For each problem name the law or laws that apply. Express the law in words and in equation. Explain how the kinetic molecular theory accounts for the principle used in solving the problem.

1. A sample of carbon dioxide occupies 11.4 L at STP. What volume would it occupy at $25^{o}C$ and 76 cm Hg?

Your answer:

Solution: 12.4 L

First arrange the given information in table form.

Initial conditions	Final conditions
V_1 = 11.4 L	V_2 = ?
P_1 = 1 atm*	P_2 = 76 cm Hg = 1 atm (same as P_1)
T_1 + $0^{o}C$ + 273° = 273 K*	T_2 = $25^{o}C$ + 273° = 298 K

*STP means standard temperature ($0^{o}C$ or 273 K) and pressure (76.0 cm Hg or 760 torr).

At constant pressure and quantity (moles) of gas, the volume and the absolute temperature are directly proportional. This is Charles' law, and it can be expressed as $\frac{V}{T} = k$. From the combined gas law, $\frac{P_1V_1}{T_1} = \frac{P_2V_2}{T_2}$, you can cancel the pressures because they are constant, to obtain the relationship: $\frac{V_1}{T_1} = \frac{V_2}{T_2}$.

$$V_2 = \frac{V_1 \times T_2}{T_1} = \frac{11.4 \text{ L} \times 298 \text{ K}}{273 \text{ K}} = 12.4 \text{ L}$$

According to the kinetic molecular theory, the absolute temperature and the average kinetic energy of a gas sample are directly proportional. Because collisions of molecules with the container involve no loss of kinetic energy, the volume of the gas sample must increase proportionately to the absolute temperature so that the pressure will remain constant.

---

2. A sample of gas occupies 5.93 L at $50.0^{o}C$ and 1.60 atm pressure. What volume does it occupy at $200^{o}C$ and 1.00 atm pressure?

Your answer:

Solution:

Initial conditions	Final conditions
$V_1 = 5.93$ L	$V_2 = ?$
$P_1 = 1.60$ atm	$P_2 = 1.00$ atm
$T_1 = 50.0^\circ C + 273^\circ = 323$ K	$T_2 = 200^\circ C + 273^\circ = 473$ K

For a constant mass of gas, the volume of the sample is directly proportional to the absolute temperature (Charles' law) and inversely proportional to the pressure (Boyle's law). The required equation for this problem is the combined gas law equation;

$\frac{P_1V_1}{T_1} = \frac{P_2V_2}{T_2}$, from which

$$V_2 = \frac{P_1 \times V_1 \times T_2}{T_1 \times P_2} = \frac{1.60 \text{ atm} \times 5.93 \text{ L} \times 473 \text{ K}}{323 \text{ K} \times 1.00 \text{ atm}}$$

$$= 13.9 \text{ L}$$

According to the kinetic molecular theory, as the temperature of a sample of gas is increased, the molecules acquire a proportionately higher kinetic energy. Consequently the molecules move faster and either occupy more space or exert more pressure. In the system described by these data, the volume is increased by the decrease of pressure and by the increase of temperature.

---

3. A bicycle tire correctly inflated on a warm $38^\circ C$ day looks "flat" on a $2.0^\circ C$ day. If you assume that the volume of the tire is constant, by what ratio should you multiply the original pressure to calculate the pressure in the "flat" tire? (Choose from among the following: $\frac{2}{38}$, $\frac{38}{2}$, $\frac{275}{311}$, $\frac{311}{275}$, or none of these.)

Your answer:

Solution: $\frac{275}{311}$

Initial conditions	Final conditions
V_1 = constant	V_2 = constant
P_1	P_2
$T_1 = 38^\circ C + 273^\circ = 311$ K	$T_2 = 2.0^\circ C + 273^\circ = 275$ K

With the combined gas law, you can relate pressure and temperature:

$$\frac{P_1V_1}{T_1} = \frac{P_2V_2}{T_2} \Longrightarrow \frac{P_1}{T_1} = \frac{P_2}{T_2}$$

Pressure and absolute temperature of a gas sample are directly proportional when the quantity (moles) of the sample and the volume remain constant.

$$P_2 = \frac{P_1 \times T_2}{T_1} = \frac{P_1 \times 275\ K}{311\ K} = \frac{275}{311} \times P_1$$

According to the kinetic molecular theory, if the temperature decreases at constant volume, the molecules of a gas sample have less average energy (and therefore less speed), and consequently the molecules have fewer "hits" per area of the container (less pressure).

---

4. A sample of 26.5 mL of gas is collected over water at 22°C. The vapor pressure of water at 22°C is 19 torr. If the total pressure in the gas-filled container equals the barometric pressure (746 torr), what volume does this gas alone occupy at STP?

Your answer:

Solution: 23.5 mL

Initial conditions	Final conditions
V_1 = 26.5 mL	V_2 = ?
P_1 = 746 - 19 = 727 torr*	P_2 = 1 atm = 760 torr
T_1 = 22°C + 273° = 295 K	T_2 = 0°C = 273 K

*According to Dalton's law of partial pressures, the total pressure equals the sum of the partial pressures of individual gases in the mixture. In this problem, the vessel contains a mixture of the gas and water vapor. Therefore;

$$P_{gas} + P_{H_2O} = P_{total}$$

$$P_{gas} + 19\ torr = 746\ torr$$

$$P_{gas} = 727\ torr$$

According to the combined gas law,

$$V_2 = \frac{P_1 \times V_1 \times T_2}{P_2 \times T_1} = \frac{727\ torr \times 26.5\ mL \times 273\ K}{760\ torr \times 295\ K} = 23.5\ mL$$

---

5. What is the final temperature (in Celsius degrees) of a gas sample that initially has a volume of 50.0 mL, pressure of 730 mm Hg, and temperature of 25°C, if the final volume is 45.7 mL and the final pressure is 1460 mm Hg?

Your answer:

Solution: 272°C

Initial conditions	Final conditions
$V_1 = 50.0$ mL	$V_2 = 45.7$ mL
$P_1 = 730$ mm Hg	$P_2 = 1460$ mm Hg
$T_1 = 25^{\circ}C + 273^{\circ} = 298$ K	$T_2 = ?$

According to the combined gas law,

$$T_2 = \frac{P_2 \times V_2 \times T_1}{P_1 \times V_1} = \frac{1460 \text{ mm Hg} \times 45.7 \text{ mL} \times 298 \text{ K}}{730 \text{ mm Hg} \times 50.0 \text{ mL}} = 545 \text{ K}$$

T_2 in Celsius = 545 K - 273° = 272°C

II. The "Ideal" Gas Law

1. A sample of an "ideal" gas has a mass of 0.683 g and occupies a volume of 0.50 L at 0.950 atm and 373°C. Calculate the mass of 1.0 mole of this gas.

Your answer:

Solution: 76 g/mole

The ideal gas law, PV = nRT, relates volume, pressure, and temperature of a gas to moles. Moles, in turn, can be related to molecular weight or the number of particles present. To calculate moles,

$$PV = nRT$$

$$n = \frac{PV}{RT} = \frac{0.950 \text{ atm} \times 0.50 \text{ L}}{[0.082 \text{ (L} \times \text{atm)/(mole} \times \text{K)}] \times 646 \text{ K}} = 0.0090 \text{ mole}$$

$$\text{mass of 1.0 mole} = \frac{\text{mass of sample}}{\text{number of moles in sample}} = \frac{0.683 \text{ g}}{0.0090 \text{ mole}} = 76 \text{ g/mole}$$

An alternate method of calculating this molecular weight would be to calculate the volume occupied by this sample at STP using the combined gas law and relate this volume to moles and mass of 1.0 mole with Avogadro's hypothesis and the molar volume concept.

$V_1 = 0.50$ L	$V_2 = ?$
$P_1 = 0.950$ atm	$P_2 = 1.00$ atm
$T_1 = 373^{\circ}C + 273^{\circ} = 646$ K	$T_2 = 273$ K

$$V_2 = \frac{V_1 \times P_1 \times T_2}{T_1 \times P_2} = \frac{0.50\ L \times 0.950\ atm \times 273\ K}{646\ K \times 1.00\ atm} = 0.20\ L$$

$$?\ \frac{grams}{mole} = \frac{0.683\ g}{0.20\ L} \times \frac{22.4\ L}{1.0\ mole} = 76\ g/mole$$

---

2. What is the pressure exerted by 0.54 g of sulfur dioxide gas in a 152 mL container at $20^{\circ}C$? (Assume ideal gas behavior.)

Your answer:

Solution: 1.3 atm

The ideal gas law, PV = nRT, can be used in this problem because the equation relates moles (and therefore grams) of a given gas sample to the volume, temperature, and pressure of the gas.

PV = nRT can be arranged to $P = \frac{nRT}{V}$.

$P = ?$

$V = 152\ mL = 0.152\ L$

$T = 20^{\circ}C + 273^{\circ} = 293\ K$

$R = 0.082\ \frac{L \times atm}{mole \times K}$

$$n = 0.54\ g\ SO_2 \times \frac{1\ mole\ SO_2}{64.06\ g\ SO_2} = 0.0084\ mole$$

$$P = \frac{[0.0084\ mole \times 0.082\ (L \times atm/mole \times K)] \times 293\ K}{0.152\ L} = 1.3\ atm$$

---

3. What is the density of nitrogen gas (a) at STP? (b) at $30^{\circ}C$ and 740 mm Hg pressure?

Your answer:

Solution: (a) 1.25 g/L

At STP 1 mole of any gas occupies 22.4 L. Therefore the density at STP can be calculated by dividing the mass of 1 mole by the volume of 1 mole. The mass of 1 mole of nitrogen gas, N_2 is 28.0 g.

$$d_{STP} = \frac{28.0 \text{ g/mole}}{22.4 \text{ L/mole}} = 1.25 \text{ g/L}$$

(b) 1.1 g/L

To solve this problem using the ideal gas law,

$$PV = nRT$$

$$P = 740 \text{ mm Hg} \times \frac{1 \text{ atm}}{760 \text{ mm Hg}} = 0.974 \text{ atm}$$

$$n = 1.00 \text{ mole}$$

$$T = 30^{o}C + 273^{o} = 303 \text{ K}$$

$$V = \frac{nRT}{P} = \frac{[1.00 \text{ mole} \times 0.082 \text{ (L} \times \text{atm)/mole} \times \text{K)}] \times 303 \text{ K}}{0.974 \text{ atm}} = 26 \text{ L}$$

$$d = \frac{28.0 \text{ g/mole}}{26 \text{ L/mole}} = 1.1 \text{ g/L}$$

To solve this problem using the combined gas law,

$$\frac{P_1V_1}{T_1} = \frac{P_2V_2}{T_2}$$

$$V_2 = \frac{P_1 \times V_1 \times T_2}{P_2 \times T_1} = \frac{1.00 \text{ atm} \times 22.4 \text{ L} \times 303 \text{ K}}{0.974 \text{ atm} \times 273 \text{ K}} = 25.5 \text{ L}$$

$$d = \frac{28.0 \text{ g/mole}}{25.5 \text{ L/mole}} = 1.10 \text{ g/L}$$

---

8 Liquids and Solids; Intermolecular Attractions

I. States of Matter

1. Define normal boiling point. How would this differ from a boiling point that you might measure in your laboratory?

Your answer:

Solution: The normal boiling point is the temperature at which vapor pressure equals standard pressure (1 atm, the barometric reading at sea level on a clear day). The boiling point at a slightly higher altitude would be slightly lower than the normal boiling point. The boiling point of a liquid that you determine in the lab is dependent on the atmospheric pressure within the lab.

---

2. Ammonia has a normal melting point of $-78^{\circ}C$ and a normal boiling point of $-33^{\circ}C$. In what physical state is ammonia at each of the following temperatures: $65^{\circ}F$, $-33^{\circ}C$, $-100^{\circ}C$? (Assume that the container holding the ammonia is closed and is at standard pressure.)

Your answer:

Solution: $65^{\circ}F$ is 18° on the Celsius scale; this temperature is above the boiling point, so ammonia is in the gaseous state. At $-33^{\circ}C$ (the boiling point), the gas and liquid states coexist. $-100^{\circ}C$ is below the melting point (also called the freezing point), so ammonia would exist as a solid.

---

3. If two liquids, A, whose nonpolar molecules have a mass of 58 g/mole, and B, whose nonpolar molecules have a mass of 98 g/mole, are at the same temperature and pressure, would the rates of evaporation be the same or different for each of the two liquids? Explain completely.

Your answer:

Solution: Evaporation is the process of changing from the liquid phase to the gaseous phase at any temperature. At a given temperature, the average energy of A molecules and of B molecules will be the same. However, the lighter-mass molecules of A will move (on the average) faster than will the molecules of B. Therefore, more A molecules will exceed the "escape" velocity than will the B molecules, and consequently A will evaporate more rapidly than will B.

---

4. The boiling points, vapor pressures, and heats of vaporization of three substances (X, Y, and Z) are to be compared. The three substances have approximately the same molecular weights. Molecules of substance X have no dipoles. The molecules of substance Y contain -O-H bonds, and the molecules of substance Z are polar but contain no -O-H bonds.

(a) Arrange X, Y, and Z in order of increasing intermolecular attractive forces.

Your answer:

Solution: X < Z < Y

Molecules of X (for example, Cl_2) can attract each other only by weak London forces. (Temporary fluctuating dipoles cause very weak attractions among the molecules.) Molecules of Y attract each other by hydrogen bonding—the relatively strong attraction between hydrogen of one molecule and the electronegative atom (oxygen, nitrogen, or fluorine) on another molecule. For example,

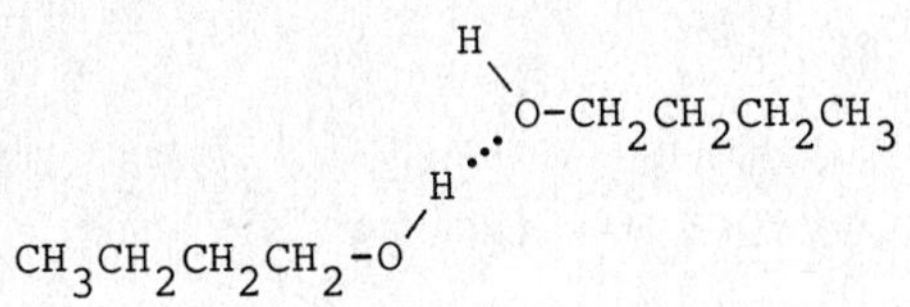

Z could be a molecule like the structure shown.

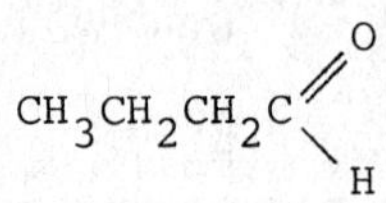

The carbon and oxygen form a polar bond and cause the molecule to have a dipole; another molecule like it will be attracted to it, and the two will align themselves so that the partially positive charge on one molecule will be attracted to the partially negative charge on the other. These attractive forces are greater than in a molecule having London forces alone, but they are much weaker than in molecules in which hydrogen bonding is possible.

(b) Arrange X, Y, and Z in order of increasing equilibrium vapor pressures and discuss the physical significance of vapor pressure.

Your answer:

Solution: Y < Z < X

The equilibrium vapor pressure is the pressure of vapor (gaseous molecules) in equilibrium with the liquid phase at a given temperature. Molecules that attract each other least will have the greatest vapor pressure.

(c) Arrange X, Y, and Z in order of increasing normal boiling points.

Your answer:

Solution: X < Z < Y

The boiling point is the temperature at which the vapor pressure of the liquid equals the atmospheric pressure. The standard or normal boiling point is the common reference temperature and is the boiling point at one standard atmosphere of pressure. Because molecules of Y attract each other much more strongly than molecules of X attract each other, more heat is required to raise the temperature of a sample of Y to its normal boiling point than is required for a sample of X.

---

5. Use Le Châtelier's principle to explain how changes in pressure affect the boiling point of a substance.

Your answer:

Solution: The volume occupied by the molecules increases greatly during a phase change from liquid to gas. liquid + heat $\rightleftharpoons$ gas (vapor) Increased pressure favors decreased volume. Therefore, increased pressure temporarily favors the reaction to the left, and more heat will be required to boil a substance (shift the reaction from left to right). Decreased pressure (at high altitudes) favors formation of the gas and permits boiling to occur at a lower temperature.

---

6. Consider an ice cube in a glass of water that is thoroughly insulated from its environment. Write an equation to illustrate the equilibrium that is established. How is this equilibrium affected by a decrease in temperature or by an increase in pressure?

Your answer:

Solution: ice + heat $\rightleftharpoons$ water (liquid)

A decrease in temperature will favor the reverse reaction; some of the liquid water will form additional ice. Because ice is less dense than liquid water, an increase in the pressure will favor the forward reaction (the more dense form of water).

---

7. In lab experiments you often need to cool a substance in an ice bath. Write an equation representing the equilibrium between ice and liquid water in the ice bath. Explain in terms of Le Châtelier's principle what happens to the quantities of ice and liquid water as the substance is cooled.

Your answer:

Solution: ice + heat $\rightleftharpoons$ water (liquid)
Assume that the ice and water are initially at equilibrium at $0^{o}C$. Heat from the substance to be cooled can be envisioned as a stress that upsets the equilibrium and that causes the reaction to shift to the right to relieve the stress. Each gram of ice that melts absorbs 80 cal of heat (called the heat of fusion) from the substance that is being cooled. The quantity of solid ice gradually decreases and the quantity of liquid water therefore increases. However, a new point of equilibrium would not be established, since heat from the room is constantly available to melt the ice also.

---

8. Write an equation to illustrate the equilibrium between $BaSO_4(s)$ and the dissociated ions in a solution. How will the addition of $BaCl_2$ affect the equilibrium?

Your answer:

Solution: $BaSO_4\ (s) + H_2O \rightleftharpoons Ba^{2+}(aq) + SO_4^{\ 2-}(aq)$
Addition of $BaCl_2$ adds extra Ba^{2+} ions to the barium sulfate solution. To diminish this stress, the reaction shifts to the left, and a new equilibrium is established in which more $BaSO_4$ precipitate is present than is present initially. Once the new equilibrium is established, the rate of $BaSO_4$ dissolving (the reaction to the right) is equal to the rate of $BaSO_4$ precipitating (the reaction to the left).

II. Heating and Cooling of Gases, Liquids, and Solids

Use Figure 8.1 and this information about water to answer the following questions.

$\Delta H_{fus} = 80.\ cal/g$
$\Delta H_{vap} = 540.\ cal/g$

Specific heats are

$H_2O(s) = 0.49\ cal/g \times {}^{o}C$
$H_2O(l) = 1.000\ cal/g \times {}^{o}C$
$H_2O(g) = 0.48\ cal/g \times {}^{o}C$

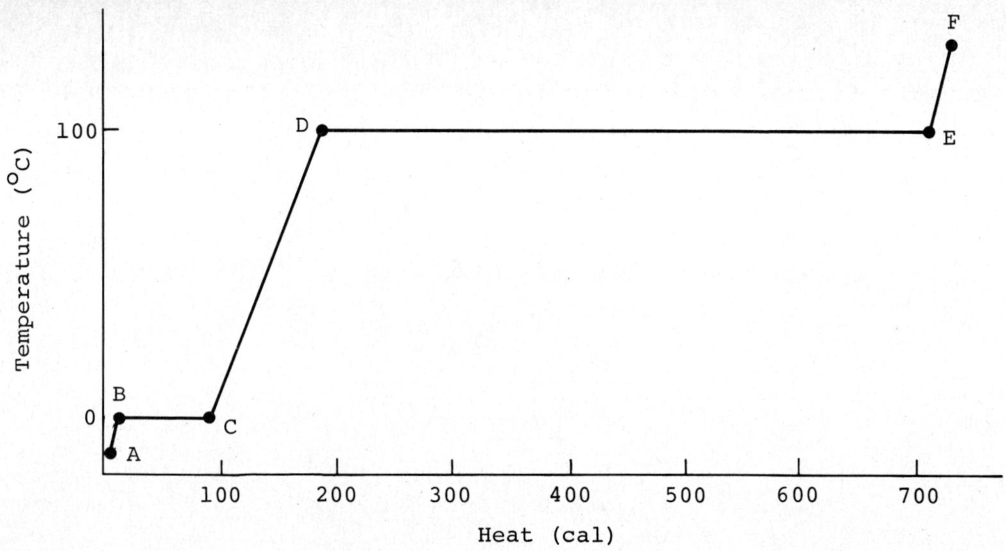

Figure 8.1 Heating curve for water

1. Consider the calories of heat needed to change 23 g of liquid water at 15°C to liquid water at 100°C.

(a) Which segment of the heating curve applies to this change?

Your answer:

Solution: Liquid water is heated over part of segment CD. Notice that no phase change occurs.

(b) How many calories are required for this process?

Your answer:

Solution:

Change of temperature = 100°C - 15°C = 85°C

$$? \text{ cal} = \frac{1.000 \text{ cal}}{\text{g} \times {}^{\circ}\text{C}} \times 85^{\circ}\text{C} \times 23 \text{ g} = 2.0 \times 10^{3} \text{ cal}$$

---

2. Consider the change indicated by segment DE of the heating curve.

(a) Describe what occurs over segment DE. In what physical state (phase) are the water molecules at E?

Your answer:

Solution: No temperature change occurs over segment DE, but a phase change occurs in which all liquid water molecules become gaseous molecules at E.

(b) Calculate the heat required to perform this conversion at 100°C for 23 g of liquid water.

Your answer:

Solution:

$$? \text{ cal} = \Delta H_{vap} \times \text{mass of } H_2O$$

$$= \frac{540 \text{ cal}}{\text{g}} \times 23 \text{ g} = 1.2 \times 10^4 \text{ cal}$$

---

3. How many calories of heat are needed to change 23 g of water at 15°C to gaseous water (steam) at 100°C?

Your answer:

Solution: Calculate the calories required for each of the two steps (represented by segment CD and DE), and add the results.

CD (as given in question 1(b)): 2.0×10^3 cal

DE (as given in question 2(b)): 1.2×10^4 cal

$$? \text{ cal} = (2.0 \times 10^3 \text{ cal}) + (12 \times 10^3 \text{ cal}) = 14 \times 10^3 \text{ cal}$$

$$= 1.4 \times 10^4 \text{ cal}$$

---

4. How many calories of heat must be removed from 125 g of steam at 125°C to cool it to ice at 0°C? Show each step of your calculation.

Your answer:

Solution: 9.2×10^4 cal

A heating curve can be read from right to left as a cooling curve.

(1) segment FE—cooling steam from 125°C to 100°C

$$? \text{ cal} = \frac{0.48 \text{ cal}}{\text{g} \times {}^\circ\text{C}} \times (125^\circ\text{C} - 100^\circ\text{C}) \times 125 \text{ g} = 1.5 \times 10^3 \text{ cal}$$

(2) segment ED—phase change from $H_2O(g)$ to $H_2O(l)$ at 100°C

$$? \text{ cal} = \frac{540.\text{ cal}}{\text{g}} \times 125 \text{ g} = 6.8 \times 10^4 \text{ cal}$$

(3) segment DC—cooling liquid water from 100°C to 0°C

$$? \text{ cal} = \frac{1.000 \text{ cal}}{\text{g} \times {}^\circ\text{C}} \times (100^\circ\text{C} - 0^\circ\text{C}) \times 125 \text{ g} = 1.25 \times 10^4 \text{ cal}$$

(4) segment CB—phase change from $H_2O(l)$ to $H_2O(s)$ at 0°C

$$? \text{ cal} = \frac{80.\text{ cal}}{\text{g}} \times 125 \text{ g} = 1.0 \times 10^4 \text{ cal}$$

Total calories for the cooling of steam at 125°C (F) to ice at 0°C (B):

$$(1.5 \times 10^3) + (68 \times 10^3) + (12.5 \times 10^3) + (10. \times 10^3) = 92 \times 10^3 \text{ cal} = 9.2 \times 10^4 \text{ cal}$$

9 Solution Chemistry

I. Solution Concentrations Expressed as Weight/Volume Percent

1. How much sucrose, $C_{12}H_{22}O_{11}$, in grams, is present in 1.00 mL of a 5.00% (w/v) sucrose solution?

Your answer:

Solution: 0.0500 g sucrose

5.00% means that 5.00 g of solute is present in each 100. mL of solution. To determine the number of grams of solute per mL, simply divide 5.00 by the 100 mL.

$$? \frac{g}{mL} = \frac{5.00\ g}{100.\ mL} = 0.0500\ g \text{ sucrose in } 1.00\ mL \text{ solution}$$

---

2. What is the w/v % of a solution containing 32.56 g NaCl in 500. mL of solution?

Your answer:

Solution: 6.51%

To get w/v % from mass of solute and volume of a given solution, divide the grams of solute by the volume in mL and multiply by 100 (to convert the fraction to percent).

$$? \frac{w}{v}\ \% = \frac{g \text{ solute}}{mL \text{ solution}} \times 100 = \frac{32.56\ g}{500.\ mL} \times 100 = 6.51\%$$

---

3. How many grams of NaCl would be needed to prepare 3.0 L of a 2.50% (w/v) NaCl solution?

Your answer:

Solution: 75 g NaCl

2.50% means that there is 2.50 g solute per 100. mL solution.

$$? \text{ g NaCl} = \frac{2.50 \text{ g}}{100 \text{ mL}} \times \frac{1000 \text{ mL}}{1 \text{ L}} \times 3.0 \text{ L} = 75 \text{ g NaCl}$$

---

4. What volume of a 15.0% solution can be made from 10.0 g $CaCl_2$? from 10.0 g NaCl? from 10.0 g of any soluble solute?

Your answer:

Solution: 66.7 mL solution

15.0% solution means $\frac{15.0 \text{ g solute}}{100 \text{ mL solution}}$. This relationship can be inverted and used as a conversion factor to calculate the volume of the solution.

$$? \text{ mL} = \frac{100 \text{ mL}}{15.0 \text{ g}} \times 10.0 \text{ g} = 66.7 \text{ mL solution}$$

You should realize that 10.0 g $CaCl_2$, or NaCl, or any solute dissolved in 66.7 mL of solution will produce a 15.0% (w/v) solution. Each of these would contain, of course, a different number of moles of solute.

II. Moles of Solute, Volume of Solution, and Molarity

1. Define molarity. Write a mathematical expression for molarity. Rewrite the equation to solve for the number of moles of solute.

Your answer:

Solution: Molarity is expressed as the number of moles of solute in a solution divided by the number of liters of solution.

$$M = \frac{\text{moles of solute}}{\text{volume in liters}} = \frac{n}{V}$$

(n represents the number of moles; V represents the volume in liters)

$$n = M \times V$$

---

2. How many moles of $FeCl_3$ are contained in a mass of 8.11 g $FeCl_3$?

Your answer:

Solution: 0.0500 mole $FeCl_3$

$$? \text{ moles } FeCl_3 = 8.11 \text{ g} \times \frac{1 \text{ mole } FeCl_3}{162.2 \text{ g}} = 0.0500 \text{ mole } FeCl_3$$

---

3. How many moles of $FeCl_3$ are in 25.0 mL of a 0.500 M solution of $FeCl_3$?

Your answer:

Solution: 0.0125 mole $FeCl_3$

$$25.0 \text{ mL} \times \frac{1 \text{ L}}{1000 \text{ mL}} = 0.0250 \text{ L}$$

$$? \text{ moles } FeCl_3 = M \times V = \frac{0.500 \text{ mole} \times 0.0250 \text{ L}}{\text{L}} = 0.0125 \text{ mole } FeCl_3$$

---

4. How many moles of Fe^{3+} ions are in 10.0 mL of a 0.0250 M solution of $FeCl_3$?

Your answer:

Solution: 2.50×10^{-4} mole Fe^{3+}

$$10.0 \text{ mL} \times \frac{1 \text{ L}}{1000 \text{ mL}} = 0.0100 \text{ L}$$

$$\text{moles } Fe^{3+} \text{ ions} = \text{moles } FeCl_3$$

$$= M \times V = \frac{0.0250 \text{ mole} \times 0.0100 \text{ L}}{\text{L}}$$

$$= 0.00025 \text{ mole} = 2.50 \times 10^{-4} \text{ mole } Fe^{3+}$$

---

5. What is the molarity of each of the following solutions?

(a) 2.92 g NaCl is dissolved in water to make a solution with a total volume of 250. mL.

Your answer:

Solution: 0.200 M

$$\text{moles NaCl} = 2.92 \text{ g} \times \frac{1 \text{ mole NaCl}}{58.44 \text{ g}} = 0.04997 \text{ mole} = 0.0500 \text{ mole}$$

$$M = \frac{\text{moles}}{\text{L}} = \frac{0.0500 \text{ mole}}{0.250 \text{ L}} = 0.200 \text{ M}$$

(b) 1.8228 g HCl is contained in a solution with a volume of 50.0 mL.

Your answer:

Solution: 1.000 M

$$? \text{ moles HCl} = 1.8228 \text{ g HCl} \times \frac{1 \text{ mole HCl}}{36.46 \text{ g}} = 0.04999 \text{ mole HCl}$$

$$M = \frac{\text{moles}}{\text{L}} = \frac{0.04999 \text{ mole}}{0.0500 \text{ L}} = 0.9999 = 1.00 \text{ M}$$

III. Calculations Involving Molarity and Percent Concentration

1. A solution is 1.253 M and contains 22.35 g of sodium carbonate, Na_2CO_3. Calculate (a) the number of moles of Na_2CO_3 in the solution, (b) the volume of the solution, and (c) the percent concentration of the solution.

Your answers:

Solution:

(a) $$? \text{ moles } Na_2CO_3 = 22.35 \text{ g} \times \frac{1 \text{ mole } Na_2CO_3}{105.99 \text{ g}} = 0.2109 \text{ mole } Na_2CO_3$$

(b) $$M = \frac{\text{moles}}{\text{L}};$$

$$V = \frac{\text{moles}}{\text{molarity}}$$

$$= \frac{0.2109 \text{ mole}}{1.253 \text{ moles/L}} = 0.1683 \text{ L} = 168.3 \text{ mL}$$

(c) $$? \frac{w}{v}\,\% = \frac{\text{g solute}}{\text{mL solution}} \times 100 = \frac{22.35 \text{ g}}{168.3 \text{ mL}} \times 100 = 13.28\%$$

---

2. For a 16.00 L NaOH solution with a concentration of 0.4935 M, calculate the following.
   (a) How many moles of NaOH are in the solution?
   (b) How many grams of NaOH are present?
   (c) What is the w/v % concentration of this solution?

Your answers:

Solution:

(a) $? \text{ moles} = M \times V = \dfrac{0.4935 \text{ mole} \times 16.00 \text{ L}}{\text{L}} = 7.896 \text{ moles}$

(b) $? \text{ grams} = 7.896 \text{ moles} \times \dfrac{40.00 \text{ g}}{\text{mole}} = 315.8 \text{ g}$

(c) $? \dfrac{w}{v} \% = \dfrac{\text{g solute}}{\text{mL solution}} \times 100 = \dfrac{315.8 \text{ g}}{16.00 \text{ L}} \times \dfrac{1 \text{ L}}{1000 \text{ mL}} \times 100$

$= 1.974\%$

Notice that the volume in liters is converted to milliliters.

---

3. If 15.0 g of glucose $C_6H_{12}O_6$ is dissolved in water to make exactly 1000.00 mL of solution, calculate (a) the number of moles of solute, (b) the molarity of the solution, and (c) the w/v % concentration of the solution.

Your answers:

Solution:

(a) $? \text{ moles glucose} = 15.0 \text{ g} \times \dfrac{1 \text{ mole glucose}}{180.16 \text{ g}} = 0.0833 \text{ mole glucose}$

(b) $M = \dfrac{\text{moles}}{\text{L}} = \dfrac{0.0833 \text{ moles}}{1.00 \text{ L}} = 0.0833 \text{ M}$

(c) $? \dfrac{w}{v} \% = \dfrac{\text{g solute}}{\text{mL solution}} \times 100 = \dfrac{15.0 \text{ g}}{1000 \text{ mL}} \times 100 = 1.50\%$

---

4. Contrast 0.900% solutions of NaCl and KCl with the following questions.
   (a) What is the molarity of a 100 mL solution of 0.900% NaCl?
   (b) What is the molarity of a 100 mL solution of 0.900% KCl?
   (c) Is the KCl solution hypotonic, hypertonic, or isotonic to a physiological saline solution?

Your answers:

Solution:

(a) 100 mL of 0.900% NaCl solution contains 0.900 g NaCl.

$$? \text{ moles NaCl} = 0.900 \text{ g} \times \frac{1 \text{ mole NaCl}}{58.45 \text{ g}} = 0.0154 \text{ mole NaCl}$$

$$M = \frac{\text{moles}}{\text{L}} = \frac{0.0154 \text{ mole}}{0.100 \text{ L}} = 0.154 \text{ M NaCl}$$

(b) 100 mL of 0.900% KCl solution contains 0.900 g KCl.

$$? \text{ moles KCl} = 0.900 \text{ g} \times \frac{1 \text{ mole KCl}}{74.55 \text{ g}} = 0.0121 \text{ mole}$$

$$M = \frac{\text{moles}}{\text{L}} = \frac{0.0121 \text{ mole}}{0.100 \text{ L}} = 0.121 \text{ M KCl}$$

(c) Physiological saline solution is 0.900% NaCl (0.154 M). The KCl solution contains fewer moles of ions per unit volume of solution than does physiological saline solution and is therefore hypotonic (less concentrated) to physiological saline.

---

5. Which solution is more concentrated in the specified ion?
(a) Hydroxide ion in a 2.0 M or a 2.0% solution of NaOH.

Your answer:

Solution: You should change the given information to a common reference point, moles NaOH per milliliter, for example.

$$2.0 \text{ M NaOH:} \quad ? \frac{\text{moles}}{\text{mL}} = 2.0 \text{ M} = \frac{2.0 \text{ moles NaOH}}{1000 \text{ mL}}$$

$$= \frac{0.0020 \text{ mole NaOH}}{\text{mL}}$$

$$= \frac{0.0020 \text{ mole } OH^-}{\text{mL}}$$

$$2.0\% \text{ NaOH:} \quad ? \frac{\text{moles}}{\text{mL}} = 2.0\% = \frac{2.0 \text{ g}}{100 \text{ mL}} \times \frac{1 \text{ mole NaOH}}{40.00 \text{ g}}$$

$$= \frac{0.00050 \text{ mole NaOH}}{\text{mL}}$$

$$= \frac{0.00050 \text{ mole } OH^-}{\text{mL}}$$

The 2.0 M solution contains four times as many OH^- ions as does the 2.0% solution.

(b) Chloride ion in a 1.0 M solution of $FeCl_3$ or in a 3.0% solution of KCl.

Your answer:

Solution:

$$1.0\ M\ FeCl_3:\quad ?\ \frac{\text{moles}}{\text{mL}} = 1.0\ M = \frac{1.0\ \text{mole}\ FeCl_3}{1000\ \text{mL}} = \frac{0.0010\ \text{mole}\ FeCl_3}{\text{mL}}$$

One mole of $FeCl_3$ furnishes 3 moles of Cl^-.

$$\frac{0.0010\ \text{mole}\ FeCl_3}{\text{mL}} \times \frac{3\ \text{moles}\ Cl^-}{1\ \text{mole}\ FeCl_3} = \frac{0.0030\ \text{mole}\ Cl^-}{\text{mL}}$$

$$3.0\%\ KCl:\quad ?\ \frac{\text{moles}}{\text{mL}} = 3.0\% = \frac{3.0\ g}{100\ \text{mL}} \times \frac{1\ \text{mole}\ KCl}{74.55\ g}$$

$$= \frac{0.00040\ \text{mole}\ KCl}{\text{mL}}$$

$$= \frac{0.00040\ \text{mole}\ Cl^-}{\text{mL}}$$

The 1.0 M $FeCl_3$ solution has 7.5 times as many Cl^- ions as does the 3.0% KCl solution.

IV. Dilution Problems

For solutions that are diluted with water, you can assume that the volumes are additive.

1. What is the molarity of NaOH in a solution made by diluting 5.0 mL of 6.0 M NaOH solution with 10.0 mL of water?

Your answer:

Solution: 2.0 M

It is important to recognize that moles solute in initial solution = moles solute in final (diluted) solution.

M_1V_1 = moles solute in initial solution

M_2V_2 = moles solute in final solution

Therefore:

$$M_1V_1 = M_2V_2$$

$$\frac{6.0\ \text{moles} \times 5.0\ \text{mL}}{L} = M_2 \times (5.0\ \text{mL} + 10.0\ \text{mL})$$

$$M_2 = \frac{6.0\ M \times 5.0\ \text{mL}}{15.0\ \text{mL}} = 2.0\ M$$

Although you would expect V_1 and V_2 to be expressed in liters to be consistent with the definition of molarity (moles/liter), it is permissible to express V_1 and V_2 in milliliters because the units eventually cancel.

---

2. 25.0 mL of 6.0 M H_2SO_4 solution is to be diluted to 2.0 M H_2SO_4. Calculate the amount of water needed and describe the procedure necessary to make the dilution.

Your answer:

Solution: 50.0 mL water needed

First calculate the total volume required for the 2.0 M solution.

$$M_1V_1 = M_2V_2$$

$$6.0\ M \times 25.0\ mL = 2.0\ M \times V_2$$

$$V_2 = \frac{6.0\ M \times 25.0\ mL}{2.0\ M} = 75.0\ mL$$

The volume of water required is 75.0 mL - 25.0 mL = 50.0 mL. Acid must always be added to water (and not the reverse), so you should measure 50.0 mL of H_2O and add the 25.0 mL of 6.0 M H_2SO_4 to it.

---

3. How much 18.0 M H_2SO_4 solution must be added to water to prepare 3.0 L of 3.0 M H_2SO_4 solution?

Your answer:

Solution: 0.50 L

$$M_1V_1 = M_2V_2$$

$$18.0\ M \times V_1 = 3.0\ M \times 3.0\ L$$

$$V_1 = \frac{3.0\ M \times 3.0\ L}{18.0\ M} = 0.50\ L$$

---

4. Often a solution is prepared by mixing two other solutions. When this is done, the volume of one solution dilutes the solute in the other solution. For example, if 10.0 mL of 0.534 M NaOH solution is mixed with 5.0 mL of another aqueous solution, what is the molarity of NaOH in the resulting solution? Assume that no reaction occurs between the components of the solutions.

Your answer:

Solution: 0.356 M

$$M_1V_1 = M_2V_2$$

$$0.534\ M \times 10.0\ mL = M_2 \times (5.0\ mL + 10.0\ mL)$$

$$M_2 = \frac{0.534\ M \times 10.0\ mL}{15.0\ mL} = 0.356\ M$$

---

5. If the solution added to the NaOH solution in Exercise 4 contains a compound that would react with the NaOH, you can still calculate the initial concentration of the NaOH (or of the Na^+ ions or the OH^- ions) in the total volume of the mixture. This type of calculation is important in equilibrium experiments. For example, if 5.0 mL of 0.026 M $FeCl_3$ is mixed with 5.0 mL of 0.002 M NaSCN,
   (a) What is the initial concentration of Fe^{3+} ions in the mixture?
   (b) What is the initial concentration of SCN^- ions in the mixture? Assume that the solutions mix but that the ions do not react initially.

Your answer:

Solution: (a) 0.013 M in Fe^{3+}; (b) 0.001 M in SCN^-

(a) Let M_2 = molarity of Fe^{3+} ions in the final solution (after mixing, before reacting).

$$0.026\ M \times 5.0\ mL = M_2 \times 10.0\ mL$$

$$M_2 = \frac{0.026\ M \times 5.0\ mL}{10.0\ mL} = 0.013\ M$$

(b) Let M_2 = molarity of SCN^- ions in the final solution (after mixing, before reacting).

$$0.002\ M \times 5.0\ mL = M_2 \times 10.0\ mL$$

$$M_2 = \frac{0.002\ M \times 5.0\ mL}{10.0\ mL} = 0.001\ M$$

V. Stoichiometric Problems Involving Solutions

1. How many grams of aluminum react with 10.0 mL of 6.00 M H_2SO_4? The products of the reaction are $Al_2(SO_4)_3$ and H_2 gas.

Your answer:

Solution: 1.08 g Al

The balanced equation for the reaction is

$$2\ Al + 3\ H_2SO_4 \longrightarrow Al_2(SO_4)_3 + 3\ H_2$$

The unit of moles is the "bridge" used in all stoichiometric problems to convert from one component of a reaction to another. Therefore, convert the solution units to moles of reactant; then you can relate the moles of the reactant to the other component in the reaction.

$$?\ \text{moles } H_2SO_4 = 6.00\ M \times 10.0\ mL \times \frac{1\ L}{1000\ mL} = 0.0600\ \text{mole } H_2SO_4$$

$$?\ g\ Al = 0.0600\ \text{mole } H_2SO_4 \times \frac{2\ \text{moles Al}}{3\ \text{moles } H_2SO_4} \times \frac{26.98\ g\ Al}{1\ \text{mole Al}} = 1.08\ g\ Al$$

---

2. A certain solution of nitric acid is 49% HNO_3 (w/v %).

(a) How many moles of HNO_3 are available in a 100 mL sample of this solution?

Your answer:

Solution: 0.78 mole HNO_3

49% (w/v %) means that there are 49 g HNO_3 in 100 mL of solution.

$$?\ \text{mole } HNO_3 = 49\ g\ HNO_3 \times \frac{1\ \text{mole } HNO_3}{63.02\ g\ HNO_3} = 0.78\ \text{mole } HNO_3$$

(b) Will the 100 mL of 49% HNO_3 in part (a) be sufficient to react with 100 g of lead in the following reaction?

$$3\ Pb + 4\ HNO_3 \longrightarrow 3\ PbO_2 + 4\ NO + 2\ H_2O$$

Your answer:

Solution: yes

$$?\ \text{moles } HNO_3 = 100\ g\ Pb \times \frac{1\ \text{mole Pb}}{207.2\ g\ Pb} \times \frac{4\ \text{moles } HNO_3}{3\ \text{moles Pb}}$$

$$= 0.6435\ \text{mole } HNO_3 \text{ required to react with 100 g Pb}$$

Therefore the 0.78 mole HNO_3 in 100 mL of 49% HNO_3 solution is sufficient for reaction with 100 g Pb.

---

3. The reaction between aluminum and NaOH (lye) is

$$2\ Al + 2\ NaOH + 2\ H_2O \longrightarrow 3\ H_2 + 2\ NaAlO_2.$$

(a) If 15.0 g of aluminum reacts with 12.0 M NaOH solution, what volume of NaOH solution is required for the (exact) reaction?

Your answer:

Solution: 0.0463 L or 46.3 mL NaOH solution

$$?\ \text{moles NaOH} = 15.0\ \text{g Al} \times \frac{1\ \text{mole Al}}{26.98\ \text{g Al}} \times \frac{2\ \text{moles NaOH}}{2\ \text{moles Al}}$$

$$= 0.556\ \text{mole NaOH}$$

$$\text{volume NaOH solution} = \frac{0.556\ \text{mole}}{12.0\ \text{M}} = 0.0463\ \text{L}$$

(b) How many grams of hydrogen gas can be theoretically produced by the reacting quantities in part (a)?

Your answer:

Solution: 1.68 g H_2

$$?\ \text{g}\ H_2 = 0.556\ \text{mole NaOH} \times \frac{3\ \text{moles}\ H_2}{2\ \text{moles NaOH}} \times \frac{2.016\ \text{g}\ H_2}{1\ \text{mole}\ H_2} = 1.68\ \text{g}\ H_2$$

or

$$?\ \text{g}\ H_2 = 15.0\ \text{g Al} \times \frac{1\ \text{mole Al}}{26.98\ \text{g Al}} \times \frac{3\ \text{moles}\ H_2}{2\ \text{moles Al}} \times \frac{2.016\ \text{g}\ H_2}{1\ \text{mole}\ H_2} = 1.68\ \text{g}\ H_2$$

---

10 Equilibrium Reactions

I. Le Châtelier's Principle

1. What is Le Châtelier's principle?

Your answer:

Solution: Le Châtelier's principle states that if a stress (a change) is applied to a system already in equilibrium, the system will shift (readjust), if possible, to reduce the stress.

For the following exercises, consider this exothermic reaction.

$$2\ NO(g) + O_2(g) \rightleftharpoons 2\ NO_2(g)$$

Assume that the system is initially at equilibrium. For each stress imposed on the system, evaluate (by answering the questions) the effect that the stress will have on the equilibrium.

2. Imposed stress: increase in the concentration of O_2

(a) How does this stress affect the position of the equilibrium?

Your answer:

Solution: The reaction to the right is favored relative to the reverse reaction because the forward reaction consumes the excess O_2 and consequently diminishes the stress. When the system again reaches equilibrium, the concentration of NO_2 has increased and the concentration of NO has decreased relative to their initial concentrations. The O_2 concentration is greater than it was initially but less than it was just after the extra was added. Because the concentration of the product has increased relative to the concentrations of the reactant, the "position" of the equilibrium is said to have shifted to the right.

(b) Will the value of the equilibrium constant (K_{eq}) be affected by the shift in the position of the equilibrium? If yes, will the constant increase or decrease?

Your answer:

Solution: The changes in the concentrations, indicated in the answer to part (a), have occurred in such a way that the equilibrium constant remains the same value as it was before. Only changes in temperature can change the value of the equilibrium constant.

---

3. Imposed stress: decrease in the concentration of NO_2

(a) How does this stress affect the position of the equilibrium?

Your answer:

Solution: The effect of the stress can be diminished by "making more NO_2"; the reaction shifts to the right in the direction of the NO_2. Consequently, the concentrations of both NO and O_2 will decrease.

(b) Will the value of the equilibrium constant (K_{eq}) be affected by the shift? If yes, will the constant increase or decrease?

Your answer:

Solution: When the new equilibrium is achieved, the concentration of NO_2 will be smaller than its initial value, but larger than its value just after the stress was imposed. The concentrations of NO and O_2 will be lower than their original values. Consequently, the constant will be unchanged.

---

4. Imposed stress: increase in the temperature of the system (The reaction is exothermic.)

(a) How does this stress affect the position of the equilibrium?

Your answer:

Solution: The exothermic reaction is written

$$2\ NO(g) + O_2(g) \rightleftharpoons 2\ NO_2(g) + heat$$

An increase in temperature increases the rate of all chemical reactions. In an equilibrium reaction, an increase in temperature will upset the equilibrium; the reaction that uses heat (the reverse reaction in this example) will have a faster rate than the reaction in the other direction. The position of this equilibrium is shifted to the left, increasing the concentrations of NO and O_2 and decreasing the concentration of NO_2.

(b) Will the value of the equilibrium constant increase, decrease, or remain unchanged?

Your answer:

Solution: When the new equilibrium is established, the concentration of the products is smaller than its initial value, and the concentrations of the reactants are larger than their initial values; therefore the K_{eq} value decreases.

---

5. Imposed stress: decrease in the volume of the system
(a) How does this stress affect the position of the equilibrium?

Your answer:

Solution: In an equation involving gaseous molecules, the coefficients can be interpreted as volumes.

$$2 \text{ volumes NO} + 1 \text{ volume } O_2 \rightleftharpoons 2 \text{ volumes } NO_2$$

which simplifies to

$$3 \text{ volumes reactants} \rightleftharpoons 2 \text{ volumes products}$$

If the volume containing the entire reaction is decreased, the reaction will shift to the right because the products occupy less volume than the reactants.

(b) Will the value of the K_{eq} increase, decrease or remain unchanged as a result of the shift in the position of the equilibrium?

Your answer:

Solution: A new equilibrium is established in the new, smaller volume, but because the temperature is unchanged, the value of the constant is also unchanged.

---

6. Imposed stress: increase in the pressure of the system
   (a) How is the position of the equilibrium affected by increasing the pressure of the system?

Your answer:

Solution: Increasing the pressure of the system can have the same effect as decreasing the volume of the system because pressure of a gas is inversely proportional to the volume of a gas at the same temperature. Therefore the equilibrium position will shift to the right as it does in exercise 5.

(b) Will the value of the constant increase, decrease or remain unchanged?

Your answer:

Solution: The K_{eq} will remain constant.

---

7. Imposed stress: decrease in the temperature of the system
   (a) How does this stress affect the position of the equilibrium?

Your answer:

Solution: Because the reaction is exothermic,

$$2\ NO(g) + O_2(g) \rightleftharpoons 2\ NO_2(g) + \text{heat}$$

a decrease in the temperature (which will slow the rate of both forward and reverse reactions) will cause the position of the equilibrium to shift in a direction that produces heat (to the right), and consequently, the concentration of NO_2 increases and the concentrations of NO and O_2 decrease.

(b) Will decreasing the temperature of the reaction change the value of the equilibrium constant?

Your answer:

Solution: When the new equilibrium is established, the concentration of the product is increased above its initial value, and the concentrations of the reactants are decreased below their initial values: therefore, the K_{eq} value increases.

---

8. Imposed stress: addition of a catalyst to the reaction vessel
(a) How does the addition of a catalyst affect the position of the equilibrium?

Your answer:

Solution: The addition of a catalyst increases the rate of both the forward and the reverse reactions, but the position of the equilibrium is unchanged.

(b) Will the value of the K_{eq} increase, decrease or remain unchanged?

Your answer:

Solution: The K_{eq} remains unchanged because no relative change occurs in the concentrations of the products or reactants.

---

9. For the reaction $2\ NO(g) + O_2(g) \rightleftharpoons 2\ NO_2(g)$ + heat, is nitrogen dioxide, NO_2, more stable at 1000 K or at 500 K?

Your answer:

Solution: If the temperature of the reaction is raised from 500 K to 1000 K, the reaction will shift in the reverse direction (to the left), which will result in the decomposition of NO_2. Therefore, NO_2 is more stable at the lower temperature (500 K).

---

10. For the reaction

$$H_2O(g) + CO(g) \rightleftharpoons H_2(g) + CO_2(g) + \text{heat}$$

which of the following stresses will cause a change in the equilibrium composition of the reaction?
(a) Addition of a catalyst.
(b) Increase in pressure of the system.
(c) Increase in concentration of CO.
(d) Decrease in concentration of CO_2.
(e) Decrease in temperature of the system.

Your answer:

Solution: (c), (d), and (e) will all cause the equilibrium to shift to the right. In each case, the concentrations of products will increase, and the concentrations of reactants will decrease. (Because moles of reactants equals moles of products, no change in relative concentrations will occur when stress (b) is imposed on the reaction.)

---

11. Which of the five stresses listed in exercise 10 will alter the K_{eq} value?

Your answer:

Solution: The K_{eq} value is altered only by a change in temperature; (e) will increase the K_{eq} for the reaction.

II. Quantitative Aspects of Equilibrium

1. Consider the reaction

$$N_2(g) + 3\ H_2(g) \rightleftharpoons 2\ NH_3(g)$$

At equilibrium there are 6.0 moles of NH_3, 3.0 moles of N_2, and 4.0 moles of H_2 coexisting in a 10.0 L flask at some (unstated) temperature. Calculate the equilibrium constant for this reaction (at this temperature).

Your answer:

Solution: $K_{eq} = 19\ M^{-2}$

Solve the problem with the following steps.

(1) Write the equilibrium expression.

$$K_{eq} = \frac{[NH_3]^2}{[N_2]\ [H_2]^3}$$ ([ ] means concentration in moles per liter)

To calculate the K_{eq} value, you will substitute the equilibrium concentrations of NH_3, N_2, and H_2 into the equilibrium expression.

(2) Calculate the concentrations in molarity.

$$\frac{6.0 \text{ moles } NH_3}{10.0 \text{ L}} = 0.60 \text{ M } NH_3$$

$$\frac{3.0 \text{ moles } N_2}{10.0 \text{ L}} = 0.30 \text{ M } N_2$$

$$\frac{4.0 \text{ moles } H_2}{10.0 \text{ L}} = 0.40 \text{ M } H_2$$

(3) $K_{eq} = \frac{(0.60 \text{ M})^2}{(0.30 \text{ M})(0.40 \text{ M})^3} = 19 \text{ M}^{-2} \text{ or } \frac{19}{\text{M}^2}$

Note: in equilibrium constant calculations, the units are sometimes omitted.

---

2. For the reaction

$$H_2(g) + I_2(g) \rightleftharpoons 2\ HI(g)$$

equilibrium is established at a given temperature when the following concentrations of the reaction components are present: $[H_2]$ = 0.10 moles/liter, $[I_2]$ = 0.10 moles/liter, and $[HI]$ = 1.8 moles/liter.

(a) Calculate the value of the equilibrium constant at this temperature.

Your answer:

Solution: $K_{eq} = 3.2 \times 10^2$ (no units)

(1) Write the equilibrium expression.

$$K_{eq} = \frac{[HI]^2}{[H_2]\ [I_2]}$$

(2) The equilibrium concentrations of the reaction components are provided in molarity. Substitute these values into the equilibrium expression.

$$K_{eq} = \frac{(1.8 \text{ M})^2}{(0.10 \text{ M})(0.10 \text{ M})} = 3.2 \times 10^2$$

(b) 2.0 moles of hydrogen and 1.0 mole of iodine are put into a 5.0 L flask at a different temperature. After equilibrium is established, 0.90 mole of hydrogen iodide is present. Calculate the K_{eq} at this new temperature.

Your answer:

Solution: $K_{eq} = 0.95$

(1) Write the equilibrium expression.

$$K_{eq} = \frac{[HI]^2}{[H_2]\ [I_2]}$$

(2) Calculate the initial molarity of H_2 and I_2 and the equilibrium molarity of HI.

$$\frac{2.0 \text{ moles } H_2}{5.0 \text{ L}} = 0.40 \text{ M } H_2 \text{ (initially)}$$

$$\frac{1.0 \text{ mole } I_2}{5.0 \text{ L}} = 0.20 \text{ M } I_2 \text{ (initially)}$$

$$\frac{0.90 \text{ mole HI}}{5.0 \text{ L}} = 0.18 \text{ M HI (equilibrium)}$$

(3) Construct a table as shown.

	$[H_2]$	$[I_2]$	$[HI]$
initial concentration	0.40	0.20	0[a]
reaction change	$-x$[b]	$-x$[b]	$+2x$[b]
equilibrium concentration	$0.40 - x$[c]	$0.20 - x$[c]	$0 + 2x$[c]

[a] Assume that no HI is present initially.

[b] Some H_2 and I_2 must react to form HI (in order to establish equilibrium). Use "x" to indicate the reacting concentrations. The reaction will occur stoichiometrically, implying that for every mole of H_2 that reacts (the negative sign denotes "reacts"), a mole of I_2 will also react to form two moles of HI (the positive sign denotes formation).

[c] Sum the components of each column to determine the relative concentrations at equilibrium. However, from the original problem you know the $[HI]_{eq} = 0.18$ M. Therefore $2x = 0.18$, or $x = 0.09$ M. Knowing x, you can calculate

$$[H_2]_{eq} = 0.40 - 0.09 = 0.31 \text{ M}$$

$$[H_2]_{eq} = 0.20 - 0.09 = 0.11 \text{ M}$$

(4) Calculate K_{eq}.

$$K_{eq} = \frac{(0.18)^2}{(0.31)(0.11)} = 0.95 \text{ or } 9.5 \times 10^{-1}$$

---

4. For a gaseous reaction $2\text{ A} + \text{B} \rightleftharpoons 2\text{ C}$, 3.00 moles of A and 3.00 moles of B are mixed in a 1.00 L flask. The concentration of C at equilibrium is found to be 0.150 M. Calculate the K_{eq} for the reaction.

Your answer:

Solution: $K_{eq} = 9.49 \times 10^{-4}\ M^{-1}$

(1) Write the equilibrium expression.

$$K_{eq} = \frac{[C]^2}{[A]^2\ [B]}$$

(2) Calculate the initial molarity of A and B.

$$\frac{3.00 \text{ moles A}}{1.00 \text{ L}} = 3.00 \text{ M A (initially)}$$

$$\frac{3.00 \text{ moles B}}{1.00 \text{ L}} = 3.00 \text{ M B (initially)}$$

Notice that the concentration of C is provided in molarity units: 0.150 M.

(3) Construct a table.

	[A]	[B]	[C]
initial concentration	3.00	3.00	0[a]
reaction change	$-2x$[b]	$-x$[b]	$+2x$[b]
equilibrium concentration	$3.00 - 2x$[d]	$3.00 - x$[e]	$0 + 2x$[c]

[a] Assume that no C is present initially.

[b] Some A and B must react to form C (in order to establish equilibrium). Because the reaction to establish equilibrium occurs stoichiometrically, 2x of A will react with 1x of B to form 2x of C.

[c] $[C]_{eq}$: $0.150 = 2x$. Therefore $x = 0.0750$ M.

[d] $[A]_{eq}$: $3.00 - 2x = 3.00 - 2 \times (0.0750) = 2.85$.

[e] $[B]_{eq} = 3.00 - x = 3.00 - 0.0750 = 2.925 = 2.92$.

(4) Calculate K_{eq}.

$$K_{eq} = \frac{(0.150\ M)^2}{(2.85\ M)^2 (2.92\ M)} = 9.49 \times 10^{-4}\ M^{-1}$$

---

5. At some (unstated) temperature, the equilibrium constant for the reaction $X(g) + Y(g) \rightleftharpoons 2\ Z(g)$ is 100. What are the equilibrium concentrations of X, Y, and Z if the initial concentrations of X and Y are each 0.05 M?

Your answer:

Solution: $[X]_{eq} = 0.01\ M$; $[Y]_{eq} = 0.01\ M$; $[Z]_{eq} = 0.08\ M$

(1) Write the equilibrium expression.

$$K_{eq} = \frac{[Z]^2}{[X]\ [Y]} = 100$$

(2) Construct a table:

	[X]	[Y]	[Z]
initial concentration	0.05	0.05	0
reaction change	-a	-a	+2a
equilibrium concentration	0.05 - a	0.05 - a	0 + 2a

(3) Insert the equilibrium values into the equilibrium expression.

$$\frac{(2a)^2}{(0.05 - a)(0.05 - a)} = 100$$

To solve for a, take the square root of both sides of the equation.

$$\frac{2a}{0.05 - a} = 10$$

$$2a = 10 \times (0.05 - a) = 0.50 - 10a$$

$$12a = 0.5$$

$$a = 0.04$$

(4) Substitute the value of a as shown to determine the equilibrium concentrations of X, Y, and Z.

$$[X]_{eq} = 0.05 - a = 0.05 - 0.04 = 0.01\ M$$

$$[Y]_{eq} = 0.05 - a = 0.05 - 0.04 = 0.01\ M$$

$$[Z]_{eq} = 2a = 2(0.04) = 0.08\ M$$

---

6. Consider the reaction $A(g) + 2\ B(g) \rightleftharpoons C(g)$. If you initially have 3.0 moles of C in a 5.0 L flask, and at equilibrium find that you have 0.80 mole of C remaining, what is the K_{eq} for the reaction? Assume that the temperature remains constant.

Your answer:

Solution: $K_{eq} = 0.47\ M^{-2}$

(1) Write the equilibrium expression.

$$K_{eq} = \frac{[C]}{[A]\ [B^2]}$$

(2) Construct a table.

	[A]	[B]	[C]
initial concentration	0	0	$\frac{3.0}{5.0} = 0.60$
reaction change	+x	+2x	-x
equilibrium concentration	0 + x	0 + 2x	$\frac{0.80}{5.0} = 0.16$

(3) It is apparent that if you have 0.60 M C initially and 0.16 M C when equilibrium is established, then 0.44 M C must have reacted (0.60 - x = 0.16; x = 0.44). If x = 0.44 M, then at equilibrium you have

[A] = x = 0.44 M

[B] = 2x = 0.88 M

[C] = 0.16 M

(4) Calculate the equilibrium expression.

$$K_{eq} = \frac{(0.16\ M)}{(0.44\ M)(0.88\ M)^2} = 0.47\ M^{-2}$$

11 Acids and Bases

I. Definitions and Concepts

A. ACIDS

1. Write an equation for the reaction of nitric acid with water and explain how the nitric acid is consistent with the Brønsted-Lowry definition of an acid.

Your answer:

Solution: $HNO_3 + H_2O \longrightarrow H_3O^+ + NO_3^-$

According to the Brønsted-Lowry definition of an acid, nitric acid is a "proton donor" because it gives an H^+ ion to the water molecule. An earlier concept of acids (Arrhenius) states that an acid "can produce H_3O^+ ions in aqueous solution." Because many of your lab experiments are performed in water solutions, you will sometimes use the Arrhenius "language" concerning acids and bases.

---

2. Write an equation showing the acidic behavior of a bicarbonate ion in water.

Your answer:

Solution: $HCO_3^- + H_2O \rightleftharpoons H_3O^+ + CO_3^{2-}$

Why is this reaction represented as an equilibrium reaction, whereas the dissociation of nitric acid is represented as "going virtually to completion" (to the right)? You will study this principle in Section II.

B. BASES

1. Write an equation for the reaction of ammonia with water. Explain why ammonia is consistent with the Brønsted-Lowry definition of a base.

Your answer:

Solution: $NH_3 + H_2O \rightleftharpoons NH_4^+ + OH^-$

An ammonia molecule accepts a proton (H^+) from the water molecule in a reversible (equilibrium) reaction. Ammonia is also consistent with the Arrhenius definition of a base in that it "increases the hydroxide ion concentration in water."

---

2. Write an equation for the neutralization of acetic acid, $HC_2H_3O_2$, in vinegar with sodium hydroxide solution.

Your answer:

Solution: $HC_2H_3O_2 + NaOH \rightleftharpoons H_2O + Na^+ + C_2H_3O_2^-$

The hydroxide ion (of sodium hydroxide) accepts the proton (H^+) from acetic acid. Sodium hydroxide solution is a common laboratory source of base.

---

3. When used in baking, baking soda (sodium bicarbonate) reacts with acids from vinegar, sour milk, or fruit juices. Write an equation for the neutralization of bicarbonate ion with hydronium ion. Will this reaction result in an equilibrium mixture?

Your answer:

Solution: $HCO_3^- + H_3O^+ \rightleftharpoons H_2O + H_2CO_3 \longrightarrow 2\ H_2O + CO_2\ (g)$

Carbonic acid is unstable and decomposes into water and carbon dioxide gas. The CO_2 gas escapes from the solution, so few H_2CO_3 molecules remain in solution (in an open container); thus the equilibrium is "upset" and the reaction is "pulled" to the right.

C. CONJUGATE ACID/BASE PAIRS

1. In the equation given, identify the acid and base on the reactant side of the equation and the corresponding conjugate base and conjugate acid from the product side of the equation.

$$HC_2H_3O_2 + H_2O \rightleftharpoons H_3O^+ + C_2H_3O_2^-$$

Your answer:

Solution: $HC_2H_3O_2$ is an acid, and $C_2H_3O_2^-$ is its conjugate base. H_2O is a base, and H_3O^+ is its conjugate acid. Conjugates "belong together" in that members of a conjugate pair differ only by a proton (H^+).

---

2. Identify the conjugates of the acids or bases listed.
   (a) The conjugate base of H_2SO_4.
   (b) The conjugate acid of NH_3.
   (c) The conjugate base of HCl.
   (d) The conjugate acid and the conjugate base of HCO_3^-.

Your answers:

Solution:

(a) H_2SO_4/HSO_4^-, acid/conjugate base

(b) NH_3/NH_4^+, base/conjugate acid

(c) HCl/Cl^-, acid/conjugate base

(d) HCO_3^-/H_2CO_3, base/conjugate acid

HCO_3^-/CO_3^{2-}, acid/conjugate base

D. AMPHIPROTIC (AMPHOTERIC) SUBSTANCES

1. Water molecules may be called amphiprotic or amphoteric. Explain.

Your answer:

Solution: An amphiprotic (amphoteric) substance such as water can either donate or accept protons (H^+). For example,

$H_2O + NH_3 \rightleftharpoons NH_4^+ + OH^-$ (water acting as an acid)

$H_2O + HCl \longrightarrow H_3O^+ + Cl^-$ (water acting as a base)

---

2. What are some other examples of amphiprotic substances?

Your answer:

Solution:

$$HCO_3^- + H_3O^+ \rightleftharpoons H_2O + H_2CO_3 \longrightarrow CO_2(g) + 2\ H_2O$$

$$HCO_3^- + OH^- \rightleftharpoons H_2O + CO_3^{2-}$$

$$HSO_4^- + H_3O^+ \rightleftharpoons H_2SO_4 + H_2O$$

$$HSO_4^- + OH^- \rightleftharpoons H_2O + SO_4^{2-}$$

II. Relative Strengths of Acids and Bases

A. STRONG AND WEAK ACIDS

Acids can be categorized as strong or weak according to their relative tendencies to donate protons to water, that is, their tendency to ionize water. (Do not confuse the terms strong and weak with the terms concentrated and dilute, which refer to the relative proportions of solute present in solution.)

1. Write equations for the reactions of perchloric, sulfuric, hydrochloric, and nitric acids with water. Are these acids weak or strong?

Your answer:

Solution:

$$HClO_4 + H_2O \longrightarrow H_3O^+ + ClO_4^-$$

$$H_2SO_4 + H_2O \longrightarrow H_3O^+ + HSO_4^-$$

$$HCl + H_2O \longrightarrow H_3O^+ + Cl^-$$

$$HNO_3 + H_2O \longrightarrow H_3O^+ + NO_3^-$$

All are strong acids. You may determine this from a table of relative acid strengths, although it is recommended that you memorize the common strong acids. Each of these acids has a greater tendency to give up a proton (H^+) than H_3O^+ does, and because each of these acids is essentially completely dissociated in water solution, you should draw the arrow to the right (in the forward direction).

---

2. Write an equation and an equilibrium expression for each of these weak acids: $HC_2H_3O_2$, acetic acid, and HCN, hydrocyanic acid.

Your answer:

Solution:

$$HC_2H_3O_2 + H_2O \rightleftharpoons H_3O^+ + C_2H_3O_2^-$$

$$K_a = \frac{[H_3O^+]\ [C_2H_3O_2^-]}{[HC_2H_3O_2]}$$

(K_a is often referred to as K_d, the dissociation constant, or K_i, the ionization constant)

$$HCN + H_2O \rightleftharpoons H_3O^+ + CN^-$$

$$K_a = \frac{[H_3O^+]\ [CN^-]}{[HCN]}$$

---

3. At 25°C, the K_a's for several acids are

$HC_2H_3O_2$	1.8×10^{-5}	HF	6.6×10^{-4}
HCN	4.0×10^{-10}	HClO	3.2×10^{-8}

Rank these acids, including HCl, in order of increasing acid strength.

Your answer:

Solution: HCN < HClO < $HC_2H_3O_2$ < HF << HCl

The weakest acid has the smallest K_a.

B. STRONG AND WEAK BASES

1. Rank the conjugate bases of the acids in the previous question, including the conjugate base of water, in order of increasing base strength.

Your answer:

Solution: Cl^- < F^- < $C_2H_3O_2^-$ < ClO^- < CN^- < OH^-

The strongest acid has the weakest conjugate base. According to the Brønsted-Lowry definition, a base is a species that will accept a proton from an acid, which donates the proton. Of the ions listed, OH^- has the greatest tendency to accept a proton and thus is the strongest base present.

---

2. Why are NaOH and KOH considered to be strong bases?

Your answer:

Solution: NaOH and KOH are ionic compounds. When they dissolve in water, their ions are completely dissociated (dispersed). The hydroxide ion is the strongest base possible in a water solution; therefore, the solutions of NaOH and KOH (and of the hydroxides of the other IA and IIA elements) are all strong bases according to the Arrhenius definition.

---

3. Ammonia solution is frequently used in laboratory experiments. Write an equation for the equilibrium reaction of ammonia in water. The dissociation constant, K_b, for the reaction is 1.8×10^{-5}; is this base strong or weak?

Your answer:

Solution:

$$NH_3 + H_2O \rightleftharpoons NH_4^+ + OH^-$$

$$K_b = \frac{[NH_4^+]\ [OH^-]}{[NH_3]} = 1.8 \times 10^{-5}$$

(K_b is also referred to as K_d, the dissociation constant, or K_i, the ionization constant)

Ammonia solution is a weak base; it is apparent from the small dissociation constant that only a small percentage of the ammonia molecules abstract a proton from water. (In Arrhenius language, ammonia is said to increase the concentration of hydroxide ions in water by only a small amount and thus is called a weak base.)

III. K_w, pH, pOH

1. Write a chemical equation for the reversible ionization of water molecules. Write the equilibrium expression for water (which is often referred to as the ion product of water).

Your answer:

Solution:

$$2\ H_2O \rightleftharpoons H_3O^+ + OH^-$$

$$K_w = [H_3O^+]\ [OH^-] \quad \text{or} \quad [H^+]\ [OH^-]$$

At 25°C, K_w is 1.0×10^{-14}. You are expected to memorize this value.

---

2. What is the concentration of H_3O^+ in water (or in any neutral solution)?

Your answer:

Solution: 1.0×10^{-7}

$[H_3O^+] = [OH^-]$ in water (or in any neutral solution). Then

$$[H_3O^+]^2 = 1.0 \times 10^{-14}$$

$[H_3O^+] = 1.0 \times 10^{-7}$

and

$[OH^-] = 1.0 \times 10^{-7}$

---

3. Define pH. Calculate the pH of a solution in which $[H_3O^+] = 0.001$ M.

Your answer:

Solution: pH is a scale used to compare the relative acidity of substances. It is defined as $pH = -\log [H_3O^+]$.

If $[H_3O^+] = 0.001\ M = 1 \times 10^{-3}$ M, then pH = -(-3) = 3.

---

4. For a 0.10 M NaOH solution, calculate the molar concentration of the hydroxide ions, the pOH of the solution, and the pH of the solution. Is this solution acidic or basic?

Your answer:

Solution: NaOH is an ionic compound; in water the ions are completely dissociated. Therefore $[OH^-] = 1 \times 10^{-1}$.

$pOH = -\log [OH^-] = -\log(1 \times 10^{-1}) = -(-1) = 1.$

Since $[H_3O^+][OH^-] = 1 \times 10^{-14}$, then

$$pH + pOH = 14$$
$$pH + 1 = 14$$
$$pH = 13$$

The solution is basic (alkaline).

---

5. Complete the following table.

Solution or substance	$[H_3O^+]$	$[OH^-]$	pH	pOH	Neutral, acidic, or basic
quality distilled water	(a)	(b)	(c)	(d)	neutral
apples	1×10^{-3}	(e)	(f)	(g)	(h)
milk	(i)	(j)	6	(k)	(l)
scouring powder	(m)	1×10^{-5}	(n)	(o)	(p)

Your answers:

(a) (i)

(b) (j)

(c) (k)

(d) (l)

(e) (m)

(f) (n)

(g) (o)

(h) (p)

Solution:

(a) 1×10^{-7} M (b) 1×10^{-7} M (c) 7 (d) 7 (e) 1×10^{-11} M (f) 3 (g) 11 (h) acidic (i) 1×10^{-6} M (j) 1×10^{-8} M (k) 8 (l) acidic (m) 1×10^{-9} M (n) 9 (o) 5 (p) basic

---

6. What is the molarity of a HNO_3 solution that has a pH = 4?

Your answer:

Solution: pH = 4 means that $[H_3O^+] = 1.0 \times 10^{-4}$ M.

HNO_3 is a strong acid (virtually completely dissociated), so the original solution is 1.0×10^{-4} M HNO_3.

---

7. What is the molarity of a NaOH solution with pH = 12?

Your answer:

Solution: pH = 12, so pOH = 14 - 12 = 2; therefore $[OH^-] = 1 \times 10^{-2}$ M.

In a solution of NaOH, $[OH^-]$ is the same as the molarity of the original NaOH solution.

$[NaOH] = 1 \times 10^{-2}$ M = 0.01 M.

---

8. Is the pH of a 0.01 M NH_3 solution higher, lower, or the same as the pH of a 0.01 M NaOH solution? Explain.

Your answer:

Solution: In a 0.01 M NaOH solution $[OH^-] = 0.01$ M $= 1 \times 10^{-2}$ M and pOH = 2. Therefore pH = 14 - 2 = 12. In contrast, NH_3 is a weak base; relatively few of its molecules react with H_2O to take on a proton. (For $NH_3 + H_2O \rightleftharpoons NH_4^+ + OH^-$, $K_b = 1.8 \times 10^{-5}$.) In the 0.01 M

NH_3 solution $[OH^-]$ is much less than 0.01 M, $[H_3O^+]$ is higher than in the NaOH solution, and the pH is less than 12.

← more acidic

pH = 0 pH = 7 pH = 14
neutral

more basic →

IV. Neutralization and Titration

1. A 10.0 mL sample of HCl solution is titrated with 0.04952 M standardized solution of NaOH. To reach the endpoint of the reaction, 22.3 mL of base is used. Write the chemical equation for the reaction and calculate the molarity of the HCl solution.

Your answer:

Solution: $HCl + NaOH \longrightarrow NaCl + H_2O$

At the endpoint (when the indicator changes from its acidic color to its basic color), moles of acid = moles of base.

$$\text{moles of acid} = M_{acid} \times V_{acid}$$

$$\text{moles of base} = M_{base} \times V_{base}$$

$$M_a \times V_a = M_b \times V_b$$

$$M_a \times 10.0 \text{ mL} = 0.04952\text{M} \times 22.3 \text{ mL}$$

$$M_a = \frac{0.04952\text{M} \times 22.3 \text{ mL}}{10.0 \text{ mL}} = 0.110 \text{ M}$$

---

2. A 1.9093 g sample of a monoprotic acid is titrated with 21.6 mL of a 0.4335 M standardized solution of NaOH. Calculate the mass of 1.00 mole of the unknown acid.

Your answer:

Solution: 204 g/mole

moles of base = 0.4335 M × 0.0216 L

$$\text{moles of acid} = \frac{1.9093 \text{ g of sample}}{\text{mass of } 1.00 \text{ mole}}$$

Because the acid is monoprotic

moles of base = moles of acid

$$0.4335 \text{ mole} \times 0.0216 \text{ L} = \frac{1.9093 \text{ g of sample}}{\text{mass of } 1.00 \text{ mole}}$$

$$\text{mass of } 1.00 \text{ mole} = \frac{1.9093 \text{ g}}{0.4335 \text{ mole/L} \times 0.0216 \text{ L}} = 204 \text{ g/mole}$$

V. Hydrolysis of Salts

When dissolved in water, some salts change the pH of the solution, others do not; that is, some salts hydrolyze (react with water); others do not. When you write chemical equations for these reactions, write as ions all substances that are primarily dissociated, but those that are partially dissociated as molecules.

1. Write an equation for the reaction of NaCl with water. Indicate whether the solution is acidic, neutral, or basic as a result of dissolving the NaCl in the water.

Your answer:

Solution:

$NaCl(s) + H_2O \longrightarrow Na^+(aq) + Cl^-(aq)$

The Cl^- is such a very weak base that it cannot attract a proton from the water molecule. Na^+ also does not react with water to change the balance between H_3O^+ and OH^-. Because no hydrolysis occurs, the solution is neutral. NaCl simply dissolves and makes no change in the pH of the solution. The same type of conclusion can be drawn about the solution of any salt consisting of alkali or alkaline earth cations (IA or IIA) and anions from strong acids.

---

2. Write an equation for the hydrolysis of potassium cyanide. In what range is the pH of this solution?

Your answer:

Solution: $KCN + H_2O \rightleftharpoons HCN + K^+(aq) + OH^-(aq)$

Because CN^- is the conjugate base of a weak acid (HCN), it is a strong Brønsted-Lowry base and will pull a H^+ ion from water. Consequently, the OH^- concentration of the solution is increased, and the solution is basic (alkaline) with pH greater than 7.

---

3. Write an equation for the reaction of ammonium nitrate in water. Will the solution be acidic, basic, or neutral?

Your answer:

Solution: $NH_4NO_3 + H_2O \rightleftharpoons NH_3 + H_3O^+ + NO_3^-(aq)$

The NO_3^- is the conjugate base of a strong acid (HNO_3), so it is a weak Brønsted-Lowry base and will not extract a proton from water. However, NH_4^+ is a stronger acid than water is and so will donate a proton to water. The salt has increased the $[H_3O^+]$ in the water, so the solution is slightly acidic (pH less than 7).

VI. Acid-Base Buffers

A buffer is a solution containing a weak acid and a salt of the conjugate base of the same weak acid. The purpose of "buffering" a solution is to enable it to counteract small changes in pH.

1. Consider a buffer solution prepared from acetic acid and sodium acetate (of approximately the same concentration).
   (a) Write equations for the equilibria that exist in the solution.
   (b) Write the equation for the reaction that occurs when a small amount of H_3O^+ is added to the buffer solution.
   (c) Write the equation for the reaction that occurs when a small amount of OH^- is added to the buffer solution.

Your answers:

Solution:

(a) $HC_2H_3O_2 + H_2O \rightleftharpoons H_3O^+ + C_2H_3O_2^-$

$NaC_2H_3O_2 + H_2O \longrightarrow Na^+(aq) + C_2H_3O_2^-(aq)$

(b) H_3O^+ from outside source: $H_3O^+ + C_2H_3O_2^- \rightleftharpoons HC_2H_3O_2 + H_2O$

(c) OH^- from outside source: $OH^- + HC_2H_3O_2 \rightleftharpoons C_2H_3O_2^- + H_2O$

In the buffer the concentration of acetic acid molecules is much, much greater than the concentration of H_3O^+ ions (because acetic acid is a weak acid). The concentration of acetate ions (conjugate base of acetic acid) is approximately the same as the concentration of the acetic acid molecules. The acetate ions and the acetic acid molecules act as reservoirs to counteract the effect of an outside source of either H_3O^+ or OH^-. The $C_2H_3O_2^-$ reacts with the hydronium ion and forms (neutral) water. As long as the reservoir of $C_2H_3O_2^-$ is not exhausted, the pH of the solution will remain relatively constant. To counteract the addition of hydroxide to the solution, the acetic acid donates a proton to form (neutral) water. As long as the reservoir of $HC_2H_3O_2$ is not depleted, the pH of the solution will remain relatively constant.

---

2. One of the buffers maintaining the pH of cellular fluids is the $H_2PO_4^-/HPO_4^{2-}$ buffer. Write one equation to illustrate the reaction of this buffer system with excess acid and another equation to illustrate the reaction of this buffer system with excess base.

Your answer:

Solution:

$HPO_4^{2-} + H_3O^+ \rightleftharpoons H_2PO_4^-$

$H_2PO_4^- + OH^- \rightleftharpoons HPO_4^{2-} + H_2O$

---

3. How does each of the changes listed affect the blood buffer system? Write an equation for each circumstance.

(a) Loss of bicarbonate ions due to severe diarrhea.

Your answer:

Solution: $CO_2 + 2\ H_2O \rightleftharpoons H_2CO_3 + H_2O \rightleftharpoons HCO_3^- + H_3O^+$

If the concentration of HCO_3^- decreases (due to diarrhea), the equilibrium will shift to the right; consequently, the concentration of H_2CO_3 will be decreased and CO_2 will be removed from the blood. Because the equilibrium shifts to the right, the concentration of H_3O^+ increases, so the pH of the blood decreases.

(b) Increased carbon dioxide concentration due to poor lung ventilation resulting from emphysema or other respiratory problems.

Your answer:

Solution: $CO_2 + 2\ H_2O \rightleftharpoons H_2CO_3 + H_2O \rightleftharpoons HCO_3^- + H_3O^+$

An increase in the concentration of CO_2 in the blood will shift the equilibrium to the right, ultimately increasing the hydronium ion concentration and lowering the pH of the blood (acidosis).

Part Two
Organic Chemistry and Biochemistry

Study Suggestions for Organic Chemistry

I. Introduction

Some people say that you must memorize most of organic chemistry. Others maintain that you have only to learn the general principles and correctly apply them. Students can combine these methods with excellent results by memorizing the fundamentals daily, practicing them with the workbook questions, and applying them on quizzes and exams.

II. Study Method

A. Read the text prior to the lecture presentation of the material.

B. Soon after lecture, use your lecture notes and text to organize the pertinent material into a "file." Many organic chemistry students have found that 4" × 6" file cards are particularly convenient for this file; sheets of notebook paper lack versatility.

C. The process of organizing the material onto the file cards will help you to organize it in your brain, but this is not sufficient: you will need continually to practice the material by writing and rewriting the reactions and structures of the compounds involved in the reactions.

D. Your file will serve as a compact set of notes that you can easily review at any time. In later chapters of the text, you will rely on your understanding of the characteristics and chemical reactions of each type of organic compound studied, and you will find that your file will enable you to easily refer back to previous reactions.

III. Suggested Form for Your File

A. Develop major headings as you progress through the material such as alkanes (open chain and cycloalkanes), alkenes, and so on.

B. Within each major heading, list subtopics such as nomenclature (systematic and common names of important compounds), "characteristic" group of this series (the functional group), physical properties (compared with other types of compounds and/or trends within the series of compounds), chemical properties—reactions, and miscellaneous information about the type of compound.

C. For each reaction, include these details.
   (1) A title or a leading question
   (2) An example from the lecture or text
   (3) A "type" (generalized) reaction, in which you use R's to represent alkyls attached to the functional group
   (4) A mechanism, if given in lecture, that includes details of the reaction in both equation form and words

D. Develop a format so that you can use the cards to test your knowledge. Reactions should be aligned on the card so that you can cover the products or the starting compounds and reagents and then write the missing components of the reaction.

E. Consider this sample card

ALKENES

Chemical Properties—Reactions

1. Electrophilic Addition

a. example

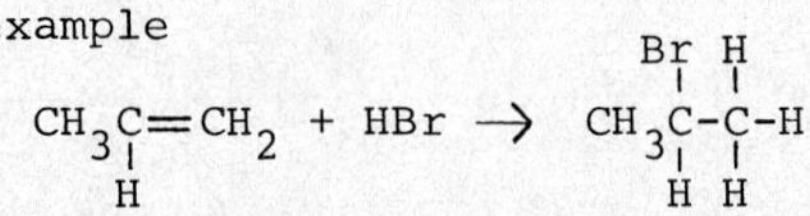

b. "type" reaction and mechanism

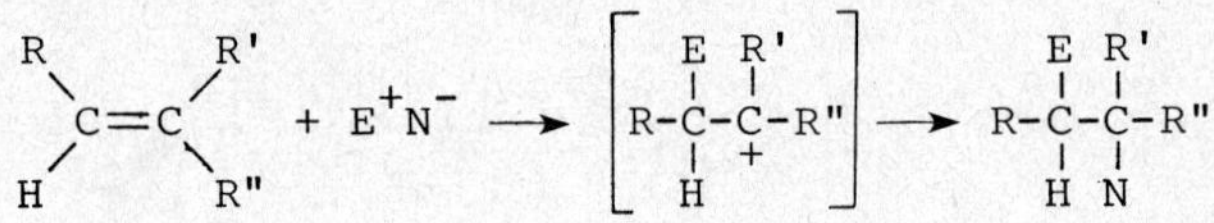

carbocation

12 Organic Chemistry—Introductory Terminology

I. Hybridization of Carbon Atoms

The fundamentals of this topic are covered in Chapter 6, "Bonding and Lewis Dot Formulas." For each molecule, draw a complete structural formula, identify the hybridization of each carbon atom and the bond angles, and indicate the geometry required for the specified type of hybridization.

1. CH_3CH_3, ethane

Your answer:

Solution: 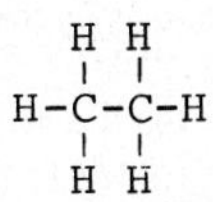

Each carbon atom is bonded to four other atoms and therefore requires four hybrid orbitals. One s orbital and three p orbitals merge to form each of four sp^3 hybrid orbitals of carbon, which overlap with the s orbitals of hydrogen to form the sigma (σ) bonds. The four pairs of bonding electrons on each carbon atom are as far apart as possible (109.5°) and generate a tetrahedral geometry.

---

2. $CH_2{=}CH_2$, ethene

Your answer:

Solution: 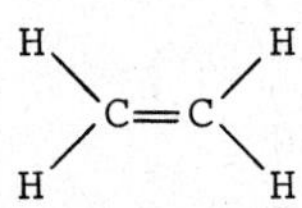

In ethene each carbon is bonded to three atoms. This necessitates three bonding orbitals or sp^2 hybridization to form the sigma (σ) bonds. Three electron pairs are farthest from each other at 120° angles in a flat plane. The electron density of the pi bond (from the sideways overlap of p orbitals) is above and below the plane of the atoms bonded to the sp^2 carbons.

---

3. $H{-}C{\equiv}C{-}CH_2CH_3$, 1-butyne

Your answer:

Solution: 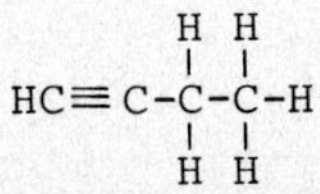

Carbons 1 and 2 are each bonded to two atoms. Each of these carbons needs two orbitals for the sigma bonds; the hybridization is sp, the bond angles are 180°, and the geometry at each of these carbons is linear. Carbons 3 and 4 are sp^3 hybridized and have 109.5° angles and tetrahedral geometry.

---

4. 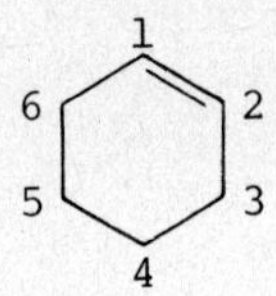

Your answer:

Solution: 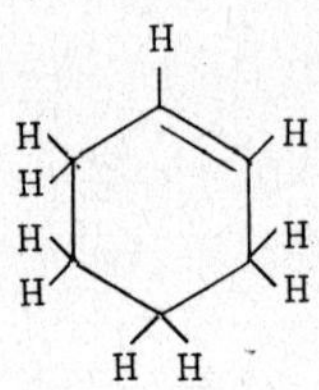

The hydrogens are often omitted in cyclic structures; however, when you do write the hydrogens on the structures, keep in mind that each carbon atom can form only four bonds. Carbons 1 and 2 are sp^2 hybridized and have 120° angles and planar geometry. Carbons 3, 4, 5, and 6 are sp^3 hybridized and have 109.5° angles and tetrahedral geometry.

II. Equivalent Hydrogens (Kinds or Types of Hydrogens) in a Molecule

Construct or visualize models of the molecules as you work the problems in this section.

1. Methane, CH_4

(a) Use a molecular model kit to construct a model of methane. Draw a sketch of the model. What are the hybridization, the angles, and the shape of the molecule?

Your answer:

Solution: 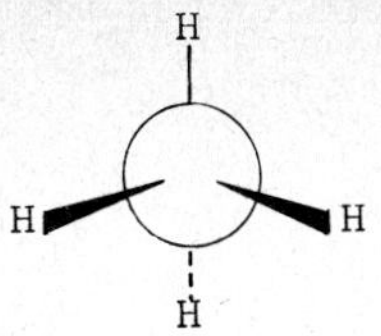 sp^3, 109.5°, tetrahedral shape

---

(b) Represent this molecule by a two-dimensional structure. How many kinds (types) of hydrogen are present in the structure?

Your answer:

Solution: 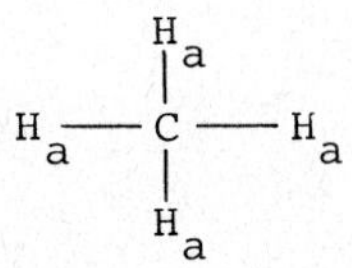

Only one kind (type) of hydrogen exists in this molecule. In other words, there are four equivalent hydrogens, all having identical chemical environments.

---

2. Propane, C_3H_8
   Draw a structure to show the bonding of each atom in this molecule. How many kinds (types) of hydrogens are there? Label them.

Your answer:

Solution: 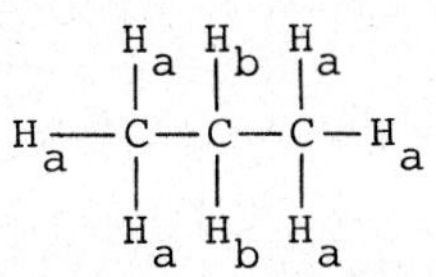

This molecule is symmetrical at the second carbon. Because there is free rotation for single (sigma) carbon-carbon bonds, all the hydrogens on the first and third carbons are alike (equivalent). The two hydrogens on the center carbon are equivalent to each other but are nonequivalent to the hydrogens on the end carbons. Propane has two types of hydrogens (two sets of equivalent hydrogens).

---

3. Pentane, $CH_3CH_2CH_2CH_2CH_3$
   Rewrite the structure and label all sets of equivalent hydrogens.

Your answer:

Solution: 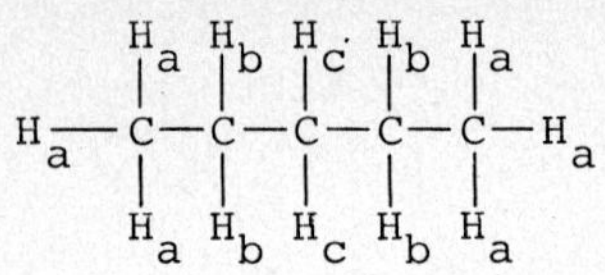

There are three sets of equivalent hydrogens. If two or more hydrogens have identical nearest-carbon-atom neighbors or have nearest-carbon neighbors that are equivalent to each other, then the hydrogens have identical chemical environments (and are equivalent). In the latter case, there is usually symmetry in the molecule.

---

4. 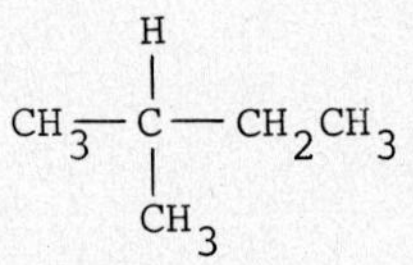

Rewrite the structure and label the different kinds of hydrogens.

Your answer:

Solution: 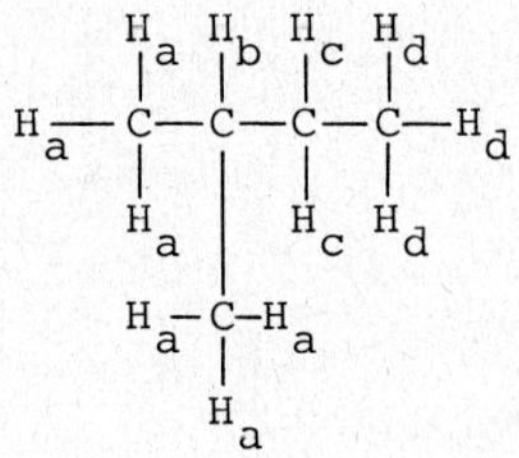

Four kinds of hydrogens (four sets of equivalent hydrogens) are indicated. Notice that the hydrogens designated as d are different from the a hydrogens. H_d is bonded to a carbon whose next-door neighbor is a $-CH_2-$ and whose second-door neighbor is the carbon that is next-door neighbor to the carbons bonded to H_a's.

---

5. 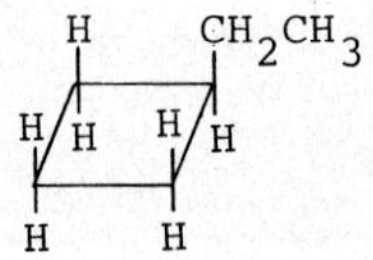 ethylcyclobutane

Rewrite the structure and label the different kinds of hydrogens.

Your answer:

Solution:

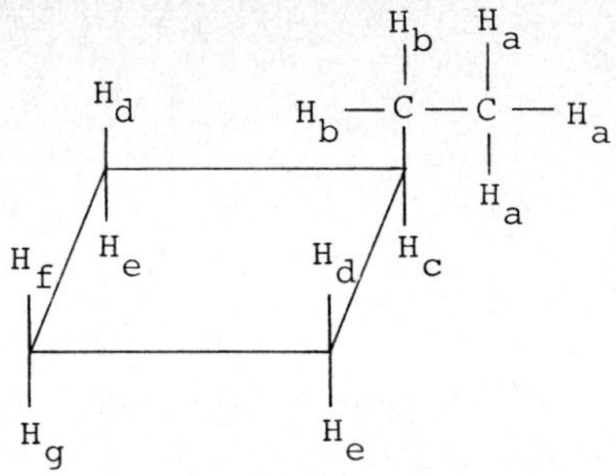

There are seven types of hydrogens (seven sets of equivalent hydrogens). Why are H_d and H_e different? The ring restricts the rotation of the single bond. H_d is on the same side of the ring as the ethyl group and thus is closer to the ethyl group than is H_e. (H_d and ethyl are cis to each other; H_e and ethyl are trans to each other).

---

6. CH_2-$CH_2CH_2CH_3$, 1-chlorobutane
   |
   Cl

   How many sets of equivalent hydrogens are there?

Your answer:

Solution:

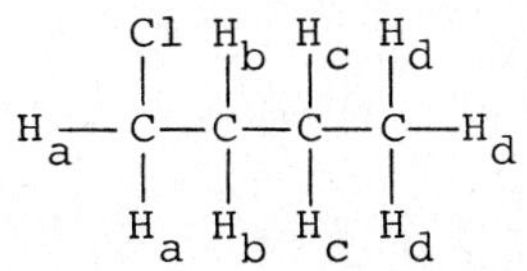

There are four sets of equivalent hydrogens.

III. Structural Isomers

Structural isomers are two or more compounds with the same molecular formulas, but with different structures, and therefore different chemical and physical properties. For each molecular formula, write structural formulas for all the possible different isomers. Identify and eliminate any duplicate structures.

1. C_4H_{10}

Your answer:

Solution:

$CH_3CH_2CH_2CH_3$, butane (n-butane)

     CH_3
     |
CH_3-CH-CH_3, 2-methylpropane (isobutane)

Nomenclature is discussed in the next chapter. If two structures have the same systematic name, they represent the same compound. Sometimes students puzzle over the statement "Draw all the possible structural isomers for C_xH_y" or the question "How many possible structural isomers are there for C_xH_y?" Consider this systematic approach to either question: to determine the number of structural isomers for C_5H_{12}, look at the different kinds of hydrogens in the isomers of C_4H_{10} (a chain of one less carbon). If you replace (in turn) each kind of hydrogen with a methyl group, you will generate four structures (each with molecular formula C_5H_{12}). Eliminate any duplicate structures (there will be one in this example), and the structures that remain will be the three structural isomers of C_5H_{12}. Try this method for the next question.

2. C_6H_{14}

Your answer:

Solution:

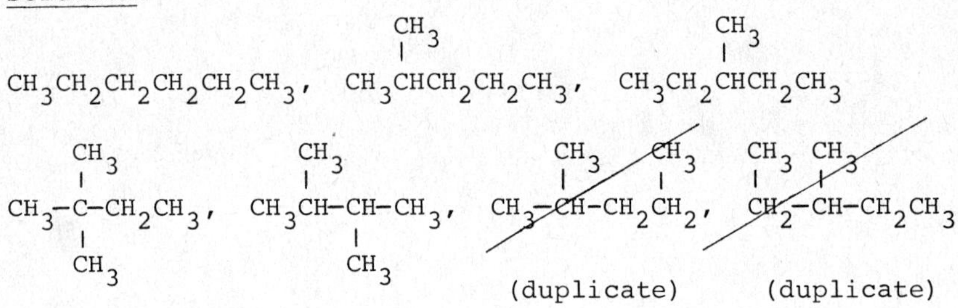

IV. Alkyl Groups

Alkyl groups are carbon branches formed by the removal of one hydrogen from an alkane. For any given alkane, the number of different alkyl groups is the same as the number of kinds of hydrogens in the alkane. Write structures for all possible alkyl groups from each alkane and memorize their names.

1. Methane, CH_4

Your answer:

Solution: CH_3-, methyl

The four hydrogens of methane are equivalent, so removing any one of them yields only one alkyl. The dash after the CH_3 represents the point of attachment of the branch to a carbon chain.

---

2. Ethane, CH_3CH_3

Your answer:

Solution: CH_3CH_2-, ethyl

Ethane has only one kind of hydrogen; consequently, there is only one ethyl group (branch).

---

3. Propane, $CH_3CH_2CH_3$

Your answer:

Solution: Two kinds of hydrogen lead to two alkyl groups: $CH_3CH_2CH_2$- (propyl) and CH_3—CH— (isopropyl).
    |
    CH_3

---

4. The butanes, $CH_3CH_2CH_2CH_3$ and CH_3CHCH_3
    |
    CH_3

Your answer:

Solution: Four kinds of hydrogen in the two butanes lead to four types of butyl branches.

$CH_3CH_2CH_2CH_2$- butyl

    CH_3
    |
CH_3CH_2CH- secondary butyl (sec-butyl)

    CH_3
    |
CH_3-C- tertiary butyl (t-butyl)
    |
    CH_3

CH_3-CH-CH_2- isobutyl
    |
    CH_3

V. Classification of Carbons and Hydrogens

From one compound to another, carbons with the same number of carbons bonded to them have some chemical similarity. We classify carbons (and the hydrogens bonded to them) as primary (1°), secondary (2°), and tertiary (3°). There are quaternary (4°) carbons, but not hydrogens.

1. What types of carbons and hydrogens are present in this molecule? Label them.

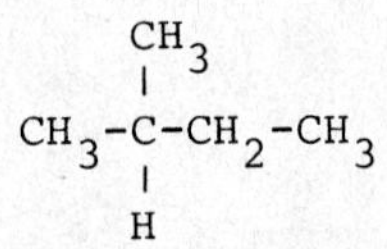

Your answer:

Solution:

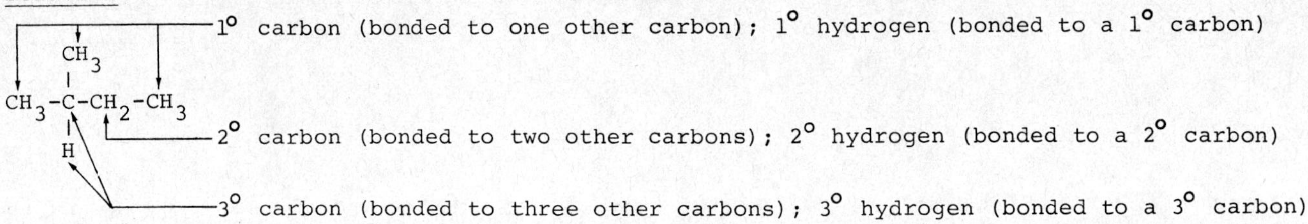

---

2. Label the types of carbons and hydrogens in this molecule.

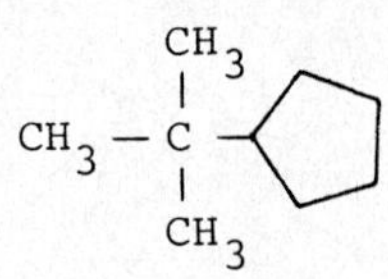

Your answer:

Solution:

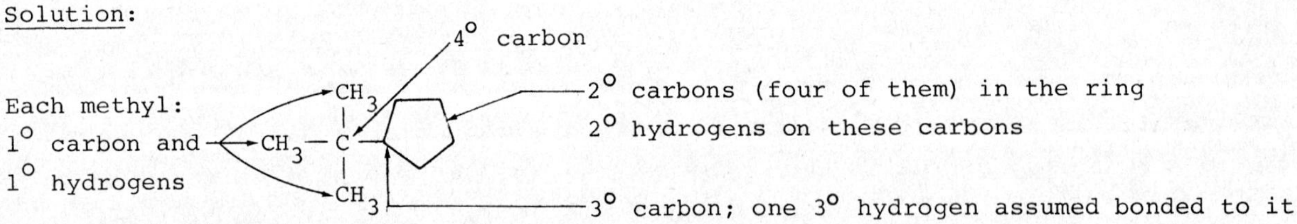

---

13 Nomenclature of Aliphatic Hydrocarbons and Halogenated Derivatives

I. Alkanes

1. Write the systematic name for each of the alkane structures shown.

(a) $CH_3CH_2CH_2\underset{\displaystyle CH_3}{\underset{|}{C}}HCH_2CH_3$

Your answer:

Solution: 3-methylhexane

The longest continuous chain in the structure is a six-carbon chain, hexane. You should number this chain from right to left to indicate the position of the one-carbon branch, methyl, with the lowest possible number.

(b) $CH_3\underset{\displaystyle \underset{\displaystyle \underset{|}{CH_2}}{} \atop \displaystyle CH_3}{\underset{|}{C}}HCH_2CH_2CH_3$

Your answer:

Solution: 3-methylhexane

Alkane chains have "freedom of rotation" around the carbon-carbon single bonds, which implies that the chains can turn or bend in three-dimensional space. Consequently, the longest continuous carbon chain need not appear linear but can be bent as in this example. The compound is identical to the compound in Exercise 1 because both have the same name.

(c) $CH_3CH_2\overset{\displaystyle CH_3}{\overset{|}{\underset{|}{\underset{\displaystyle CH_3}{C}}}}CH_2CH_2CH_3$

Your answer:

Solution: 3,3-dimethylhexane

To indicate the presence of several identical branches, you should use the prefixes di-, tri-, tetra-, penta-, hexa-, and so on. The position of each branch on the longest continuous carbon chain is indicated with a number.

(d) 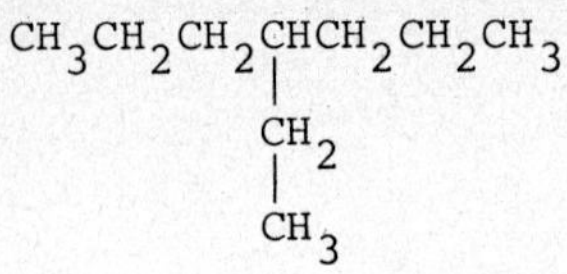

Your answer:

Solution: 4-ethylheptane

(e) 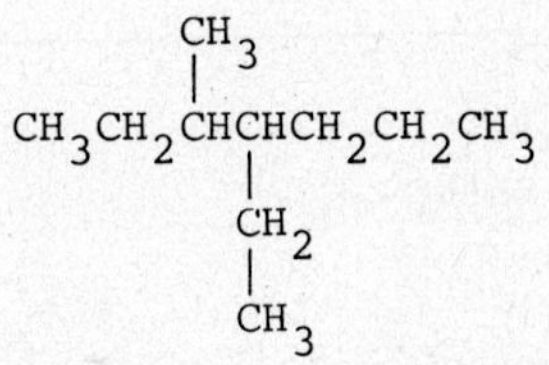

Your answer:

Solution: 4-ethyl-3-methylheptane
Alkyl branches are listed alphabetically in front of the name of the longest continuous carbon chain.

(f) 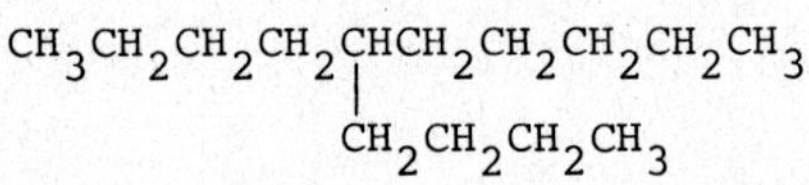

Your answer:

Solution: 5-butyldecane

(g) 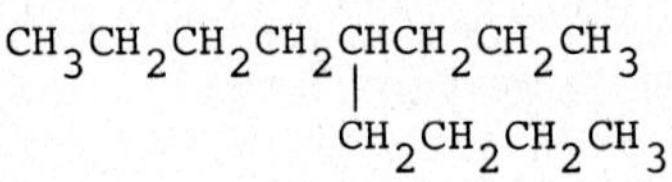

Your answer:

Solution: 5-propylnonane
In this example the longest continuous chain, which is nine carbons long, is not the linear chain.

---

2. Draw the structure of each compound named.
(a) 2,4,6-trimethyloctane

Your answer:

Solution: 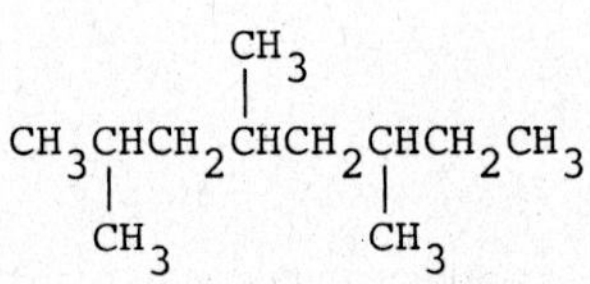

(b) 4,4-diethylnonane

Your answer:

Solution: 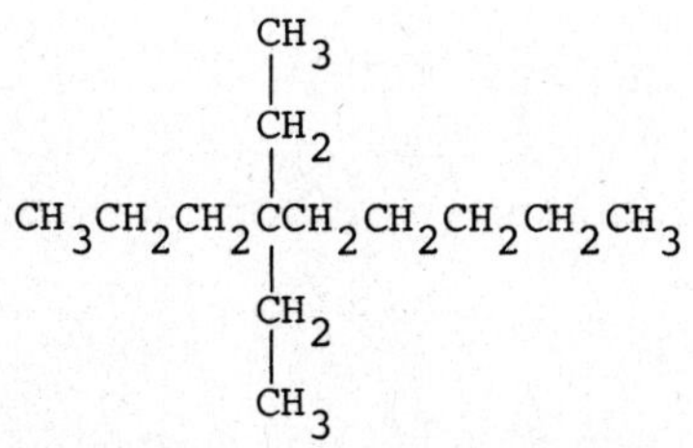

(c) 3-ethyl-2,4-dimethylheptane

Your answer:

Solution: 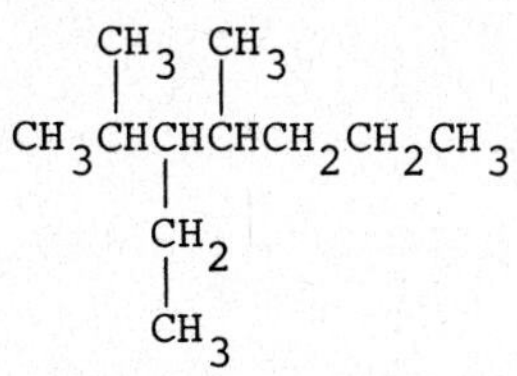

(d) 4-isopropylheptane

Your answer:

Solution: 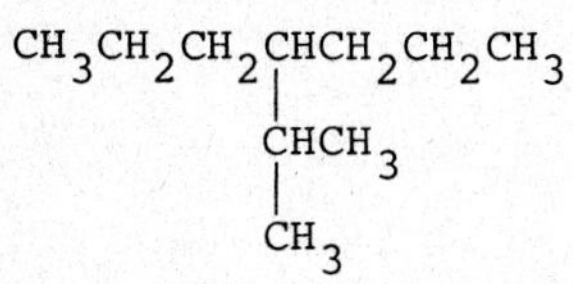

(e) 2,3,4,5-tetramethylhexane

Your answer:

Solution: 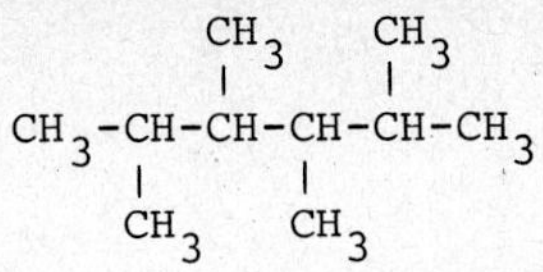

II. Haloalkanes: Open Chain and Cyclic

1. Write the systematic name for each structure shown.

(a) 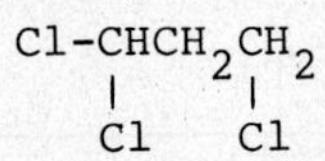

Your answer:

Solution: 1,1,3-trichloropropane

(b) 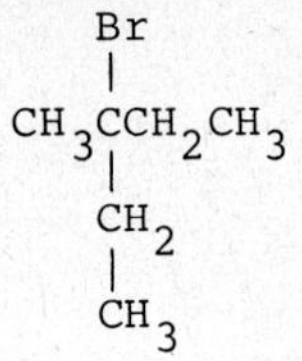

Your answer:

Solution: 3-bromo-3-methylpentane

(c) 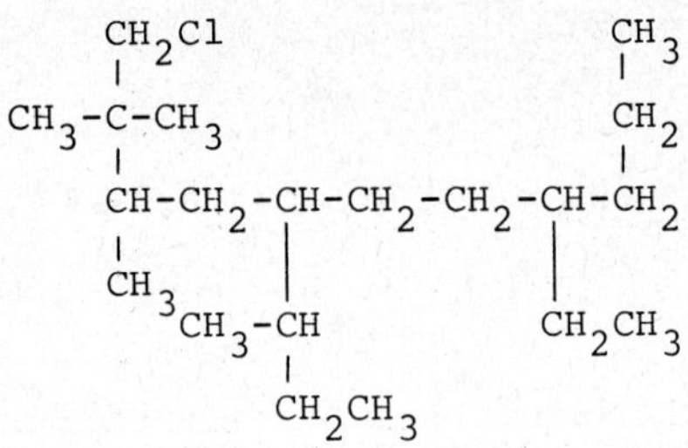

Your answer:

Solution: 1-chloro-8-ethyl-2,2,3-trimethyl-5-sec-butylundecane

---

(d) 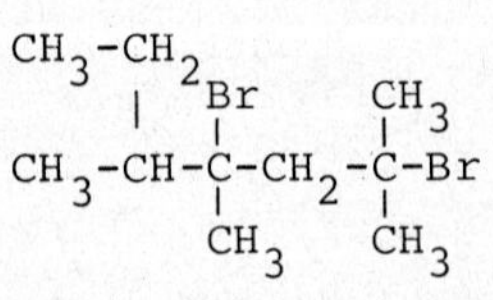

Your answer:

Solution: 2,4-dibromo-2,4,5-trimethylheptane

(e) 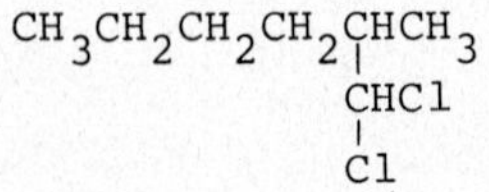

Your answer:

Solution: 1,1-dichloro-2-methylhexane

(f) 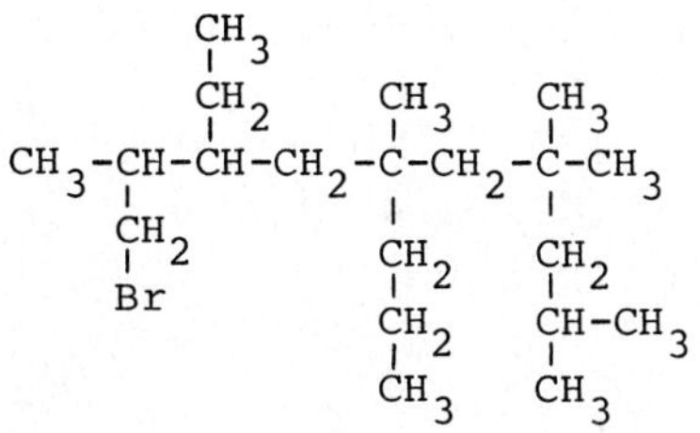

Your answer:

Solution: 1-bromo-3-ethyl-2,5,7,7,9-pentamethyl-5-propyldecane

---

(g) 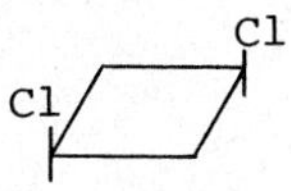

Your answer:

Solution: cis-1,3-dichlorocyclobutane

(h) 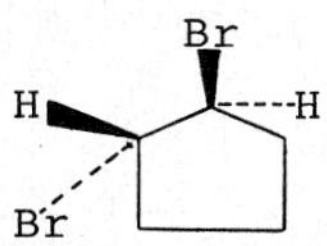

Your answer:

Solution: trans-1,2-dibromocyclopentane

---

2. Draw the structure of each compound named.

(a) 2,2,4,4-tetrabromopentane

Your answer:

Solution: 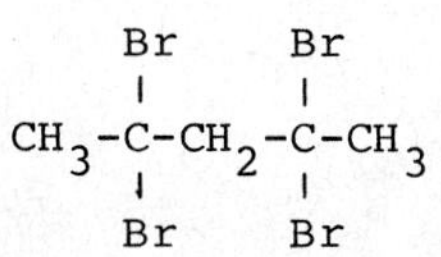

(b) 2-chloro-2-methylbutane

Your answer:

```
                      Cl
                      |
Solution:  CH3-C-CH2CH3
                      |
                      CH3
```

---

3. 2-bromo-2,3,4-trichloropentane

Your answer:

```
                      Br
                      |
Solution:  CH3-C-CH-CH-CH3
                      |   |    |
                      Cl Cl Cl
```

(d) 1,6-dibromo-2,5-dichloro-4-ethyl-3-methyloctane

Your answer:

```
                           Cl  CH3    Cl
                            |    |      |
Solution: Br-CH2-CH-CH-CH-CH-CH-CH2-CH3
                                  |       |
                                 CH2     Br
                                  |
                                 CH3
```

(e) cis-1,4-dibromocyclohexane

Your answer:

Solution:

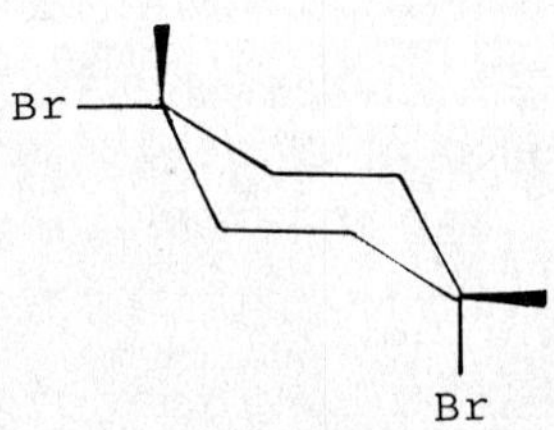

or

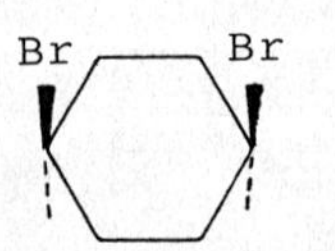

III. Aliphatic Hydrocarbons and Halogenated Derivatives

1. Write the systematic name of each compound and indicate its general molecular formula.

(a) 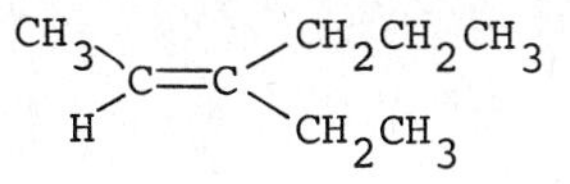

Your answer:

Solution: cis-3-ethyl-2-hexene, C_nH_{2n}

To determine if a compound should be named as "cis" or "trans" when no two groups bonded to the double bond are the same, follow the longest continuous carbon chain through the double bond. The compound is called "cis" if the longest continuous carbon chain lies completely above or completely below a line drawn through the carbons with the double bond. The "trans" isomer has the longest carbon chain crossing the line drawn through the carbons with the double bond. To determine the general molecular formula for the compound, you should subtract two hydrogens for each pi bond or ring from the general molecular formula for a saturated, open-chain alkane (C_nH_{2n+2}). The double bond of this molecule contains one pi bond, so the general molecular formula for the compound is C_nH_{2n}.

(b) 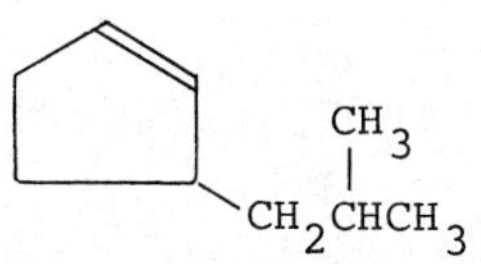

Your answer:

Solution: 3-isobutylcyclopentene, C_nH_{2n-2}

(c) H-C≡C-H

Your answer:

Solution: ethyne or acetylene (common name), C_nH_{2n-2}

A triple bond contains two pi bonds; therefore, you should subtract four hydrogens from the general molecular formula for a saturated, open-chain alkane.

(d) 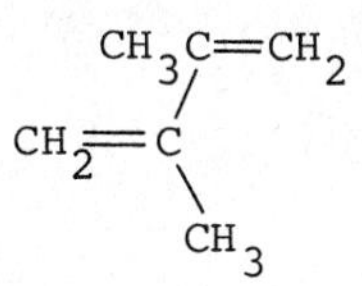

Your answer:

Solution: 2,3-dimethyl-1,3-butadiene, C_nH_{2n-2}

(e) 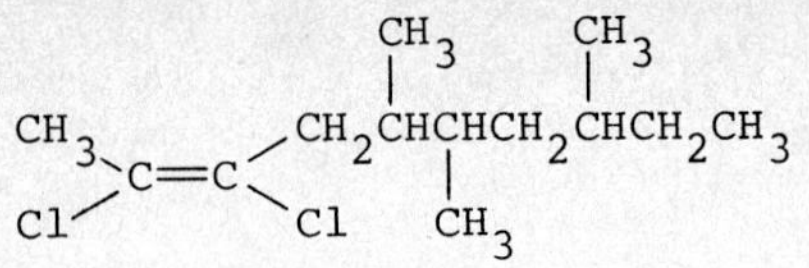

Your answer:

Solution: cis-2,3-dichloro-5,6,8-trimethyl-2-decene, $C_nH_{2n-2}Cl_2$

You should subtract two hydrogens for the pi bond (of the double bond) and two hydrogens for the two chlorines from the general molecular formula for a saturated, open-chain alkane.

---

2. Draw the structure of each compound named and determine the general molecular formula for the compound.

(a) trans-4-chloro-4-isopropyl-6-methyl-2-heptene

Your answer:

Solution: 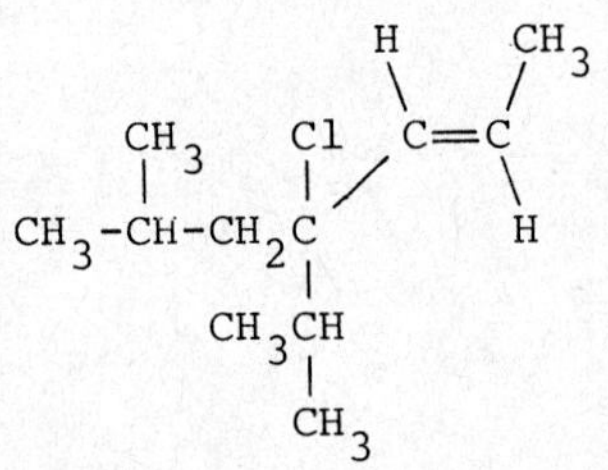 $C_nH_{2n-1}Cl$

(b) t-butylcyclohexane

Your answer:

Solution: 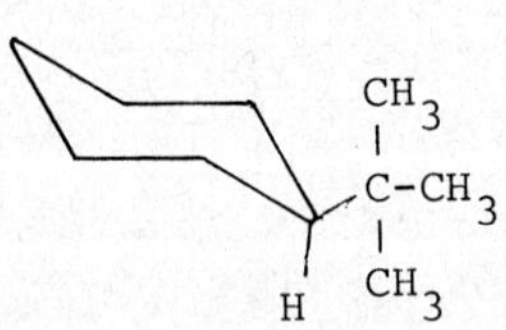 C_nH_{2n}

(c) 5-bromo-6-chloro-4-isopropyl-5-methyl-2-heptyne

Your answer:

Solution: 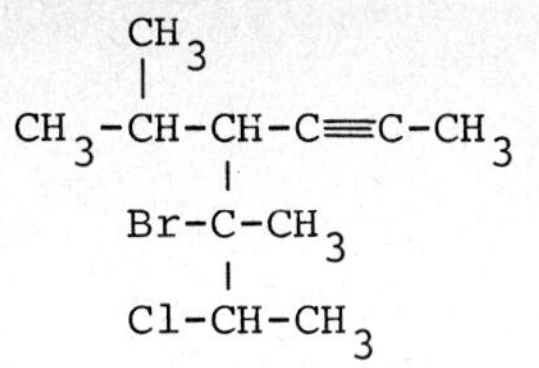 $C_nH_{2n-4}BrCl$

---

3. Write the structural formulas and the systematic names for all possible isomers of C_4H_8 that are alkenes. What types of isomers are represented by these compounds?

Your answer:

Solution: 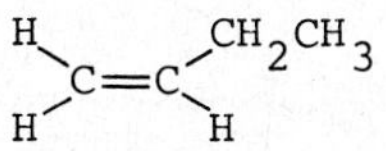 ; 1-butene (This compound has no cis or trans isomers because carbon one has two identical groups attached to it.)

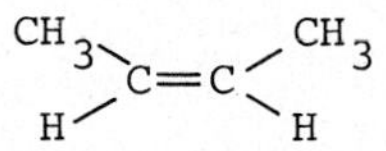 ; cis-2-butene 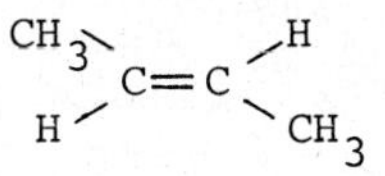 ; trans-2-butene

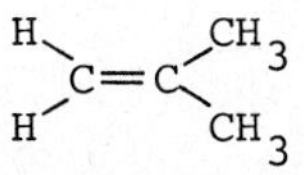 ; 2-methyl-1-propene (Methylpropene is acceptable because neither number is necessary.)

The structural isomers are 1-butene, methylpropene, and cis- or trans-2-butene. The stereoisomers (geometric isomers) are cis-2-butene and trans-2-butene.

---

4. Write the structural formulas and the systematic names for all possible isomers of C_4H_6 that are alkynes.

Your answer:

Solution: $H-C\equiv C-CH_2CH_3$, 1-butyne

$CH_3-C\equiv C-CH_3$, 2-butyne

The molecular formula C_4H_6 implies that there can be two pi bonds or one pi bond and one ring in the compound. Other isomers that are not alkynes are 1,3-butadiene, 1,2-butadiene, cyclobutene, and methylcyclopropene.

---

14 Properties and Reactions of Alkanes, Alkenes, and Alkynes

I. Alkanes

A. PHYSICAL PROPERTIES

1. If you perform an experiment involving test tube reactions of hydrocarbons, why should you not rinse the test tubes with distilled water between each reaction? How should you rinse them?

Your answer:

Solution: Because hydrocarbon molecules have little or no polarity, they are known as nonpolar solvents. (Carbon and hydrogen have similar electronegativities, and in many molecules the symmetry of the molecular shape enables cancellation of even the very small dipoles.) Water is well known as a polar solvent, and due to the polarity of the water molecule and the hydrogen bonding between molecules, water molecules exhibit a high attraction for each other as well as for other polar molecules and ions. This property is expressed in the common phrase "like dissolves like," which implies that nonpolar compounds do not dissolve well in polar solvents. (The converse is also true: polar compounds do not dissolve well in nonpolar solvents.) Consequently, when a hydrocarbon is added to a test tube rinsed with water, the water molecules will adhere to each other forming large droplets, which will obscure the hydrocarbon test results. To "clean" the test tubes between reactions, you can rinse each with an organic solvent such as acetone, followed by a final rinse with the hydrocarbon you will use in the next reaction.

---

2. Rank the following compounds in order of increasing boiling point: octane, heptane, propane, and decane.

Your answer:

Solution: propane < heptane < octane < decane
Each compound listed is an alkane. Therefore, as is typical of alkanes, each compound is nonpolar and forms no hydrogen bonds with other molecules. Consequently, size is the only factor that influences the boiling point trend: the lower the molecular weight (the smaller the chain length of straight-chain hydrocarbons), the lower the boiling point.

---

3. Rank the following compounds in order of increasing boiling point.

(a) 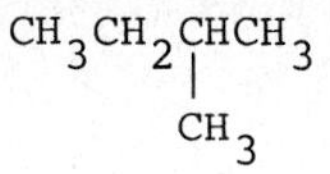 (b) 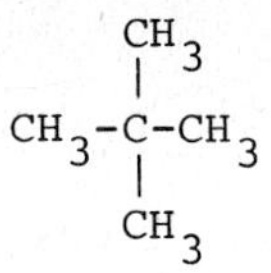 (c) $CH_3CH_2CH_2CH_2CH_3$ (d) $CH_3CH_2CH_2CH_2CH_2CH_3$

Your answer:

Solution: (b) $<$ (a) $<$ (c) $<$ (d)

Compounds (a), (b), and (c) are isomers because they have the same molecular formula, but they differ from each other in the degree of branching. The greater the degree of branching in a molecule, the more the molecule resembles a ball. Molecules with such a shape exhibit much less attraction for each other than do straight chains of carbons that become entangled with each other in the liquid phase. Consequently, a higher temperature is required to boil a sample of compound (c) than is required to boil a sample of compound (b). Compound (d) is not isomeric to the other three compounds, but instead, it has a higher molecular weight and therefore a higher boiling point.

B. CHEMICAL REACTIONS

1. We are dependent daily on one important reaction of alkanes. What reaction is it?

Your answer:

Solution: oxidation (also known as combustion or burning)

Common combustible alkanes include methane, which is natural gas; butane, which is used in cigarette lighters; and isomers of heptane and octane, which are found in gasoline. The chemical equation representing the complete combustion of butane is

$$2\ C_4H_{10} + 13\ O_2\ \text{(excess)} \longrightarrow 8\ CO_2 + 10\ H_2O$$

Incomplete combustion occurs when the oxygen supply is limited and results in some of the carbon deposited as soot or oxidized only to CO (carbon monoxide) rather than to CO_2 (carbon dioxide). Combustion reactions also apply to alkenes and alkynes.

---

2. Write a series of equations representing the complete chlorination of methane.

Your answer:

Solution: The reactions shown are substitution reactions in which a chlorine atom substitutes for a hydrogen atom. The other chlorine atom bonds with the hydrogen (which is removed from the methane) to form hydrogen chloride.

$$CH_4 + Cl_2 \xrightarrow[\text{or heat}]{\text{uv light or sunlight}} CH_3Cl + HCl$$

$$CH_3Cl + Cl_2 \xrightarrow[\text{or heat}]{\text{uv light or sunlight}} CH_2Cl_2 + HCl$$

$$CH_2Cl_2 + Cl_2 \xrightarrow[\text{or heat}]{\text{uv light or sunlight}} CHCl_3 + HCl$$

$$CHCl_3 + Cl_2 \xrightarrow[\text{or heat}]{\text{uv light or sunlight}} CCl_4 + HCl$$

---

3. Complete the following equations for the reactions of alkanes. If no reaction occurs, write "N.R." and explain why it does not.

(a) $CH_3\text{-}\underset{\underset{CH_3}{|}}{CH}\text{-}CH_3 + Br_2 \xrightarrow{\text{dark}}$

Your answer:

Solution: $CH_3\text{-}\underset{\underset{CH_3}{|}}{CH}\text{-}CH_3 + Br_2 \xrightarrow{\text{dark}} \text{N.R.}$

Halogenation of alkanes must be initiated by the action of sunlight or ultraviolet light on the halogen molecule (Cl_2, Br_2).

(b) $CH_3CH_3 + Br_2 \xrightarrow{\text{ultraviolet light}}$

Your answer:

Solution: $CH_3CH_3 + Br_2 \xrightarrow{\text{uv}} CH_3CH_2Br + HBr$

It is reasonable to expect the monobromoethane to react with additional molecules of Br_2 to produce dibrominated, tribrominated, etc., products.

(c) 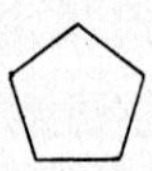 $+ O_2$ (excess) $\longrightarrow$

Your answer:

Solution:

2 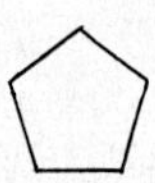 $+ 15\ O_2$ (excess) $\longrightarrow 10\ CO_2 + 10\ H_2O$

($2\ C_5H_{10}$)

"Excess" indicates that sufficient O_2 is available to enable complete combustion. Often you will see "spark" or a similar word written on the arrow to indicate that the reaction is initiated by a spark.

II. Reactions of Alkenes

1. Use the reagent indicated in each part to write the equations for the reactions with each isomer of C_4H_8 (1-butene, cis-2-butene, trans-2-butene, and methylpropene). Write a description of each reaction.

(a) Reaction with Br_2

Your answer:

Solution:

(i) 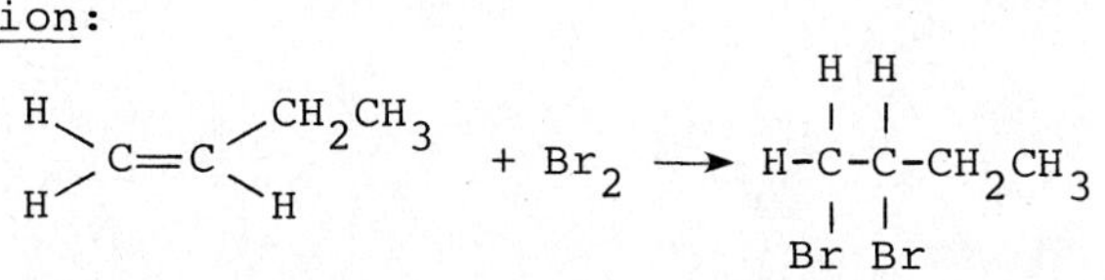

1-butene 1,2-dibromobutane

(ii) 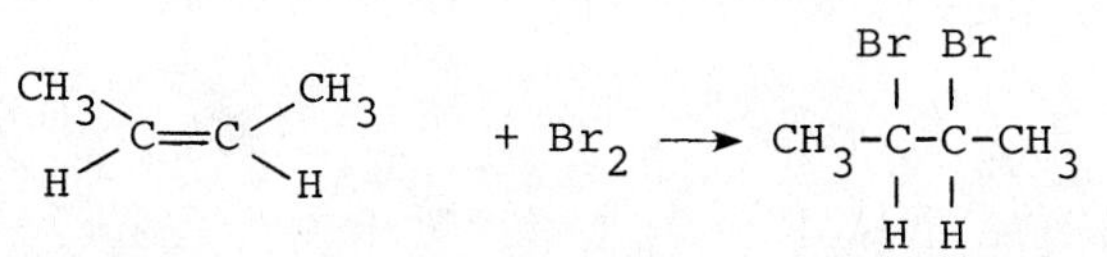

cis-2-butene 2,3-dibromobutane

(iii) 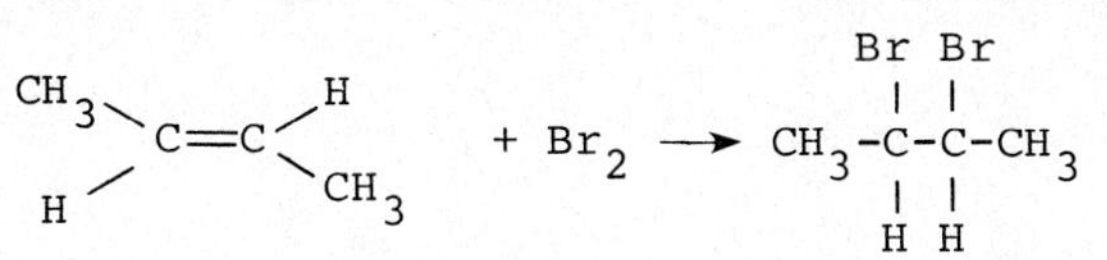

trans-2-butene 2,3-dibromobutane

(iv) 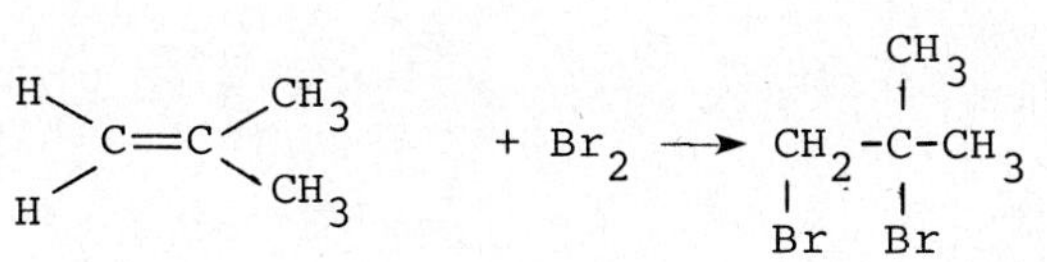

methylpropene 1,2-dibromo-2-methylpropane

In each of these addition reactions, one bromine adds to each of the carbon atoms that contains the double bond in the original alkene. The mechanism of the reaction is shown in part (b), reaction with HBr.

(b) Reaction with HBr (dry)

Your answer:

Solution:

(i)

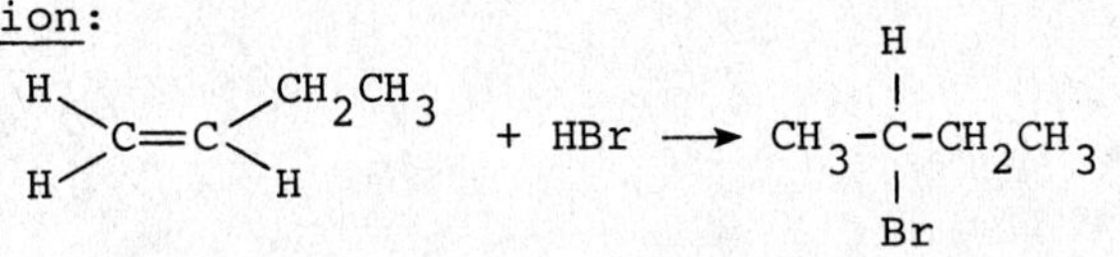

1-butene 2-bromobutane

(ii)

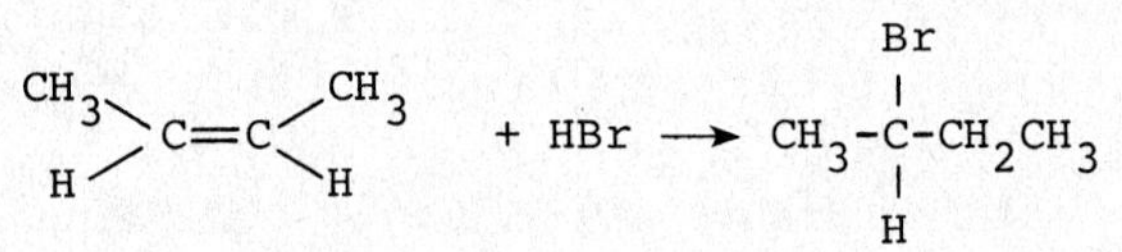

cis-2-butene 2-bromobutane

(iii)

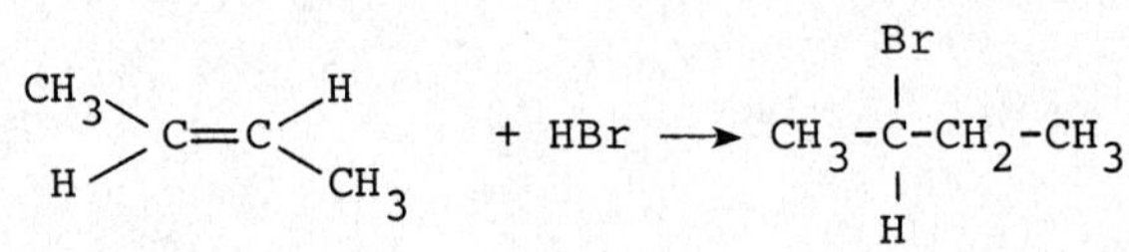

trans-2-butene 2-bromobutane

(iv)

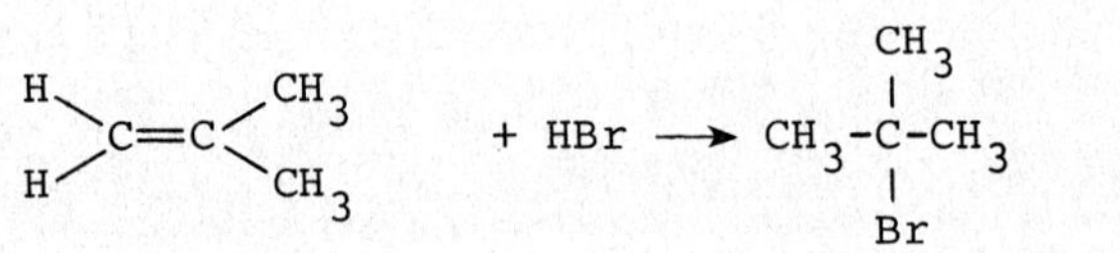

methylpropene 2-bromo-2-methylpropane

In each of these addition reactions, the bromine adds to one carbon of the double bond, and the hydrogen adds to the other carbon. The products shown are the "major" products formed, as predicted by Markovnikov's rule. According to this rule, the hydrogen of the reagent bonds to the carbon (of the carbons sharing the double bond) that already has the larger number of hydrogens bonded directly to it; the remainder of the reagent then adds to the other carbon.

Mechanism (stepwise explanation of the reaction)

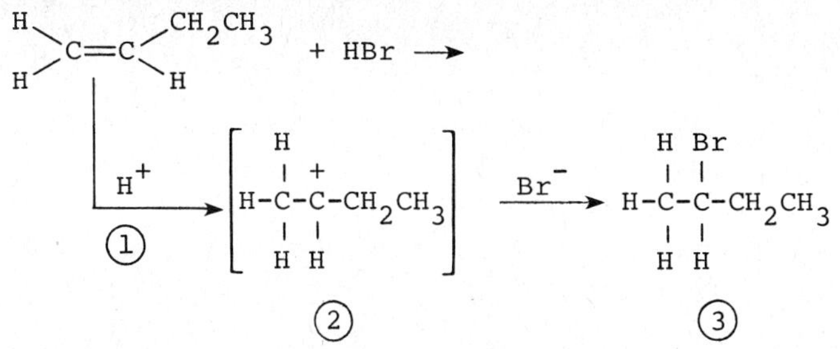

(1) The positive part of the reagent (H^+) is electron-seeking (called the electrophile). It attacks the pi electrons and uses them to bond to one of the carbons with the double bond. Which carbon? The one that would make the poorest (least stable) carbocation always gets the electrophile.

(2) Consequently, carbon 2 becomes "electron-deficient" because the electron it donated to form the pi bond is used to bond H (the electrophile) to carbon 1. Carbon 2 acquires a positive charge and is called a carbocation (the whole sepcies is also called a carbocation), which is a temporary intermediate. This particular carbocation is a secondary carbocation; if H had bonded to carbon 2, the carbocation would be a primary carbocation. The ease of formation of carbocations decreases in the order $3^\circ > 2^\circ > 1^\circ$.

(3) Br^- (the nucleophile) seeks a positive center. It forms a bond by furnishing both electrons to the positively charged carbon.

"Type" Reaction for Electrophilic Addition to Alkenes

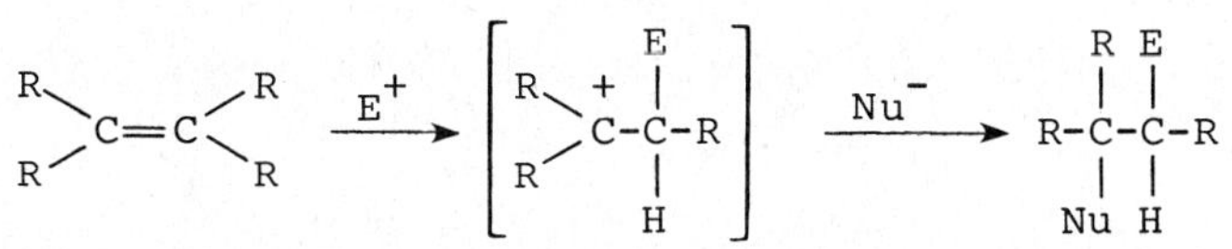

tertiary carbocation
intermediate

R's represent alkyl groups or other organic chains that are not reactive with any of these reagents.

E^+ represents the electrophile.

Nu^- represents the nucleophile.

This "type" reaction represents electrophilic addition to thousands of different alkenes and other compounds containing carbon double bonds. A similar "type" reaction can be written for compounds with triple bonds.

Note: All sigma bonds in the starting compound have remained unchanged. The starting compound and the reagent (represented by E-Nu) have added together to form a saturated compound.

(c) Reaction with H_2O (acid-catalyzed)

Your answer:

Solution:

(i) 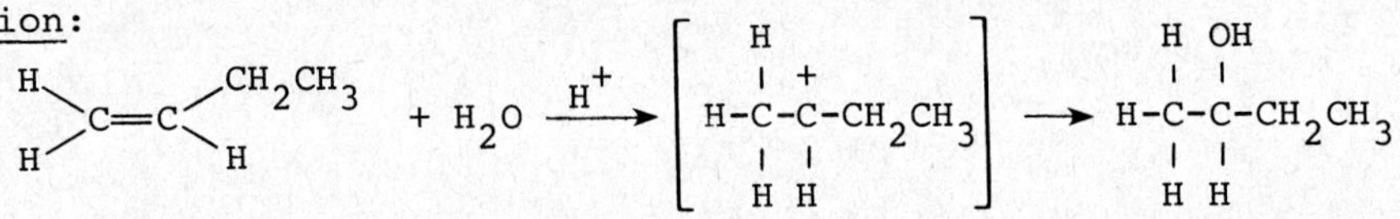

1-butene 2° carbocation 2-butanol (naming of alcohols is presented in Chapter 16)

(ii) 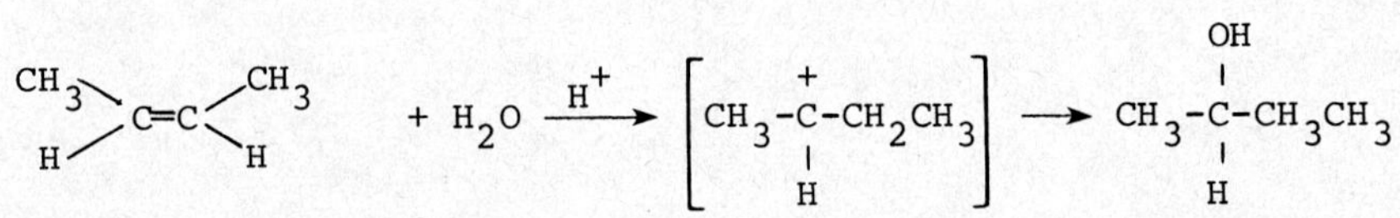

cis-2-butene 2° carbocation 2-butanol

(iii) The reaction for trans-2-butene is identical to the reaction for cis-2-butene.

(iv) 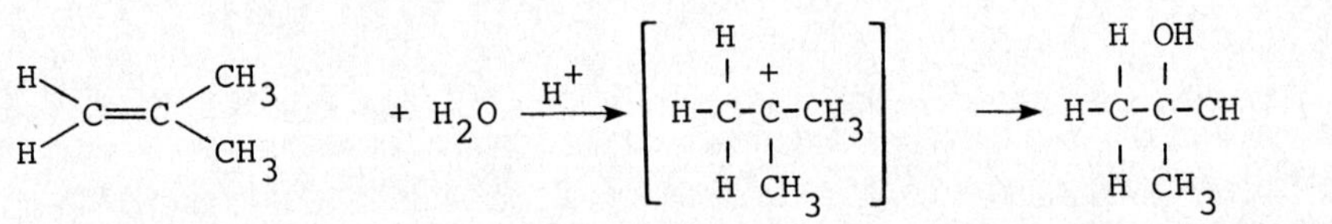

methylpropene 3° carbocation 2-methyl-2-propanol

(d) Reaction with H_2 in the presence of a metal catalyst (Ni or Pt or Pd)

Your answer:

Solution:

(i)

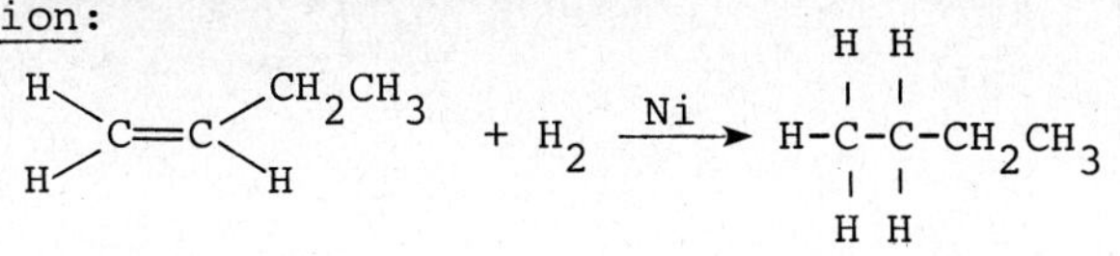

1-butene butane

(ii)

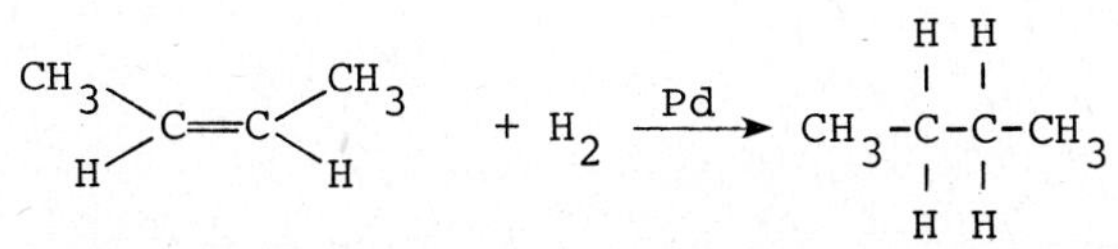

cis-2-butene butane

(iii) The reaction for trans-2-butene is identical to the reaction for cis-2-butene.

(iv)

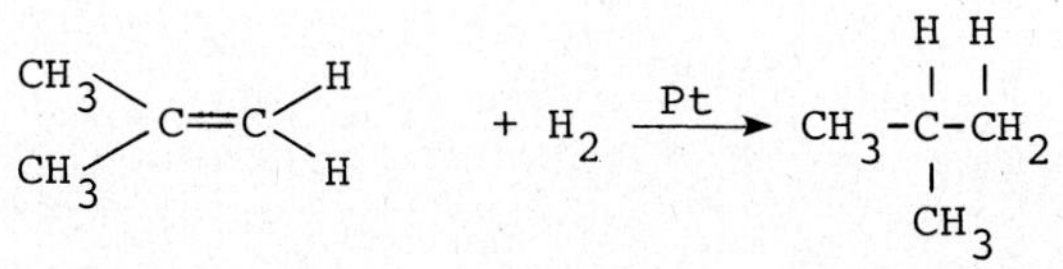

methylpropene methylpropane

"Type" reaction:

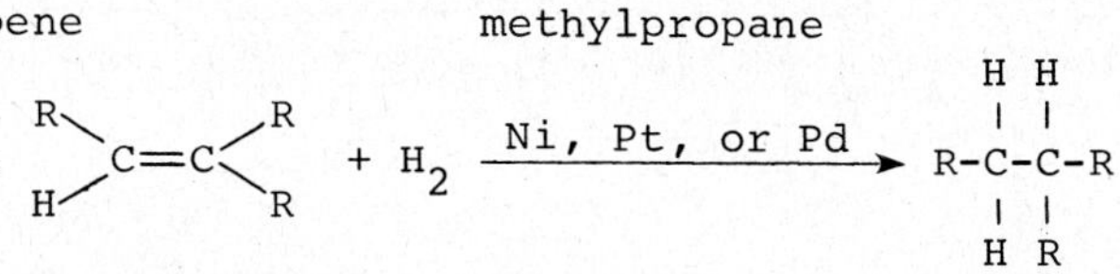

This reaction is an addition reaction, but it is not specifically an electrophilic addition.

(e) Reaction with cold, dilute $KMnO_4$, potassium permanganate solution

Your answer:

Solution

(i)

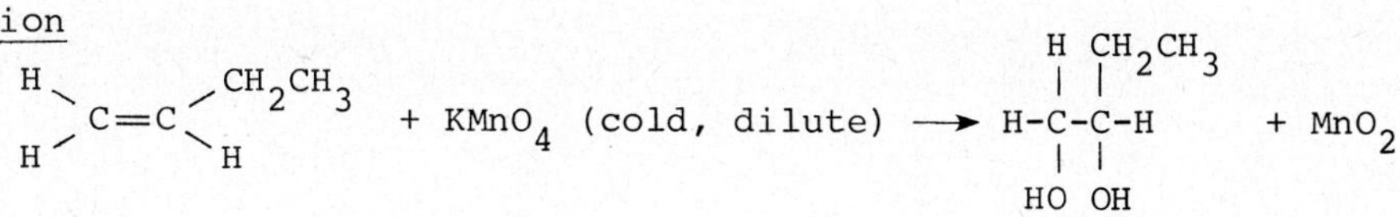

1-butene 1,2-butanediol

(ii) 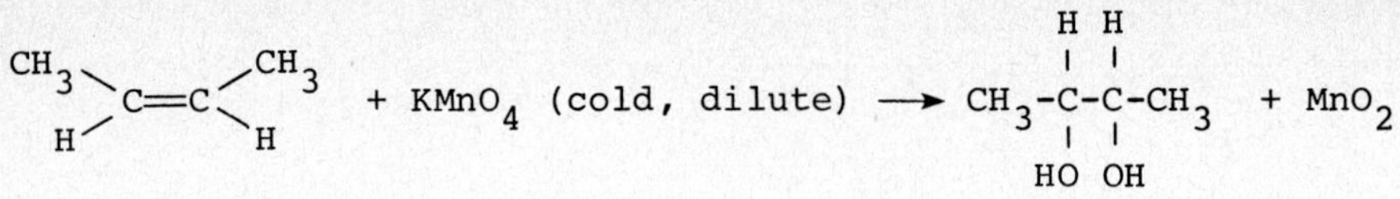

cis-2-butene 2,3-butanediol

(iii) The reaction for trans-2-butene is identical to the reaction for cis-2-butene.

(iv) 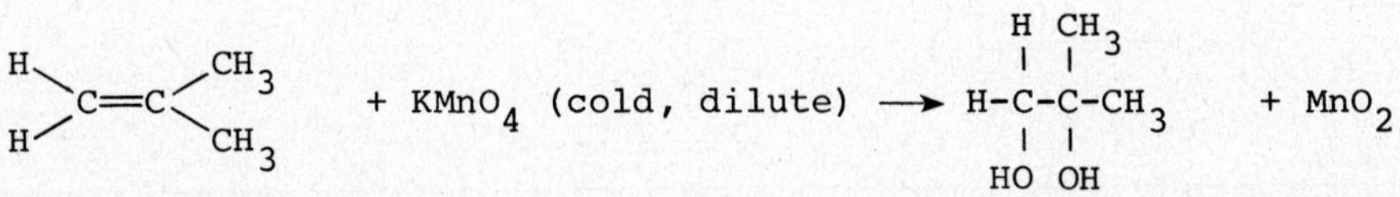

methylpropene 2-methyl-1,2-propanediol

"Type" reaction: 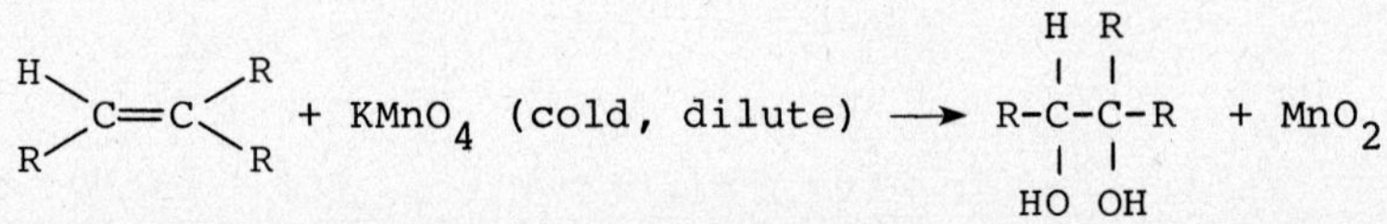

This reaction is the Baeyer test for unsaturated carbon compounds. A positive test is recognized by the occurrence of the brown precipitate of MnO_2. When a cold, dilute solution of $KMnO_4$ is used, the organic products are glycols (diols). However, when the solution is warmed, products form in which the unsaturated carbons are oxidized further to form ketones or carboxylic acids. These products form concurrently with the breakage of not only the pi bond but also the sigma bond between the unsaturated carbons.

(f) Reaction with O_3, ozone, followed by Zn and acid (called ozonization or ozonolysis)

Your answer:

Solution:

(i) 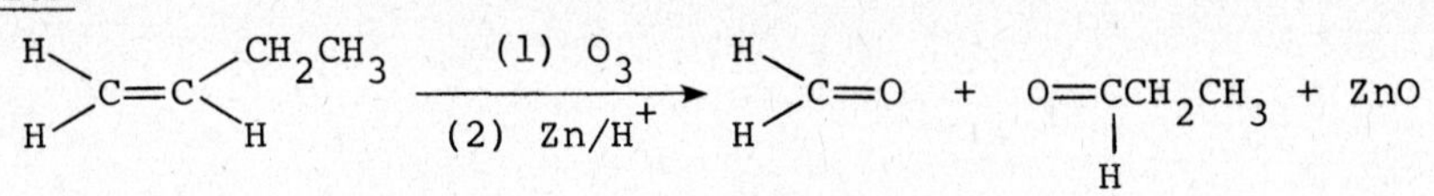

1-butene methanal propanal

(ii) 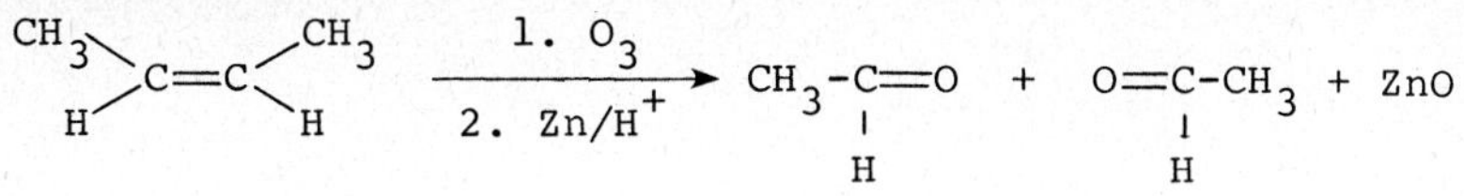

cis-2-butene ethanal ethanal

(iii) The reaction for trans-2-butene is identical to the reaction for cis-2-butene.

(iv) 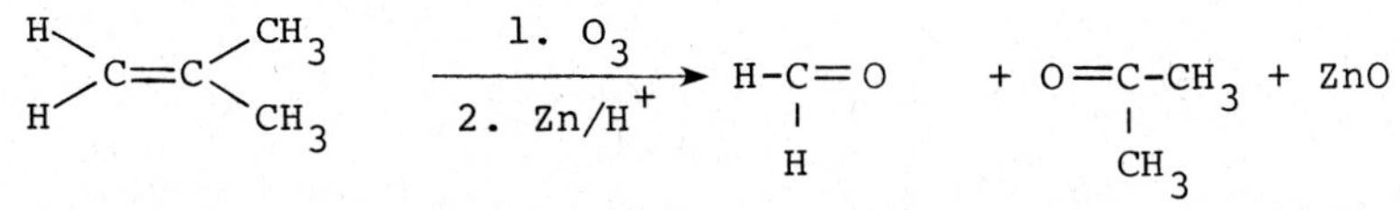

methanal propanone

"Type" reaction:

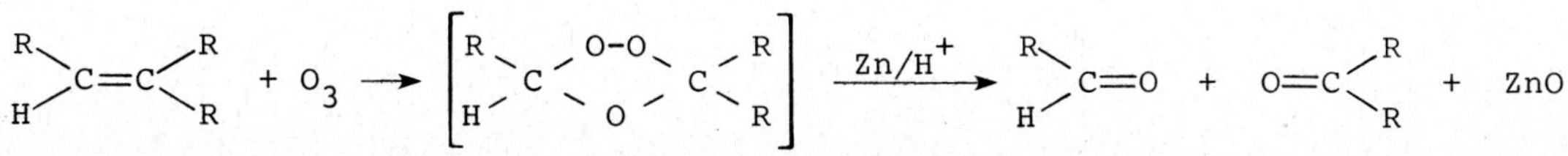

The intermediate formed is an unstable ozonide. The products formed can be an aldehyde and a ketone (as shown) or two aldehydes or two ketones; if the reactant is a cycloalkene, the product is a dialdehyde, a diketone, or a molecule containing an aldehyde and a ketone.

---

2. Write the reactions for the polymerization of four molecules of 1-butene. Write a formula to represent this polymer with n repeating units.

Your answer:

Solution

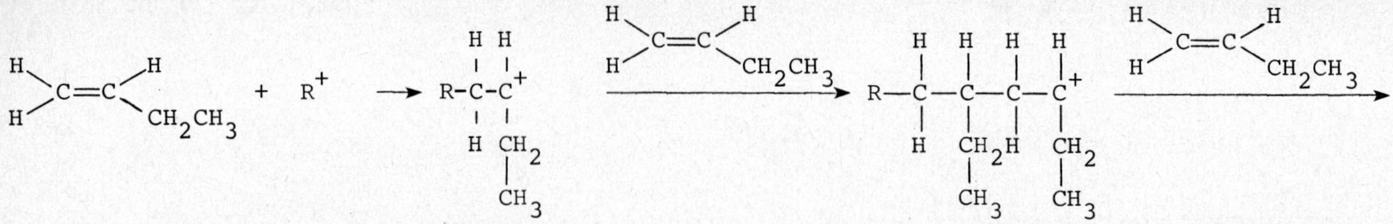

1-butene (the monomer) | initiator (required to start reaction) | carbocation (acts as electrophile when next 1-butene is added to chain)

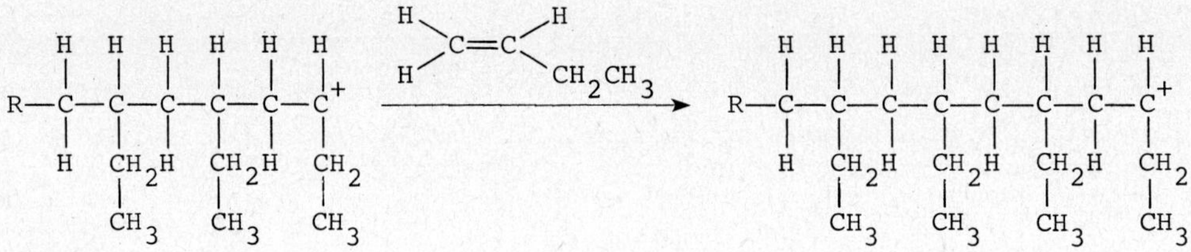

This reaction continues until a nucleophile, such as Br^-, is added to the mixture to terminate the reaction. This nucleophile is appropriately called a terminator.

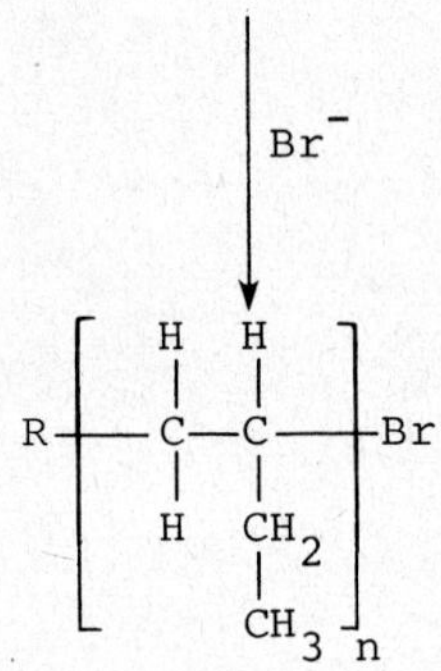

n represents the number of 1-butene units used to synthesize the polymer.

III. Reactions of Alkynes

1. Use the reagent indicated in each exercise to write the equations for the reactions with each isomer of C_4H_6, which is an alkyne (1-butyne and 2-butyne). Write a description of each reaction and name the products.

(a) Reaction with excess Cl_2 (called chlorination)

Your answer:

Solution:

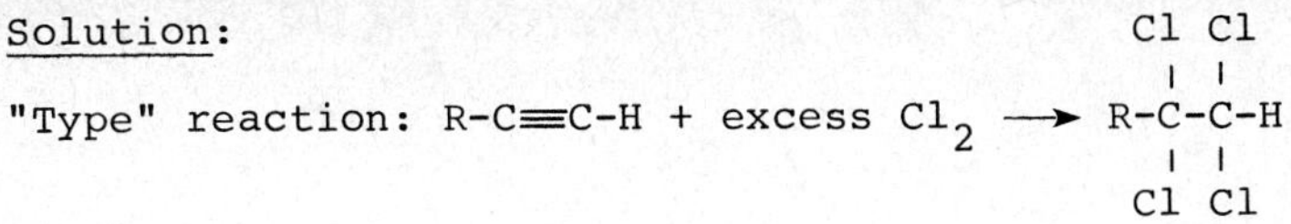

"Excess" implies that enough reagent is provided to completely saturate the triple bond. Because there are two pi bonds in a triple bond, two molecules of halogen will add to each molecule of alkyne.

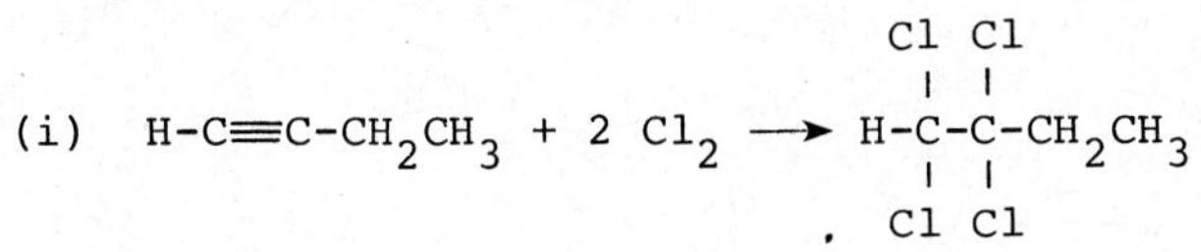

1-butyne 1,1,2,2-tetrachlorobutane

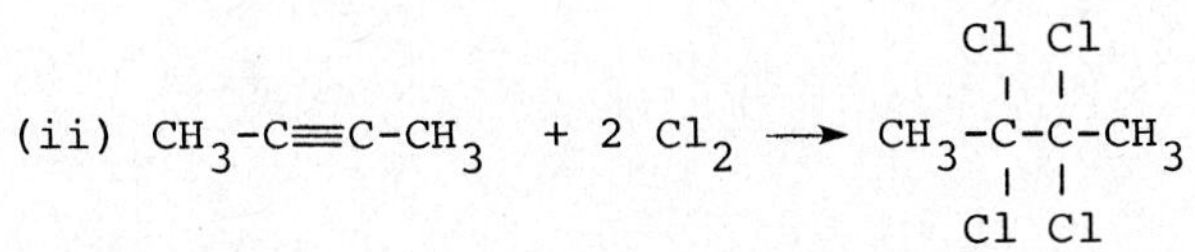

2-butyne 2,2,3,3-tetrachlorobutane

(b) Reaction with dry HCl (not solution)

Your answer:

Solution:

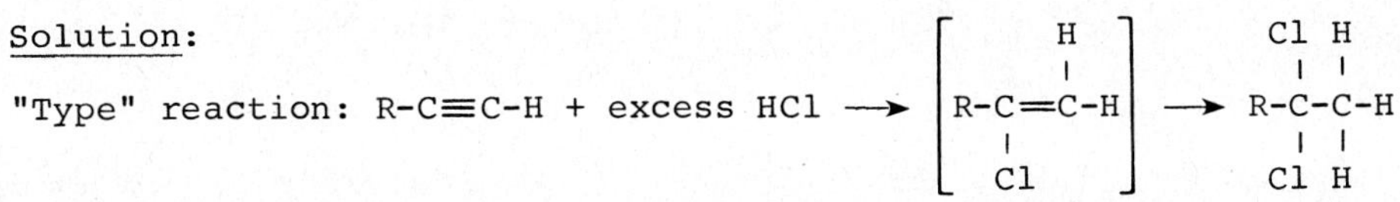

Use Markovnikov's rule to add one molecule of HCl at a time.

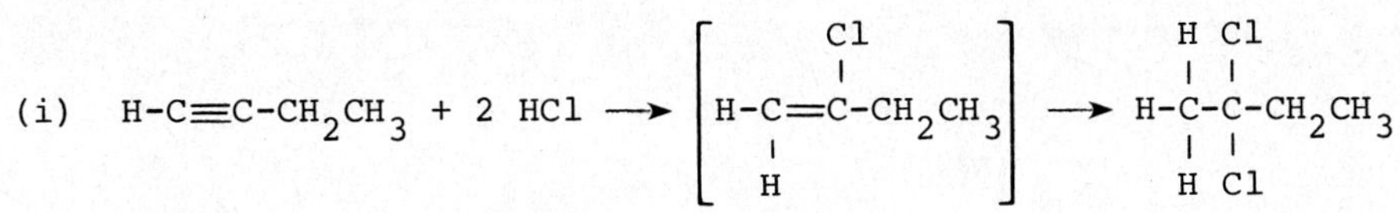

1-butyne 2,2-dichlorobutane

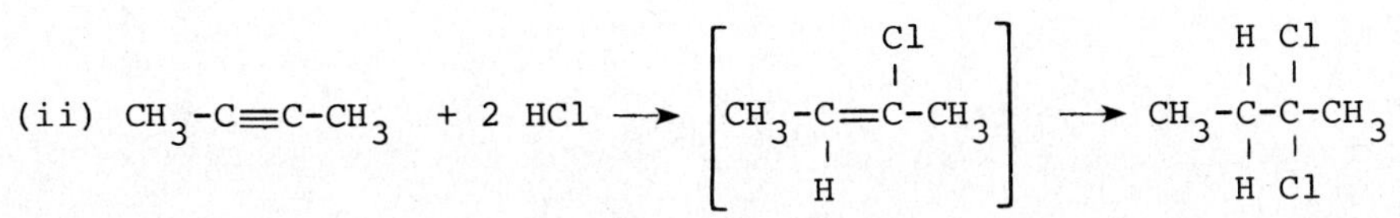

2-butyne 2,2-dichlorobutane

However, if the carbons at the double bond or triple bond each have the same number of hydrogens directly bonded to them, a mixture of products may result. Consider this example.

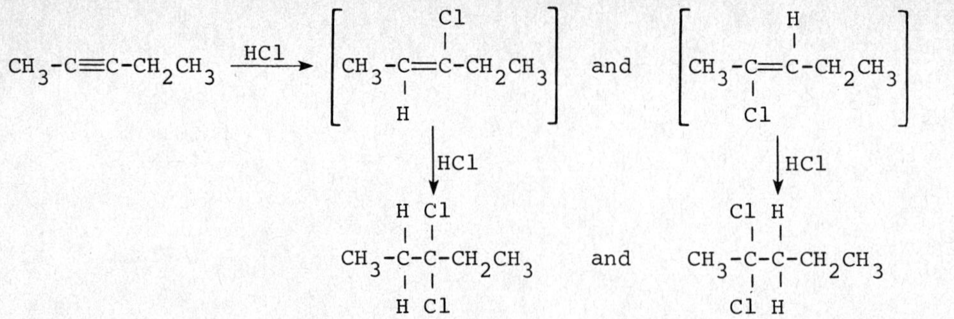

3,3-dichloropentane 2,2-dichloropentane

Each isomer will constitute approximately 50% of the product mixture.

(c) Reaction with H_2 in the presence of a metal catalyst

Your answer:

Solution:

"Type" reaction: 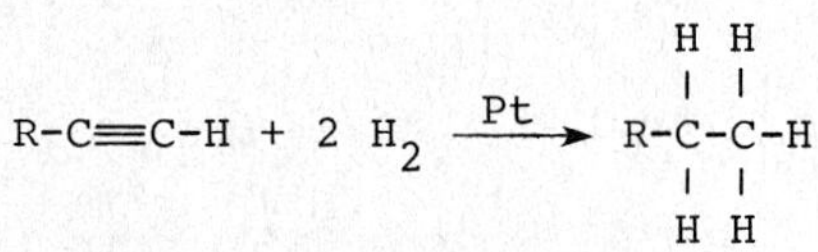

Two moles of molecular hydrogen are required to saturate 1 mole of an alkyne.

(i) 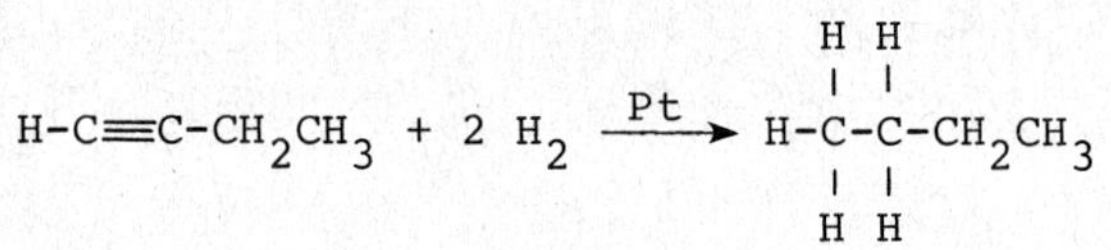

1-butyne butane

(ii) 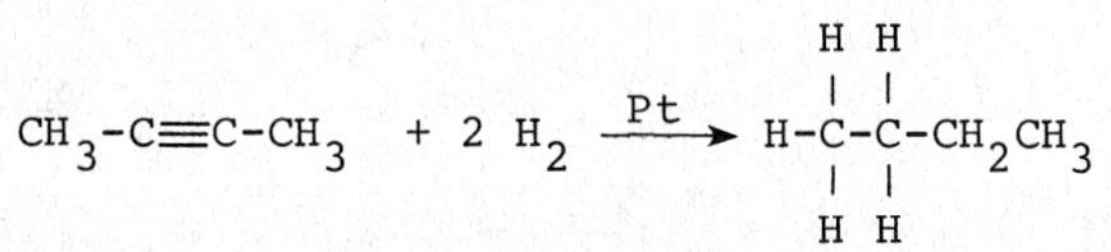

2-butyne butane

(d) Reaction with $KMnO_4$

Your answer:

Solution:

"Type" reaction: R-C≡C-H + $KMnO_4$ ⟶ oxidized products + MnO_2 (brown precipitate)

(i) $H-C\equiv C-CH_2CH_3$ + $KMnO_4$ ⟶ oxidized products + MnO_2
1-butyne

(ii) $CH_3-C\equiv C-CH_3$ + $KMnO_4$ ⟶ oxidized products + MnO_2
2-butyne

IV. Miscellaneous Reactions and Terminology Review

The following structural formulas are to be used for the questions in this section. Write both the structure and the name of the compound that best answers each question, and of any products.

A: $H-C\equiv C-\underset{\displaystyle CH_3}{\underset{|}{CH}}-CH_3$

B: $CH_3-\overset{\displaystyle CH_3}{\overset{|}{CH}}-CH_2CH_2CH_3$

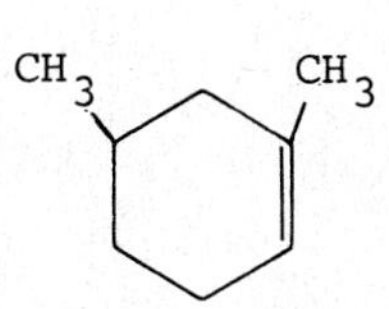

C

D: $CH_3-CH=CH-CH_2CH_3$

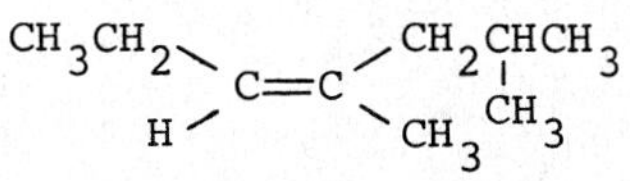

E

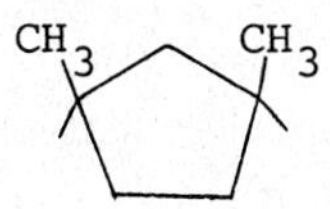

F

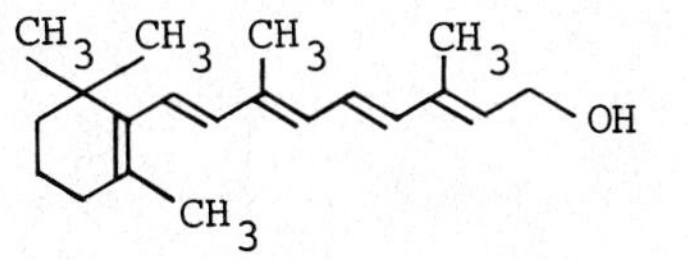

G

1. Which compounds will undergo combustion when ignited? Write a balanced equation for the complete combustion of one of them.

Your answer:

Solution: All of the compounds shown (and most of the organic compounds used in your laboratory) will burn if ignited by a spark or flame. For example,

$H-C\equiv C-\underset{\displaystyle CH_3}{\underset{|}{CH}}-CH_3$, 3-methyl-1-butyne: $C_5H_8 + 7\ O_2 \xrightarrow{\text{spark}} 5\ CO_2 + 4\ H_2O$

---

2. Write an equation for the reaction of compound C with bromine in the dark.

Your answer:

Solution:

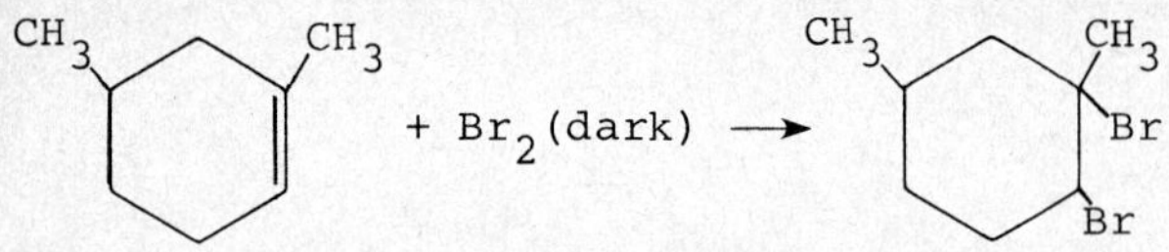

2,4-dimethylcyclohexene 1,2-dibromo-2,4-dimethylcyclohexane

This reaction represents addition of bromine to a pi bond.

---

3. Compound D reacts with HBr (dry) to give a mixture of two products. Write the equation for the reaction and explain why more than one product forms. How are the two products related?

Your answer:

Solution:

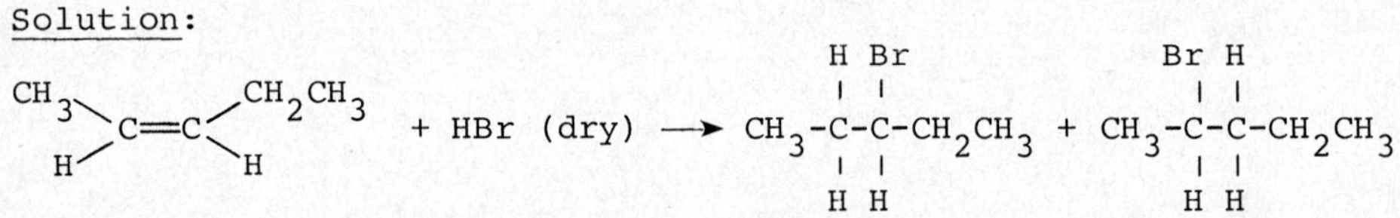

cis-2-pentene 3-bromopentane 2-bromopentane

An approximately 50%-50% mixture of the two products results. This is because the more positive hydrogen (of HBr) can add with relatively the same ease to either carbon 2 or carbon 3 of cis-2-pentene and form the same type of intermediate—a secondary carbocation.

---

4. Write an equation for the reaction of compound A with excess hydrogen in the presence of a nickel catalyst.

Your answer:

Solution:

$$H{-}C{\equiv}C{-}CH(CH_3){-}CH_3 + H_2\ (excess) \xrightarrow{Ni} CH_3{-}CH_2{-}CH(CH_3){-}CH_3$$

3-methyl-1-butyne 2-methylbutane

---

5. Suppose you perform a Baeyer test on each of the compounds. In which test tubes would the solution remain purple and in which would you see a brown sludge?

Your answer:

Solution: In the Baeyer test, $KMnO_4$ oxidizes the carbons sharing a pi bond. The $KMnO_4$ is reduced to MnO_2, a muddy brown precipitate, which is visual proof that the reaction has occurred. You should expect to see the brown sludge in test tubes containing compounds A, C, D, E, and G. The solution should remain purple (indicating no reaction) in test tubes containing compounds B and F.

---

6. Compound E reacts in a very dilute aqueous solution of HCl or H_2SO_4. Write an equation for the reaction.

Your answer:

Solution:

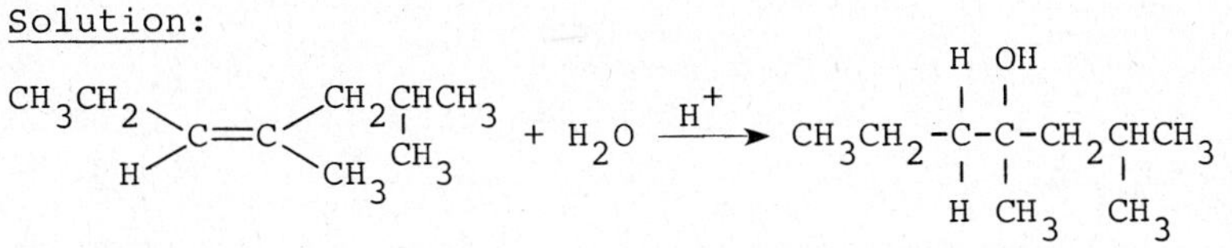

cis-4,6-dimethyl-3-heptene   2,4-dimethyl-4-heptanol

---

7. Compound G is vitamin A. Write equations for its reaction with (a) excess Br_2 (dark) and (b) O_3 (ozone) followed by Zn and acid.

Your answer:

Solution:

(a) 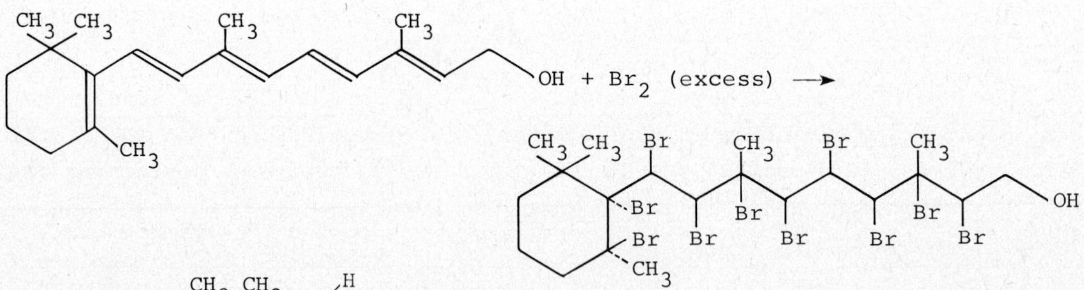

(b) 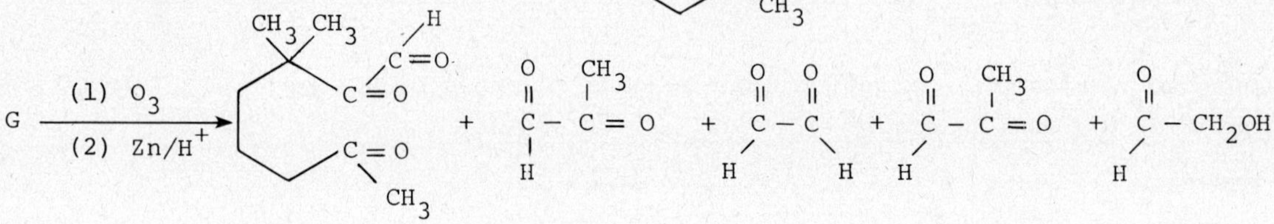

---

8. Which compounds can form haloalkanes (alkyl halides) only by a substitution reaction? In other words, which compounds cannot add chlorine, bromine, etc.? What conditions are necessary for such a reaction?

Your answer:

Solution: Compounds B and F cannot add bromine or chlorine because there are no pi bonds in these compounds (they are saturated compounds). Sunlight, heat, or ultraviolet light must be present to activate the halogen to enable it to remove a hydrogen from the alkane.

$$CH_3\text{-}\underset{\displaystyle CH_3}{\underset{|}{CH}}\text{-}CH_2CH_2CH_3 + Cl_2 \xrightarrow{\text{sunlight}} C_6H_{13}Cl + HCl$$

The principles necessary to predict which H will be substituted by the Cl are not shown here. Monosubstitution by a Cl is indicated in this reaction, although additional substitution can be expected to occur.

---

9. Which compounds will become halogenated primarily by addition of bromine or chlorine? What conditions are necessary for the reaction?

Your answer:

Solution: Compounds A, C, D, E, and G each add bromine or chlorine readily. A large supply (designated as excess bromine or chlorine) is needed to saturate compound G, vitamin A. The conditions often specified for this reaction are "dark" or "in the dark"; this indicates that you want the products for only the addition reaction and not for any substitution reactions that can occur on the saturated carbons, which would require sunlight. If these compounds react with Br_2 or Cl_2 in the sunlight, addition of the halogen occurs at the pi bond(s), followed by the slower substitution reactions, resulting in unpredictable mixtures of products.

V. Identification Problems

Identification problems are problems in which a functional group (or groups) or the structure of the whole molecule can be deduced from the information provided in the problem.

1. What functional groups are indicated for a compound giving these test results?

(a) A hydrocarbon adds 2 moles of hydrogen (over a metal catalyst) per mole of compound.

Your answer:

Solution: Addition of 2 moles of H_2 indicates that the original compound contains two pi bonds per molecule. These can be present in the molecule as two double bonds or as one triple bond.

(b) A hydrocarbon reacts with $KMnO_4$ solution and produces MnO_2, a brown sludge (precipitate).

Your answer:

Solution: This hydrocarbon contains pi bonds (as double and/or triple bonds); it is an unsaturated compound.

---

2. A sample of an unknown hydrocarbon, W, reacts with $KMnO_4$ and decolorizes bromine (in CCl_4) in the dark. When another sample of the unknown reacts with ozone followed by zinc and acid, the products are

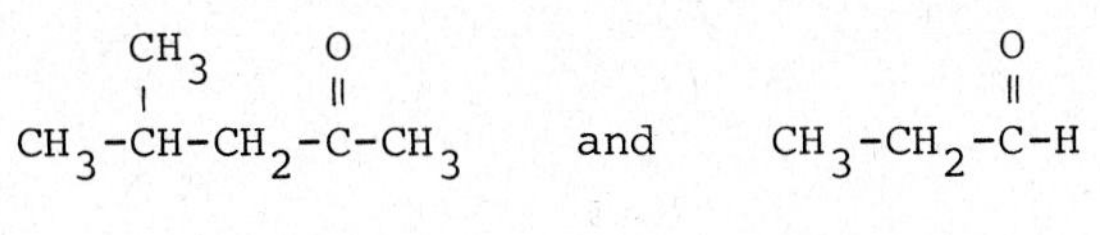

What is unknown hydrocarbon, W?

Your answer:

Solution: You should analyze an identification problem systematically. List the given clues in table form (as shown) and decide which functional groups or structures correspond to the clues.

Given information	Possible	Not possible
W + $KMnO_4$ ⟶ MnO_2 + other products	$>C=C<$ $-C\equiv C-$	alkane
W + Br_2 ⟶ colorless product	verifies double or triple bond	

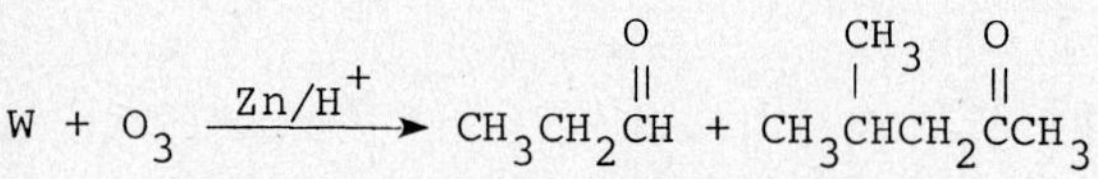

W is 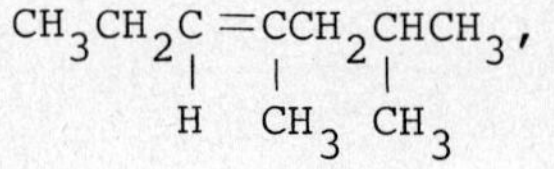, cis or trans-4,6-dimethyl-3-heptene.

---

3. Compound X has a molecular formula C_7H_{12} and reacts with excess Br_2 (in CCl_4) in the dark to form $C_7H_{12}Br_2$. After ozonolysis X yields one product.

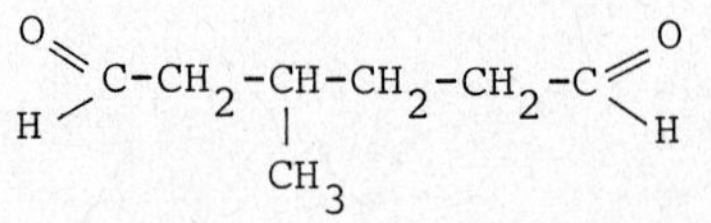

Give the structure and name of X.

Your answer:

Solution:

Given information	Possible	Not possible
X is C_7H_{12} C_nH_{2n-2} (a 4H deficiency)	2 rings, a diene, cycloalkene, alkyne	alkane
X + Br_2 (excess) ⟶ $C_7H_{12}Br_2$	cycloalkene	2 rings would not add Br_2; a diene or triple bond would add 4 bromine atoms (2 Br_2)

$$X + O_3 \xrightarrow{Zn/H^+} \text{OHC}CH_2\underset{\displaystyle CH_3}{\underset{|}{C}H}CH_2CH_2\text{CHO}$$

The carbons containing the double bond to oxygen are C=C in the original compound. Therefore X is 4-methylcyclohexene,

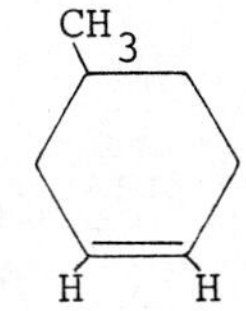

---

4. Compound Z has a molecular formula C_9H_{16}. A sample of the compound reacts with excess Br_2/CCl_4 (in the dark) to form $C_9H_{16}Br_4$. A different sample of the compound turns a solution of $KMnO_4$ to a muddy brown. Compound Z reacts with ozone followed by zinc to give two products.

$$\text{OHC-}CH_2\text{-CHO} \quad \text{and} \quad (CH_3)_2C{=}O$$

Write the structure of Z and name it.

Your answer:

Solution:

Given information	Possible	Not possible
Z is C_9H_{16} C_nH_{2n-2} (a 4H deficiency)	2 rings, a diene, cycloalkene, an alkyne	alkane
$Z + Br_2 \longrightarrow C_9H_{16}Br_4$	diene, alkyne	2 rings, cycloalkene
$Z + KMnO_4 \longrightarrow MnO_2$ + other products	diene, alkyne	

$$Z + O_3 \xrightarrow{Zn/H^+} \text{OHC-}CH_2\text{-CHO} + (CH_3)_2C{=}O$$

One molecule of each product represents a total of only six carbons. Since Z is C_9H_{16} one three-carbon unit must be used twice. Compound Z is $(CH_3)_2CHCH_2CH{=}C(CH_3)_2$, 2,6-dimethyl-2,5-heptadiene.

---

15 Aromatic Hydrocarbons

I. Nomenclature and Preparations of Simple Aromatic Hydrocarbons

Use this reference list of compounds.

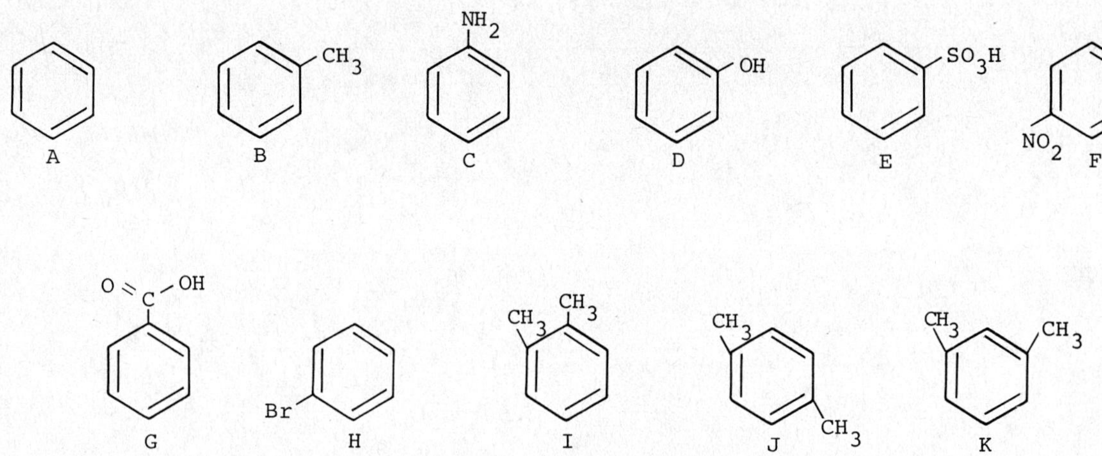

1. Give the common name of each aromatic compound shown in the reference list.

Your answer:

Solution: A, benzene B, toluene C, aniline D, phenol E, benzenesulfonic acid F, nitrobenzene G, benzoic acid H, bromobenzene I, ortho-xylene or o-xylene J, para-xylene or p-xylene K, meta-xylene or m-xylene

---

2. Write an equation for each of the following reactions. Specify the appropriate reagents, catalysts, and conditions and describe (in words) each type of reaction.

(a) A ⟶ B

Your answer:

Solution:

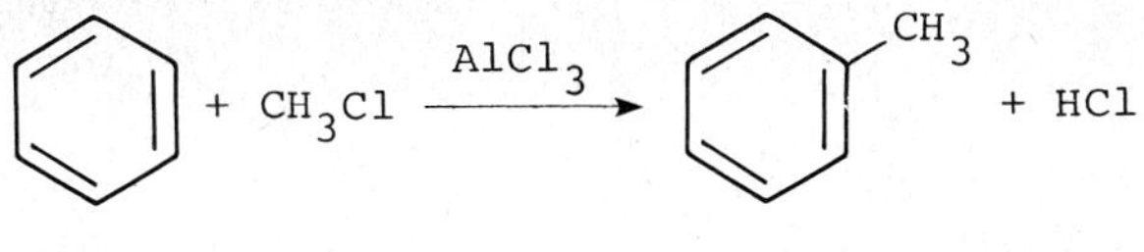

This reaction is an aromatic (electrophilic) substitution of an alkyl group for an H on the benzene ring; called alkylation or Friedel-Crafts reaction; $AlCl_3$ is the catalyst.

(b) A ⟶ F

Your answer:

Solution:

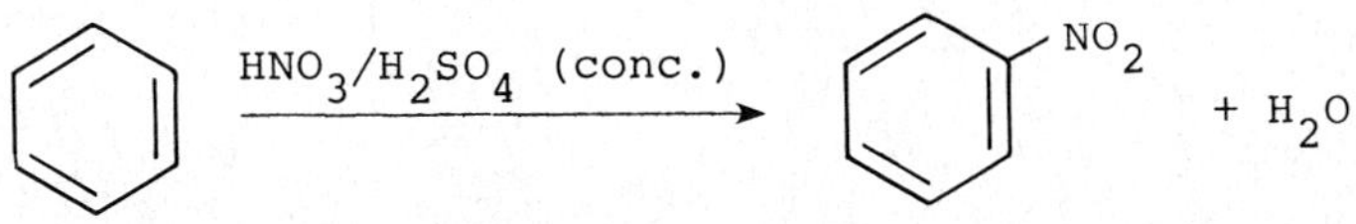

This reaction is a nitration and involves the substitution of a nitro group for H on the aromatic ring; H_2SO_4 is the catalyst.

(c) A ⟶ E

Your answer:

Solution:

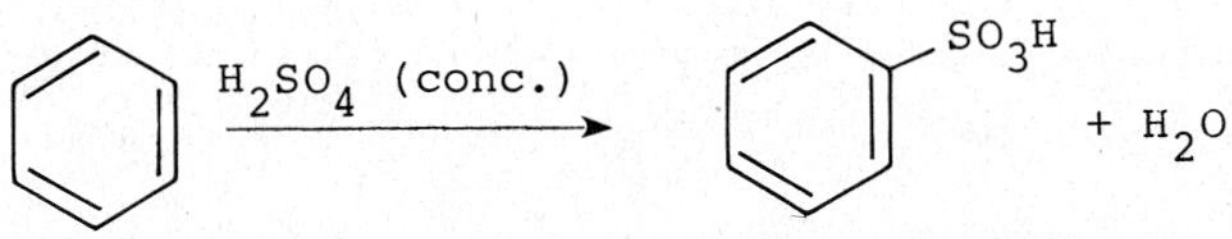

This reaction is the sulfonation of the benzene ring; H_2SO_4 acts as both the reagent and the catalyst.

(d) A ⟶ H

Your answer:

Solution:

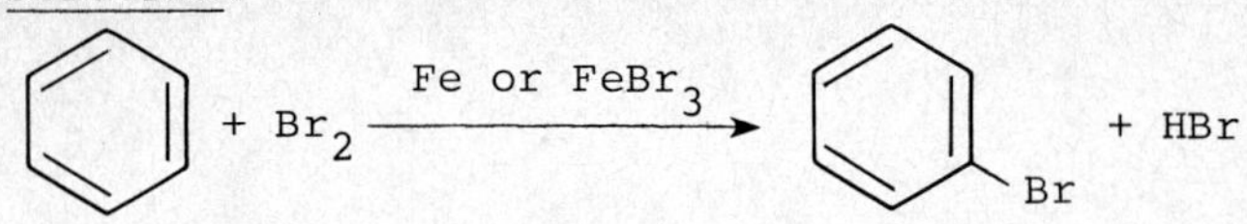

This reaction is a bromination or halogenation and is a slow substitution reaction; Fe or $FeBr_3$ acts as the catalyst. (The reaction will not proceed without the catalyst.) The HBr gas is identified by its reaction with wet litmus paper (placed over the mouth of the test tube); HBr will turn the paper red.

(e) B ⟶ G

Your answer:

Solution:

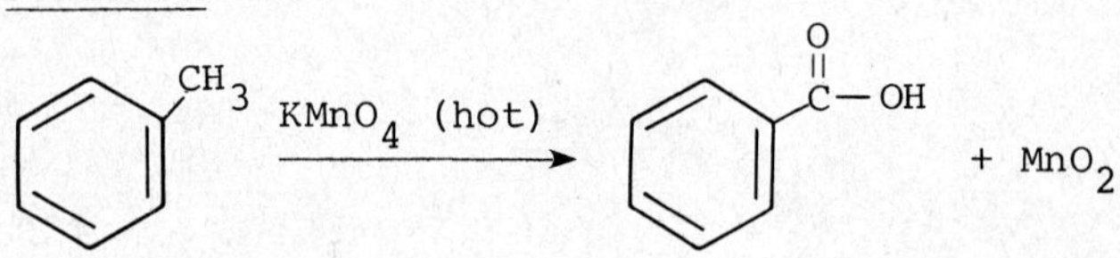

This reaction is the oxidation of an aromatic ring side chain. If the carbon chain has more than one carbon, the additional carbons will be oxidized to CO_2 and H_2O.

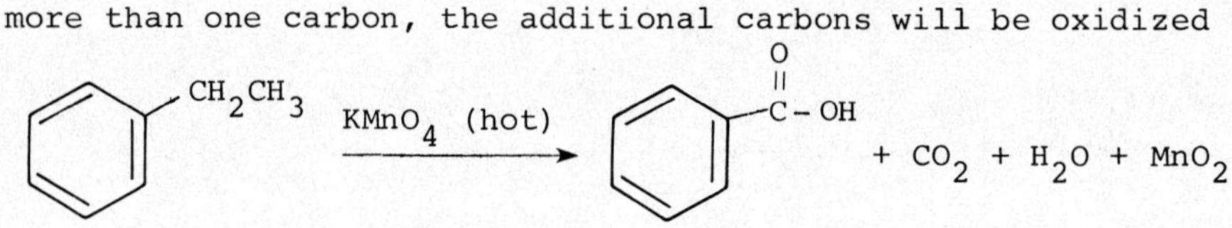

Notice that the aromatic ring is undisturbed by the $KMnO_4$, but compare this reaction with that of an alkene with $KMnO_4$.

(f) Ethylbenzene with bromine in sunlight or ultraviolet light

Your answer:

Solution:

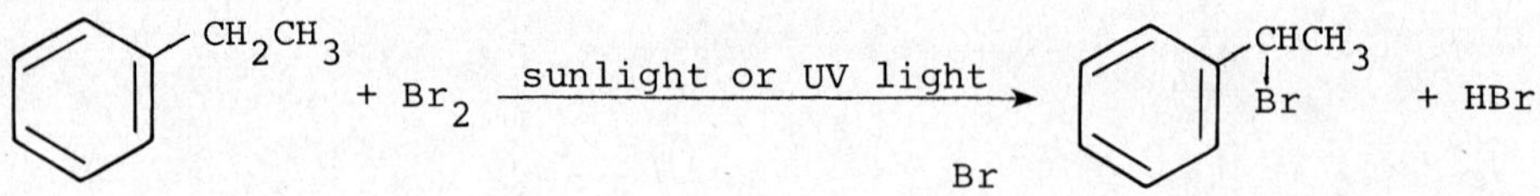

With additional substitution, 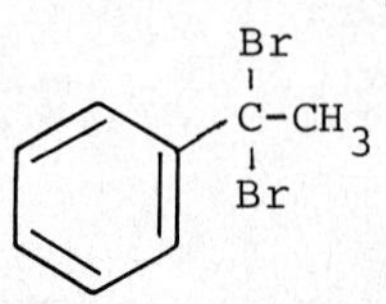 can form. A halogen activated by light attacks the hydrogen bonded to a carbon adjacent to the aromatic ring (called the benzyl carbon) and replaces the hydrogen.

II. Multiple Substitution on an Aromatic Ring

For these reactions, you should learn the following directing groups:

Ortho,para-directing groups	Meta-directing groups
alkyl (-R; methyl, ethyl, etc.)	nitro ($-NO_2$)
halogen (primarily -Br, -Cl)	carboxyl ($-CO_2H$)
amino ($-NH_2$)	sulfonic ($-SO_3H$)
hydroxyl (-OH)	

1. Complete the following equations. Name the reactants and products or supply the structure of the compound if the name is given.

(a)

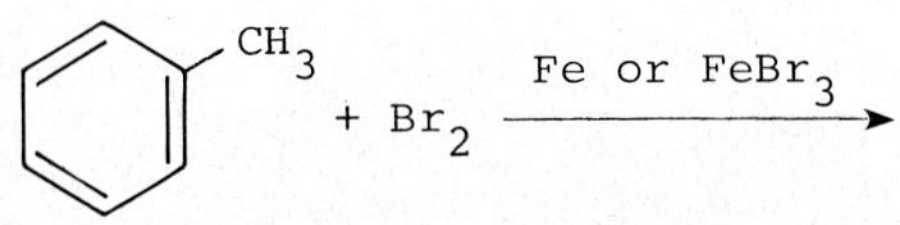

Your answer:

Solution:

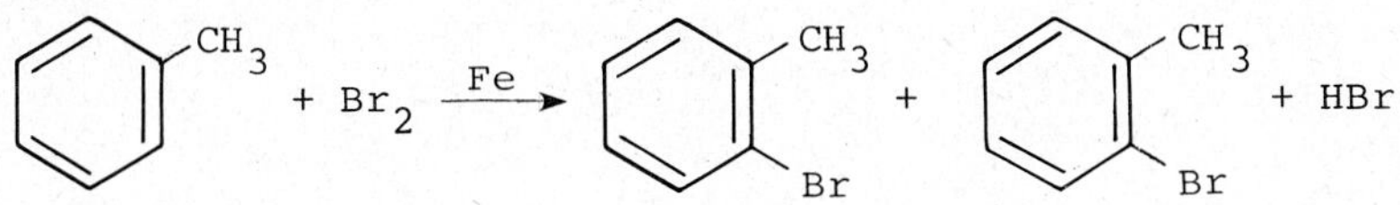

o-bromotoluene p-bromotoluene

The methyl on toluene is an ortho,para director; it "directs" the incoming bromine to the ortho or para position. The products of the reaction are a mixture of the ortho and para isomers.

(b)

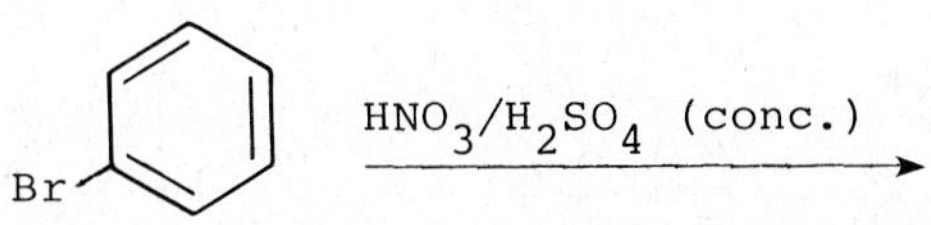

Your answer:

Solution:

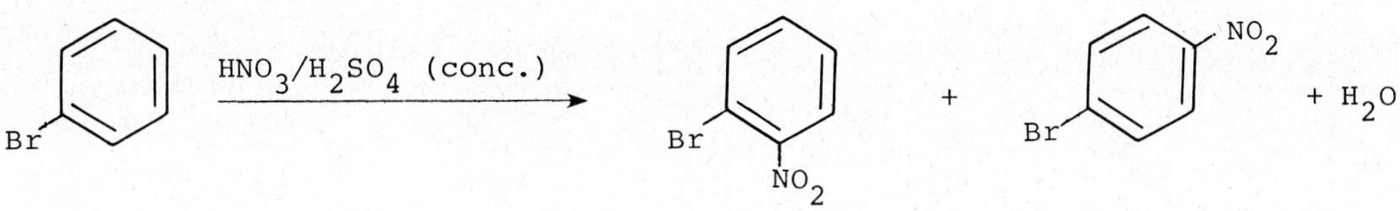

bromobenzene

o-bromonitrobenzene or 1-bromo-2-nitrobenzene

p-bromonitrobenzene or 1-bromo-4-nitrobenzene

Bromine is an ortho,para director.

(c) benzoic acid + Br_2 $\xrightarrow{\text{Fe or FeBr}_3}$

Your answer:

Solution:

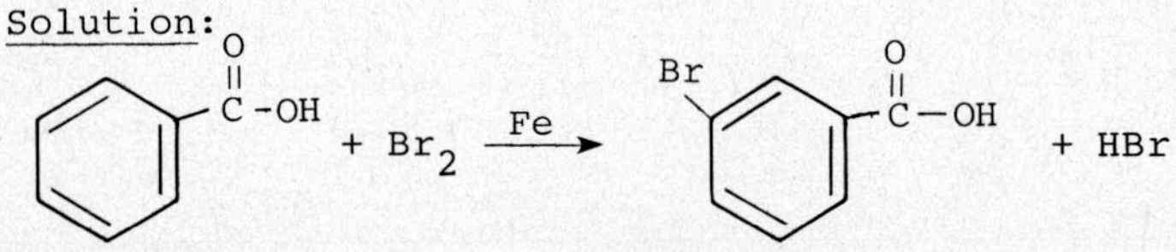

benzoic acid m-bromobenzoic acid

The carboxyl group of benzoic acid is a meta director.

(d) m-xylene $\xrightarrow{\text{KMnO}_4}$

Your answer:

Solution:

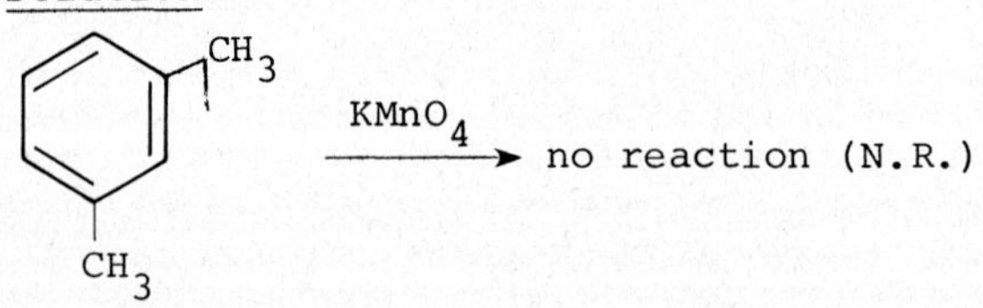

m-xylene

You can assume that $KMnO_4$ is at room temperature or below unless "hot $KMnO_4$" is specified. Compare this reaction with the one in the next part.

(e) propylbenzene $\xrightarrow{\text{hot KMnO}_4}$

Your answer:

Solution:

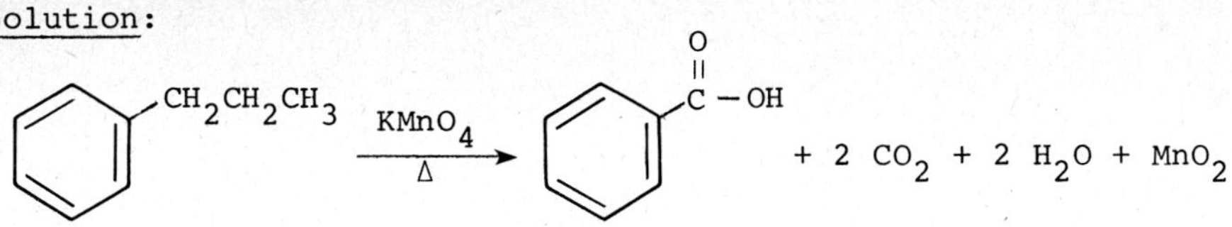

propylbenzene benzoic acid

Hot $KMnO_4$ (also indicated by Δ) oxidizes the alkyl side chain on an aromatic ring to the last carbon; the last carbon is oxidized to a carboxyl group.

(f) phenol + $\xrightarrow{HNO_3/H_2SO_4\ (conc.)}$

Your answer:

Solution:

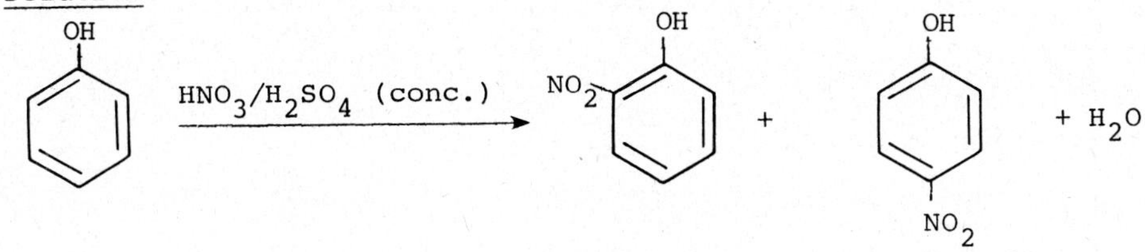

phenol o-nitrophenol p-nitrophenol

The hydroxyl group is an ortho,para director.

(g)

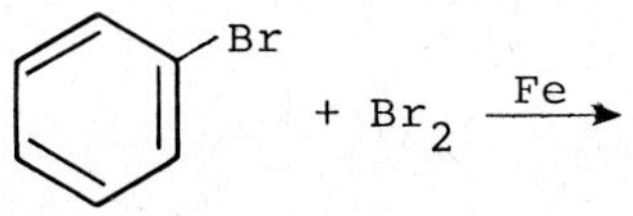

Your answer:

Solution:

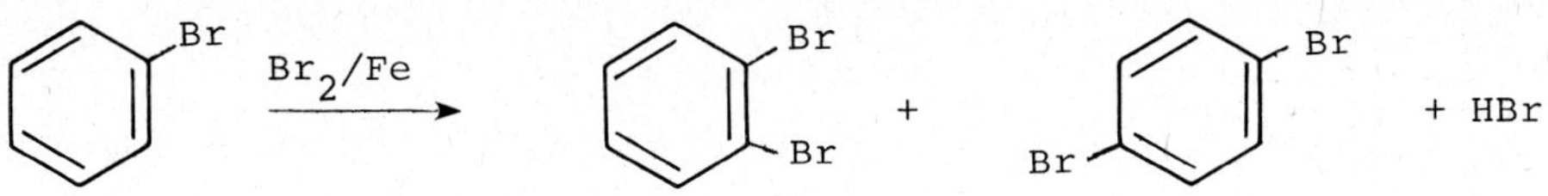

bromobenzene o-dibromobenzene p-dibromobenzene
1,2-dibromobenzene 1,4-dibromobenzene

The bromine is an ortho,para director.

(h)

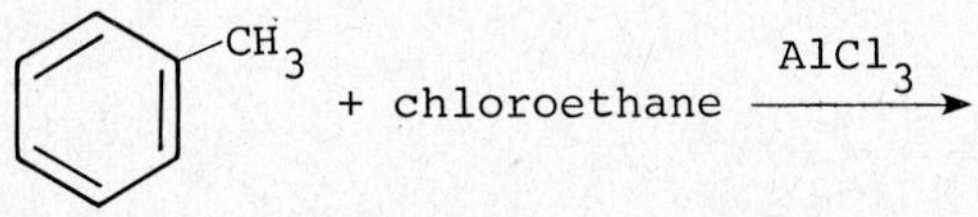

Your answer:

Solution:

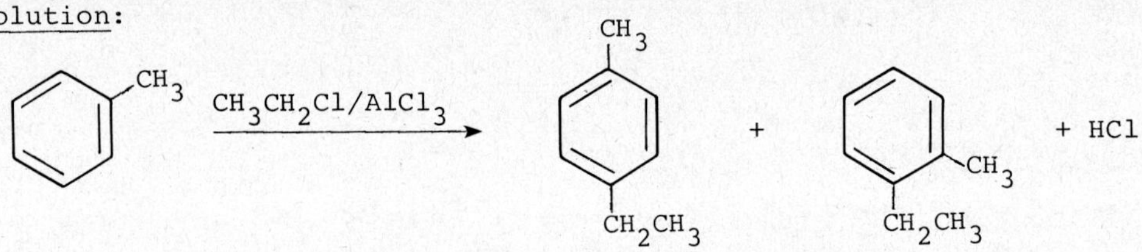

toluene p-ethyltoluene o-ethyltoluene

The methyl group of the toluene is an ortho,para director.

(i)

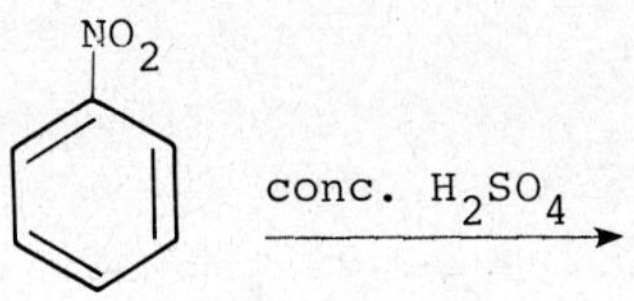

Your answer:

Solution:

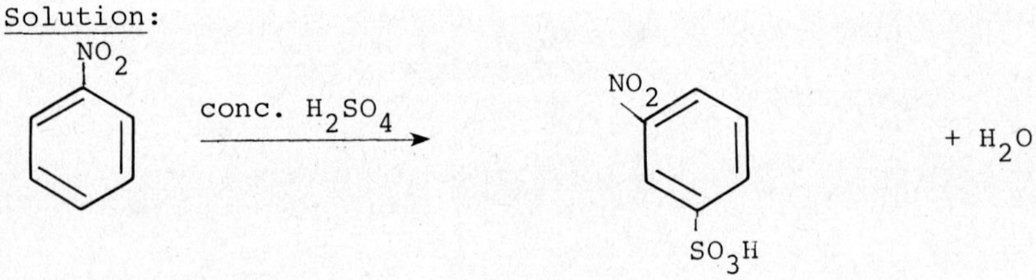

nitrobenzene m-nitrobenzenesulfonic acid

The nitro group is a meta director.

(j) propylbenzene + Br_2 $\xrightarrow{\text{sunlight or uv light}}$

Your answer:

Solution:

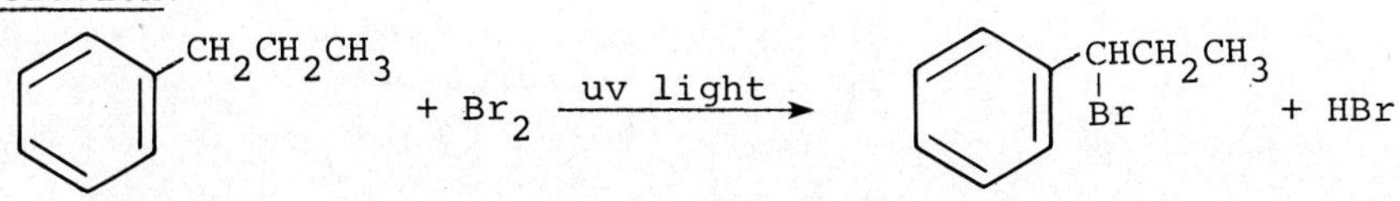

propylbenzene 1-bromo-1-phenylpropane

The hydrogens on the benzyl carbon (the carbon adjacent to the aromatic ring) are more easily replaced than are the other hydrogens. Further substitution yields

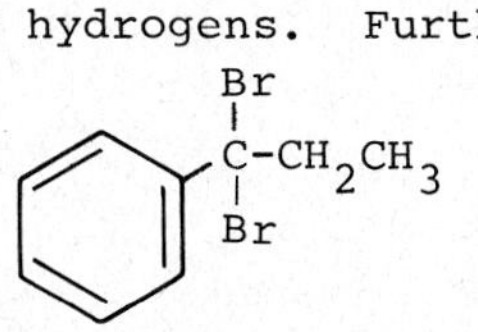

However, no substitution of a bromine for a hydrogen occurs on the aromatic ring because the required catalyst (Fe or $FeBr_3$) is absent.

III. Identification Problems (Aromatic and Aliphatic Compounds)

1. Compound A, C_7H_8, does not react with bromine except in the presence of an iron catalyst. When A is treated with hot potassium permanganate, benzoic acid forms. What is compound A?

Your answer:

Solution:

Given information	Possible	Not possible
A is C_7H_8 (C_nH_{2n-6}) (an 8 H deficiency)	4 double bonds, 1 triple and 2 double bonds, 2 rings and 2 double bonds, 1 ring and 3 double bonds (this suggests aromatic ring)	alkane, cycloalkane
$A + Br_2 \longrightarrow$ N.R.		triple bonds, nonaromatic double bonds
$A + Br_2 \xrightarrow{Fe}$ reaction	aromatic compound	

$A \xrightarrow{\text{hot } KMnO_4}$ benzoic acid

Compound A is toluene, 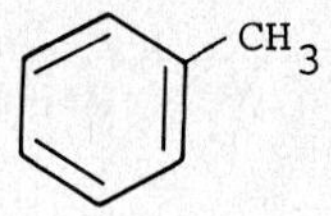

---

2. An alkene reacts with ozone followed by zinc to give two products, butanone and ethanal. Write the structural formula of the alkene. The structure of butanone is $CH_3\overset{O}{\overset{\|}{C}}CH_2CH_3$, and the structure of ethanal is $CH_3\overset{O}{\overset{\|}{C}}H$.

Your answer:

Solution: The carbons containing the double bonds to oxygen in the products are the atoms that originally were joined by a double bond in the alkene. To write the original alkene, remove the oxygens and join the two molecules with a double bond between the carbons. The original alkene is cis- or trans-3-methyl-2-pentene, $CH_3CH_2\underset{CH_3}{\underset{|}{C}}{=}CHCH_3$.

---

3. Compound D gives no reaction with bromine in the dark or in the sunlight. However, D reacts with bromine in the presence of an iron catalyst (Fe) to yield compound X. Compound X reacts with concentrated HNO_3 (with H_2SO_4 catalyst) to yield a mixture of o-bromonitrobenzene (Y) and p-bromonitrobenzene (Z). Write the structural formulas for D, X, Y, and Z.

Your answer:

Solution:

Compound Y is o-bromonitrobenzene, 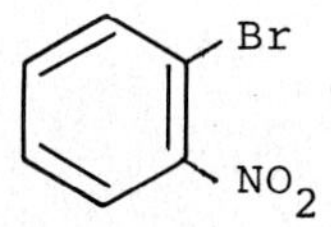

Compound Z is p-bromonitrobenzene, 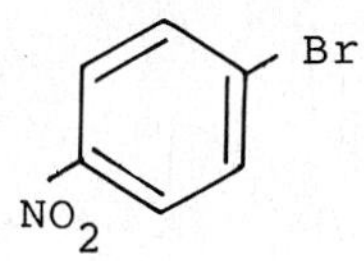

HNO_3 (with H_2SO_4 catalyst) substituted the $-NO_2$ on X to produce Y and Z. Therefore X is bromobenzene,

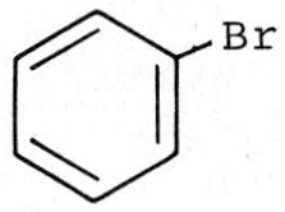

and D, which reacts with bromine to give X, is benzene, 

IV. Practice Reactions – Alkanes, Alkenes, Alkynes, and Aromatics

Write the structures and (whenever possible) names of the missing products or reactants. For each reaction, choose (from the following list) the word or words that best describe the reaction.

oxidation	addition reaction
combustion	hydrogenation
reduction	substitution reaction

1. 2,2,4-trimethylpentane + O_2 $\xrightarrow{\text{spark}}$

Your answer:

Solution: oxidation, combustion

```
     CH3  CH3
     |    |
CH3-C-CH2CHCH3
     |
     CH3
```

; $2\ C_8H_{18} + 25\ O_2 \xrightarrow{\text{spark}} 16\ CO_2 + 18\ H_2O$ + heat

---

2. 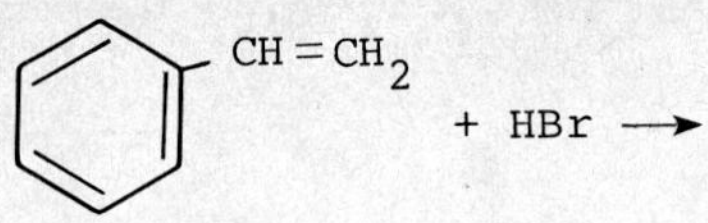

Your answer:

Solution: addition reaction

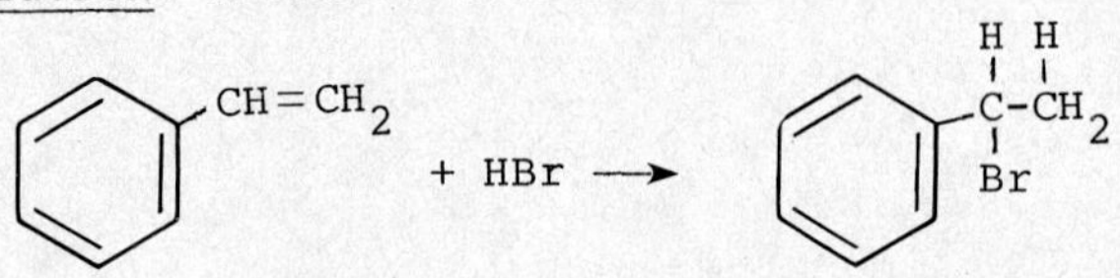

1-bromo-1-phenylethane

---

3. 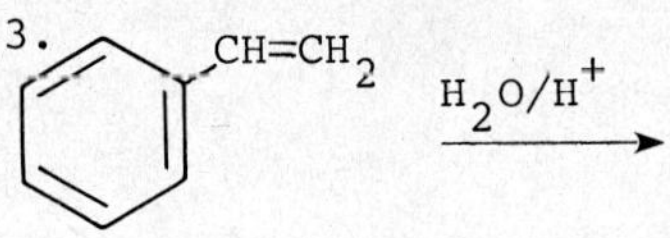

Your answer:

Solution: addition reaction

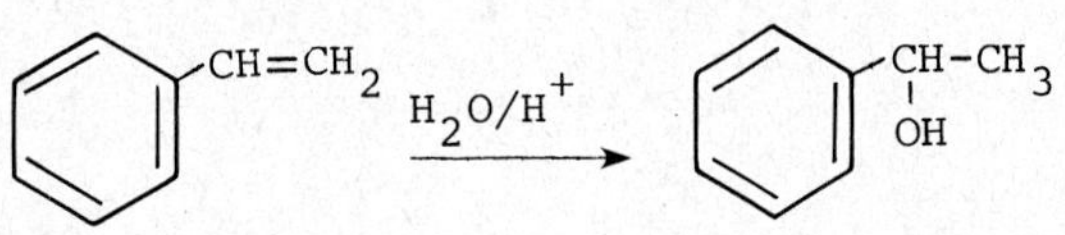

1-phenyl-1-ethanol

---

4. 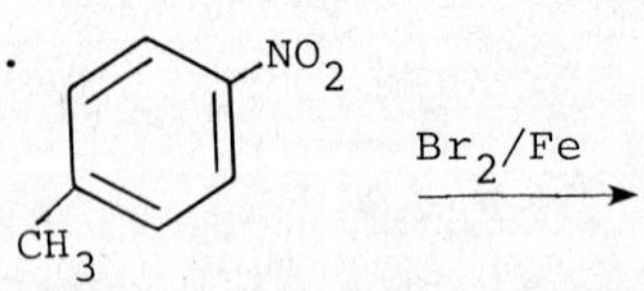

Your answer:

Solution: substitution reaction

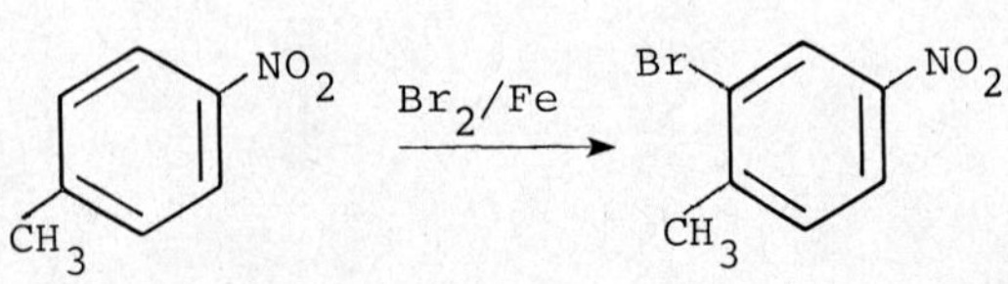

2-bromo-4-nitrotoluene

$-CH_3$ is an ortho,para director, and $-NO_2$ is a meta director; thus their directing effects in the original compound complement each other.

---

5. 2-pentyne + H_2 (excess) $\xrightarrow{Pd}$

Your answer:

Solution: addition reaction, reduction, hydrogenation

$$CH_3C{\equiv}CCH_2CH_3 \xrightarrow{H_2\text{(excess)/Pd}} CH_3\text{-}\overset{H}{\underset{H}{C}}\text{-}\overset{H}{\underset{H}{C}}\text{-}CH_2CH_3$$

pentane

---

6. 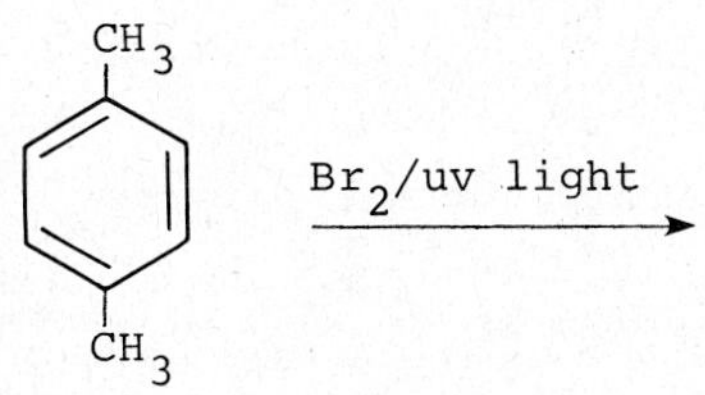

Your answer:

Solution: substitution (on side chain of aromatic ring)

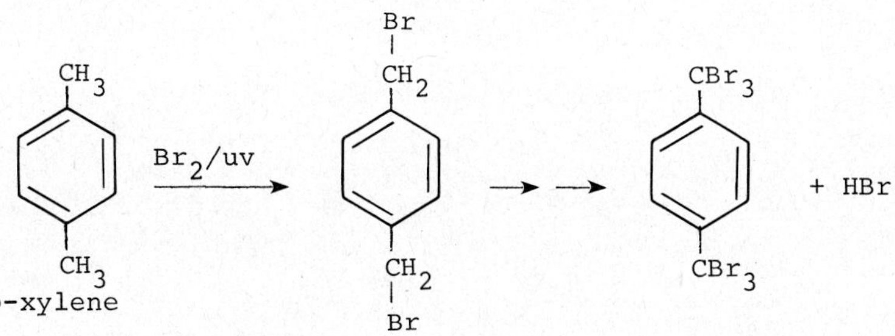

p-xylene

Bromine will substitute on the methyl; if sufficient time is permitted, all three hydrogens on each methyl can be replaced by bromines.

---

7. 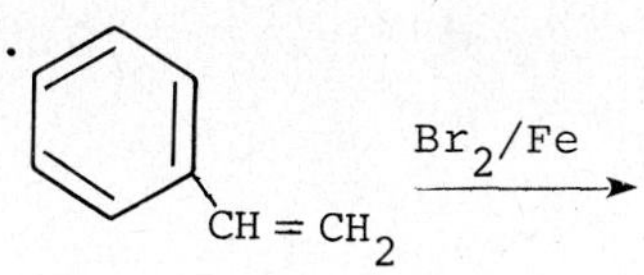

Your answer:

Solution: addition reaction (at the double bond) and substitution reaction (on the aromatic ring)

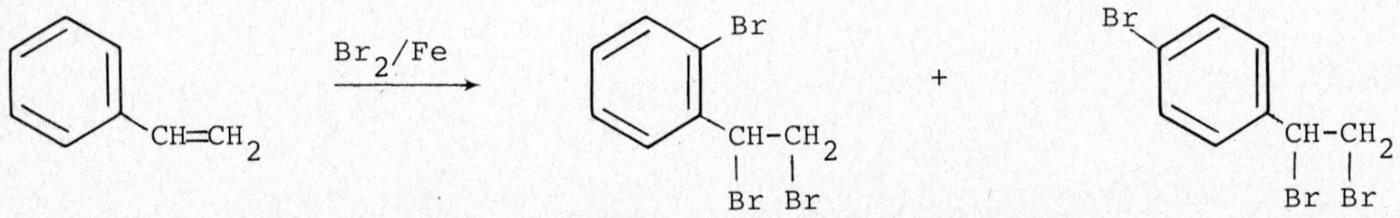

Bromine quickly adds to the double bond, and then slowly substitutes at the ortho or para positions on the ring.

---

8. benzene + ethyl chloride $\xrightarrow{AlCl_3}$? $\xrightarrow{\text{hot } KMnO_4}$?

Your answer:

Solution: The first reaction involves substitution on the aromatic ring, and the second reaction illustrates the oxidation of an alkyl substituent on an aromatic ring.

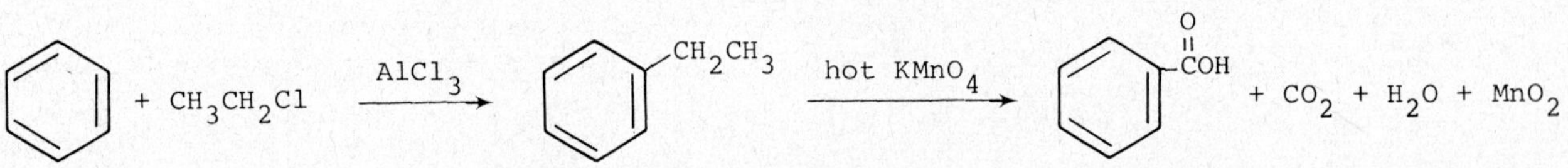

ethylbenzene benzoic acid

---

9. 1,5-hexadiene $\xrightarrow{Br_2 \text{ (excess)}/CCl_4}$

Your answer:

Solution: addition reaction

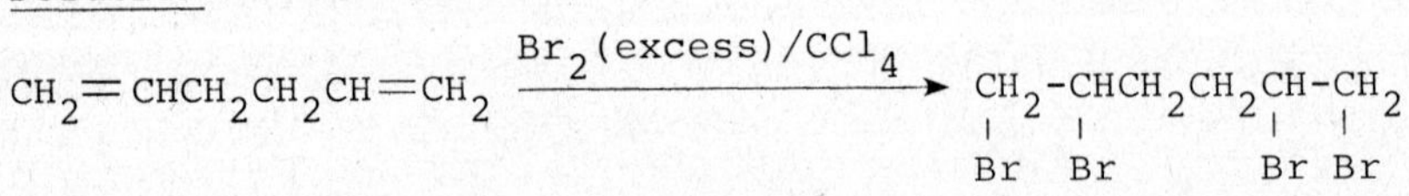

1,2,5,6-tetrabromohexane

10.

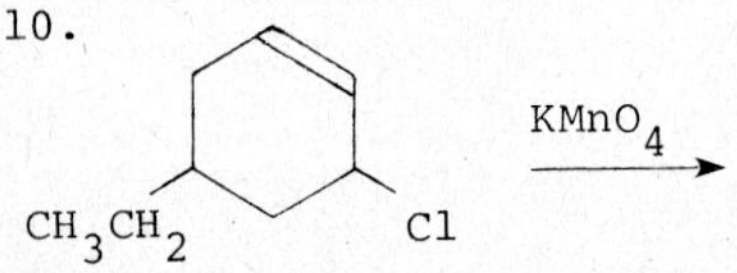

Your answer:

Solution: oxidation

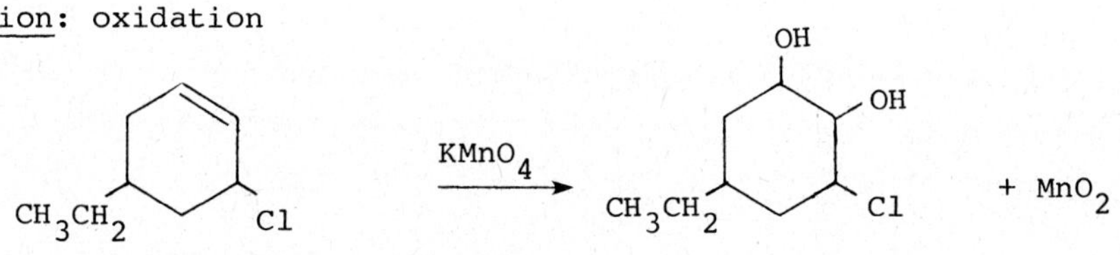

3-chloro-5-ethylcyclohexene → 3-chloro-5-ethyl-1,2-cyclohexanediol

---

11. methylpropene $\xrightarrow{HCl}$? $\xrightarrow[(2)\ \text{toluene}]{(1)\ AlCl_3\ \text{catalyst}}$

Your answer:

Solution: The first reaction is an addition reaction, and the second reaction is an aromatic substitution reaction.

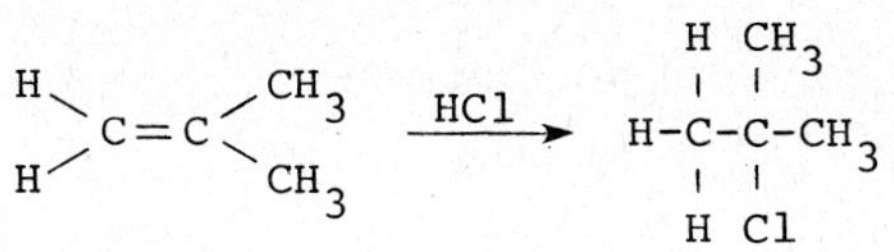

2-chloro-2-methylpropane

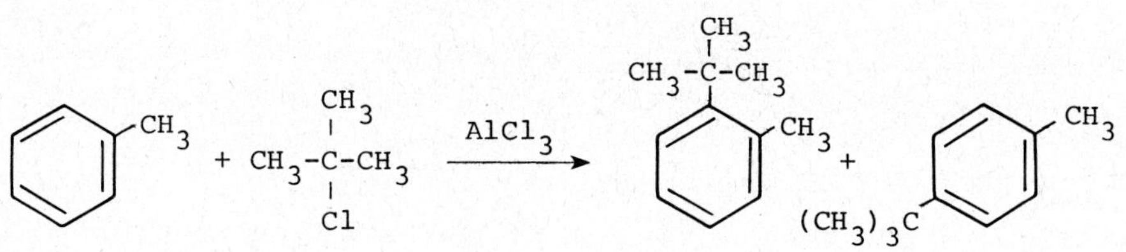

The methyl on the toluene acts as an ortho,para director.

---

16 Nomenclature of Organic Compounds Containing Oxygen and Nitrogen

Determine the name of each compound from the structure, or determine the structure from the compound named. Write the common name of the compounds marked with an asterisk, and enclose the common name in parentheses.

I. Alcohols, Phenols, and Ethers

1. $CH_3CH_2CH_2OH$ (*)

Your answer:

Solution: 1-propanol (n-propyl alcohol)
The base name of an alcohol corresponds to the longest continuous carbon chain that contains the carbon to which the hydroxyl is bonded. The hydroxyl group has priority over alkyl or halogen substituents; this implies that the hydroxyl carbon is assigned the smallest number possible. To form the name of an alcohol from the base name of an alkane, drop the "e" and add "ol." The position of the hydroxyl is identified by the number preceding the base name.

---

2. $CH_3CH(CH_3)CH_2OH$ (*)

Your answer:

Solution: 2-methyl-1-propanol (isobutyl alcohol)

---

3. 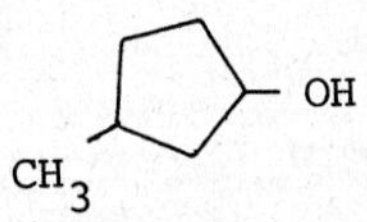

Your answer:

Solution: 3-methylcyclopentanol
No number is assigned to the position of the hydroxyl on a cyclic structure because it is assumed to be position 1.

---

4. 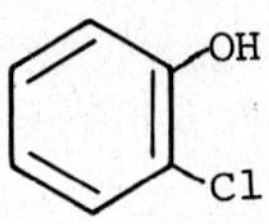

Your answer:

Solution: 2-chlorophenol or o-chlorophenol

---

5. 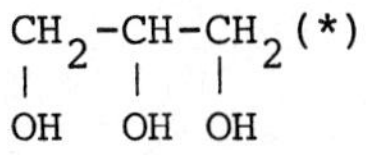

Your answer:

Solution: 1,2,3-propanetriol (glycerol or glycerine)
Notice the use of the "triol," which specified three hydroxyl groups.

---

6. CH_3-O-CH(CH_3)$_2$

Your answer:

Solution: isopropyl methyl ether

---

7. 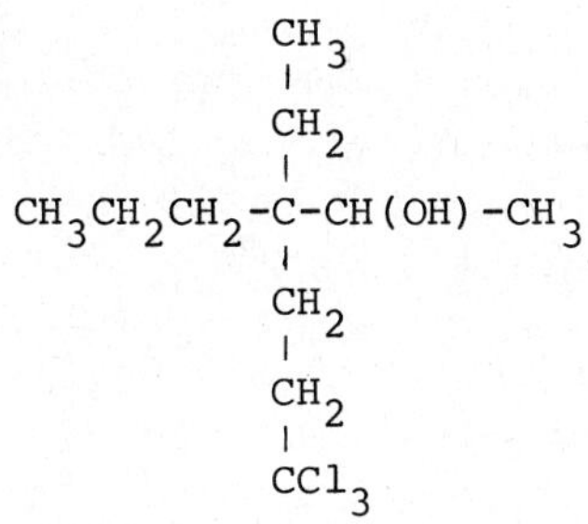

Your answer:

Solution: 6,6,6-trichloro-3-ethyl-3-propyl-2-hexanol
Notice that the six-carbon chain carrying the substituents is used to form the base name.

---

8. 4-propyl-5-hexen-2-ol

Your answer:

Solution: 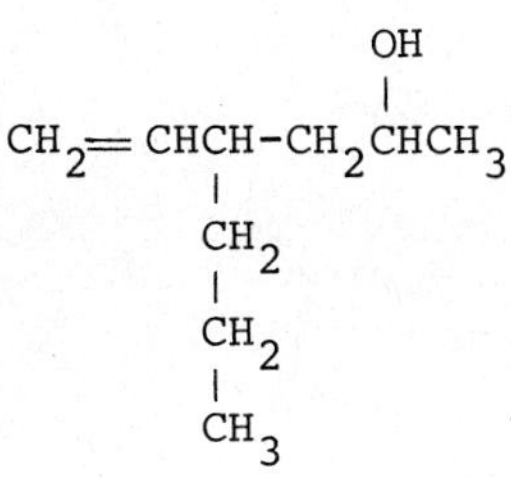

---

9. 2-methyl-2-propanol(*)

Your answer:

$$\begin{array}{c} OH \\ | \\ CH_3\text{-}C\text{-}CH_3 \\ | \\ CH_3 \end{array}$$

Solution: (t-butyl alcohol or tert-butyl alcohol)

---

10. ethyl sec-butyl ether

Your answer:

Solution: $CH_3CH_2\text{-}O\text{-}\underset{\displaystyle CH_3}{\underset{|}{C}}HCH_2CH_3$

II. Aldehydes and Ketones

1. $CH_3C\begin{smallmatrix} {=}O \\ \diagdown H \end{smallmatrix}$ (*)

Your answer:

Solution: ethanal (acetaldehyde)

The carbonyl group ($>C{=}O$) is given priority over the functional groups you have studied previously (alkyl, halogen, alcohol) and therefore is assigned the smallest number. The carbonyl is always assigned position 1 in aldehydes and is assigned the smallest number possible in ketones. From the base name of an alkane, drop the "e" and add "al" for an aldehyde and "one" for a ketone.

---

2. 3,4-dimethylpentanal

Your answer:

Solution: $CH_3\underset{\displaystyle CH_3}{\overset{\displaystyle CH_3}{\overset{|}{C}}}H\text{-}CH\text{-}CH_2C\begin{smallmatrix} {=}O \\ \diagdown H \end{smallmatrix}$

---

3. 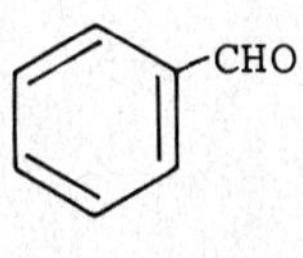 (*)

Your answer:

Solution: (benzaldehyde)

---

4. propanone (*)

Your answer:

Solution: $CH_3\overset{O}{\overset{\|}{C}}CH_3$ (acetone)

---

5. 3-methylcyclohexanone

Your answer:

Solution

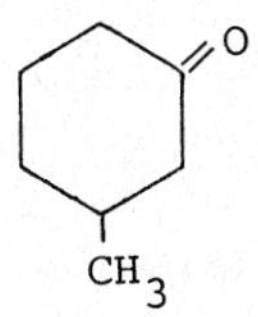

---

6. $(CH_3)_2CHCOCH_2CH_3$

Your answer:

Solution: 2-methyl-3-pentanone

---

7. 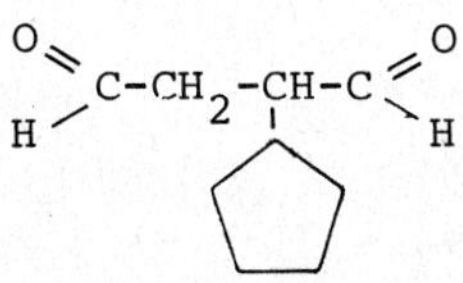

Your answer:

Solution: 2-cyclopentylbutanedial

---

8. 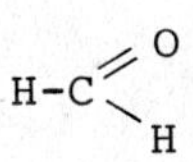 (*)

Your answer:

Solution: methanal (formaldehyde)

---

9. 4-hydroxy-2-pentanone

Your answer:

Solution: $CH_3\text{-}\underset{\displaystyle OH}{\underset{|}{CH}}\text{-}CH_2\text{-}\overset{\displaystyle O}{\overset{\|}{C}}\text{-}CH_3$

---

10. $CH_3\underset{\displaystyle CHO}{\underset{|}{CH}}CH_2CH_3$

Your answer:

Solution: 2-methylbutanal

---

11. 1,4-cyclohexanedione

Your answer:

Solution: 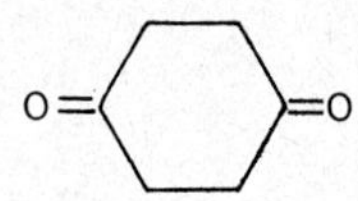

III. Carboxylic Acids and Acid Derivatives (Salts, Acyl Halides, Acid Anyhdrides, Esters, and Amides)

A. CARBOXYLIC ACIDS

1. $H\text{-}C\overset{\displaystyle /\!\!/ O}{\underset{\displaystyle \backslash OH}{}}$ (*)

Your answer:

Solution: methanoic acid (formic acid)

To form the name of the acid, drop the "e" from the base name of the alkane and add "oic acid." The carboxylic carbon is numbered as carbon 1.

---

2. $CH_3\text{-}C\overset{\displaystyle /\!\!/ O}{\underset{\displaystyle \backslash OH}{}}$ (*)

Your answer:

Solution: ethanoic acid (acetic acid)

---

3. p-methylbenzoic acid

Your answer:

Solution:

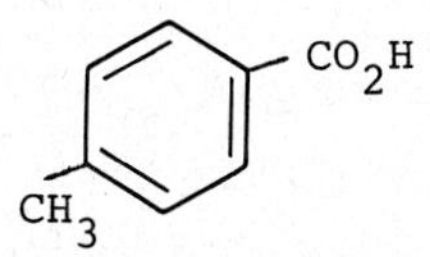

---

4. octadecanoic acid (*)

Your answer:

Solution: $CH_3(CH_2)_{16}C(=O)OH$ (stearic acid)

---

5. $HO_2C\text{-}CH_2CH_2CO_2H$ (*)

Your answer:

Solution: butanedioic acid (succinic acid)

B. SALTS OF CARBOXYLIC ACIDS

1. $CH_3C(=O)O^- \ Na^+$

Your answer:

Solution: sodium ethanoate (sodium acetate)

The cation is named first; the anion (or conjugate base) is named by dropping the "ic acid" and adding "ate" to the name of the acid.

---

2. 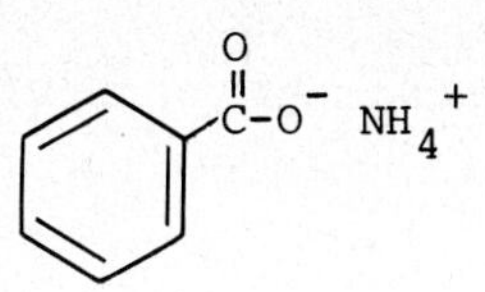

Your answer:

Solution: ammonium benzoate

C. ESTERS

1. 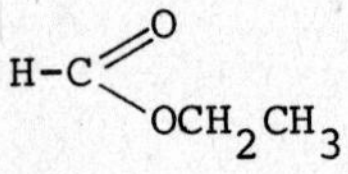 or $HCO_2CH_2CH_3$ (*)

Your answer:

Solution: ethyl methanoate (ethyl formate)

The group bonded by oxygen to the carbonyl is named first. (It is called the alcohol group in reference to the reaction during which an alcohol and an acid form an ester.) The "acid portion" of the ester is named with the same ending used for the anion or conjugate base in the salts.

---

2. 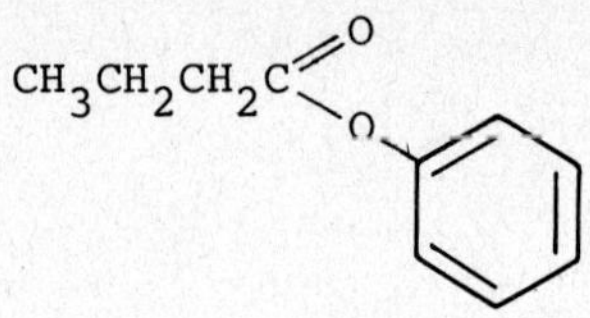

Your answer:

Solution phenyl butanoate

---

3. $(CH_3)_2CHO\underset{\underset{O}{\|}}{C}CH_2CH_3$

Your answer:

Solution: isopropyl propanoate

D. ACYL HALIDES (ACID HALIDES)

1. 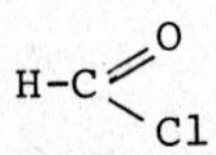 (*)

Your answer:

Solution: methanoyl chloride (formyl chloride)

$R-\overset{\overset{O}{\|}}{C}-$ is named as a substituent on the halogen. The name of the acyl halide is derived by dropping the "ic acid" and adding "yl chloride (or other halide)."

---

2. 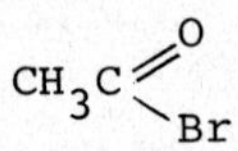 (*)

Your answer:

Solution: ethanoyl bromide (acetyl bromide)

---

3. benzoyl chloride

Your answer:

Solution:

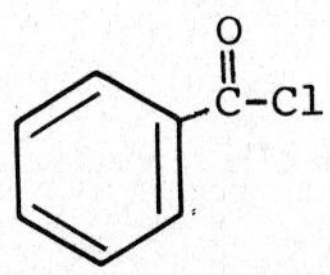

E. ACID ANHYDRIDES

1. $CH_3-\underset{\underset{O}{\|}}{C}-O-\underset{\underset{O}{\|}}{C}-CH_3$ or $(CH_3CO)_2O$ (*)

Your answer:

Solution: (acetic acid anhydride or acetic anhydride)

---

2. (benzoic acid anhydride)

Your answer:

Solution:

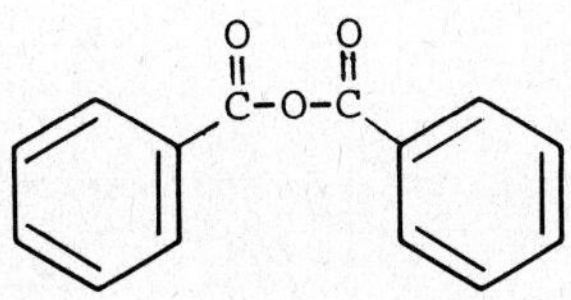

F. AMIDES

1. $H-C\overset{\displaystyle O}{\underset{\displaystyle NH_2}{\lessgtr}}$ (*)

Your answer:

Solution: methanamide (formamide)

To derive the name of an amide, drop the "oic acid" from the name of the acid and add "amide."

---

2. $CH_3CH_2NHC(=O)CH_3$ (*)

Your answer:

<u>Solution</u>: N-ethylethanamide (N-ethylacetamide)
"N-ethyl" indicates that the ethyl is bonded to the nitrogen of the amide.

---

3. N,N-dimethylbenzamide

Your answer:

<u>Solution</u>:

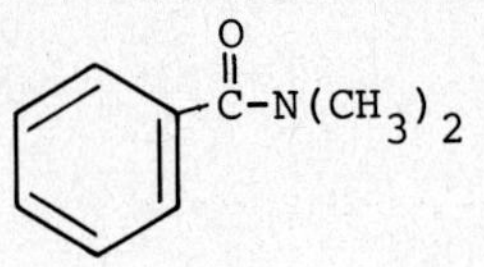

G. CARBOXYLIC ACIDS, ACID DERIVATIVES, AND REVIEW COMPOUNDS

From the given name or formula, identify the type of compound, write a "type" formula, and then write the corresponding structure or name. Also practice the questions in reverse: from the solution, can you name or draw the structure of the compound in the question?

1. methyl butanoate

Your answer:

<u>Solution</u>: ester; $R\text{-}C(=O)OR'$; $CH_3CH_2CH_2C(=O)\text{-}OCH_3$

---

2. phenyl ethanoate (*)

Your answer:

<u>Solution</u>: ester; $R\text{-}CO_2R'$; 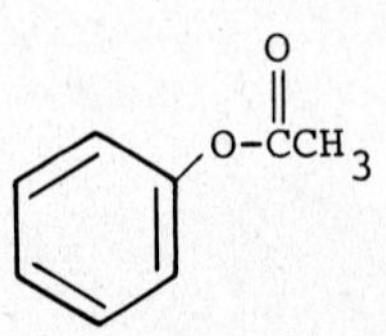 (phenyl acetate)

---

3. 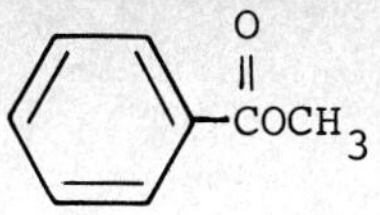

Your answer:

Solution: ester, $R\text{-}CO_2R'$; methyl benzoate

---

4. $CH_3\overset{\overset{O}{\|}}{C}\text{-}\underset{\underset{CH_3}{|}}{CH}\text{-}\overset{\overset{Cl}{|}}{C}HCH_3$

Your answer:

Solution: ketone; $R\text{-}\underset{\underset{O}{\|}}{C}\text{-}R'$; 4-chloro-3-methyl-2-pentanone

---

5. propanoic acid anhydride

Your answer:

Solution: anhydride, $R\text{-}\overset{\overset{O}{\|}}{C}\text{-}O\text{-}\overset{\overset{O}{\|}}{C}\text{-}R$, $CH_3CH_2\overset{\overset{O}{\|}}{C}\text{-}O\text{-}\overset{\overset{O}{\|}}{C}CH_2CH_3$

---

6. sodium o-chlorobenzoate

Your answer:

Solution: salt, $R\text{-}\overset{\overset{O}{\|}}{C}\text{-}O^- \; Na^+$, 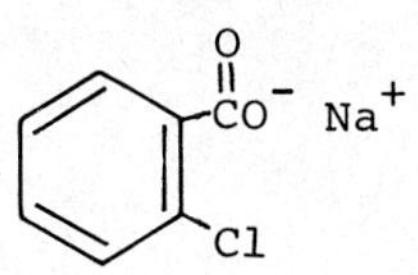

---

7. $Cl\text{-}\underset{\underset{O}{\|}}{C}\text{-}CH_2CH_2CH_3$

Your answer:

Solution: acyl (or acid) halide, $\mathrm{R{-}\overset{\overset{O}{\|}}{C}{-}Cl}$; butanoyl chloride

---

8. $\mathrm{CH_3CH_2CH_2{-}O{-}\underset{\underset{O}{\|}}{C}H}$ (*)

Your answer:

Solution: ester, $\mathrm{R{-}\overset{\overset{O}{\|}}{C}{-}OR'}$ or $\mathrm{R'O{-}\overset{\overset{O}{\|}}{C}{-}R}$; propyl methanoate (propyl formate)

---

9. $\mathrm{H_2N{-}\overset{\overset{O}{\|}}{C}{-}CH_2CH_2CH_2CH_2CH_3}$

Your answer:

Solution: amide, $\mathrm{R{-}\overset{\overset{O}{\|}}{C}{-}NH_2}$; hexanamide

---

10. hexadecanoic acid (*)

Your answer:

Solution: carboxylic acid, $\mathrm{R{-}CO_2H}$; $\mathrm{CH_3(CH_2)_{14}CO_2H}$ (palmitic acid)

---

11. trans-4-hexen-2-one

Your answer:

Solution: ketone, $\mathrm{R{-}\overset{\overset{O}{\|}}{C}{-}R'}$ (and alkene as lower priority); 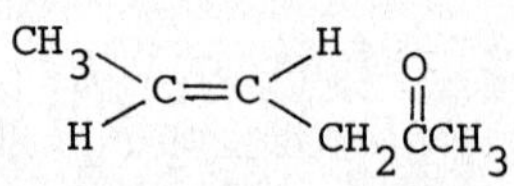

12.

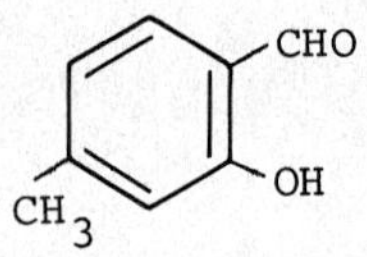

(*)

Your answer:

<u>Solution</u>: aromatic aldehyde, R-CHO (also phenol, aromatic alcohol, R-OH), (2-hydroxy-4-methylbenzaldehyde)

---

13. 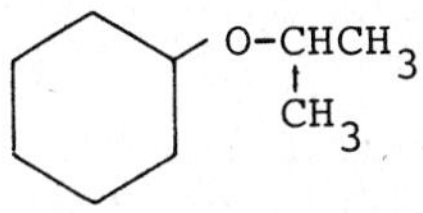

Your answer:

<u>Solution</u>: ether, R-O-R'; cyclohexyl isopropyl ether

---

14. $CH_3(CH_2)_{16}\overset{O}{\overset{\|}{C}}\text{-}OCH_2CH_3$ (*)

Your answer:

<u>Solution</u>: ester, R-C-OR'; ethyl octadecanoate (ethyl stearate)

---

15. $\underset{OH}{\underset{|}{CH_2}}\underset{NH_2}{\underset{|}{CH}}C(=O)OH$ (*)

Your answer:

<u>Solution</u>: carboxylic acid, $R\text{-}CO_2H$ (also an amine, $R\text{-}NH_2$, and an alcohol, R-OH); 2-amino-3-hydroxypropanoic acid (α-amino-β-hydroxypropionic acid or serine, an amino acid)

---

16. 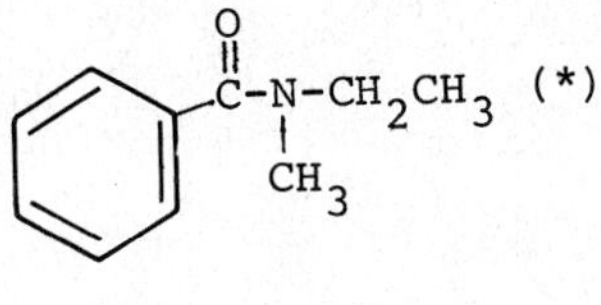

Your answer:

<u>Solution</u>: amide, $R\text{-}\overset{O}{\overset{\|}{C}}\text{-}NH_2$; (N-ethyl-N-methylbenzamide)

IV. Amines and Amine Salts

Determine the name of each compound from the structure, or determine the structure from the compound name. Write the common name of the compounds marked with an asterisk, and enclose the common name in parentheses.

```
1. CH3CHNHCH3  (*)
       |
       CH3
```

Your answer:

Solution: 2-(N-methylamino)propane (isopropylmethylamine)
Name each group attached to the nitrogen and add the word "amine" to the name.

2.

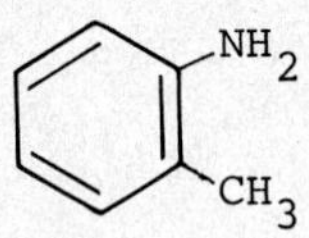

Your answer:

Solution: o-methylaniline

```
3. CH3CH2CHCH2CHCH3
           |     |
           NH2   CH3
```

Your answer:

Solution: 3-amino-5-methylhexane
The amino group has a higher priority than does the alkyl group, so the carbon to which the amino group is bonded is assigned the lower number.

```
                O
                ||
4. CH3CH2CHC-OH
           |
           N-H
           |
           CH3
```

Your answer:

Solution: 2-(N-methylamino)butanoic acid
Throughout the series of organic compounds you have studied, the newest functional group has been assigned the highest priority (and consequently receives the lowest number in nomenclature). The amino group is an exception to this generalization: it ranks in priority between the alkynes and the alcohols, and so in this problem, the acid group ranks above the amino group.

```
5. CH3-CH-CH-CH3
       |  |
       OH N-CH3
          |
          CH3
```

Your answer:

Solution: 3-(N,N-dimethylamino)-2-butanol

---

6. $CH_3\underset{|}{C}HNH\overset{}{C}CH_2CH_3$

$\quad\;\; CH_3 \quad\; \| \; O$

(CH beneath bears CH_3; C bears =O)

Your answer:

Solution: N-isopropylpropanamide

---

7. $(CH_3)_3\overset{+}{N}-H$

Your answer:

Solution: trimethylammonium ion

---

8. o-methyl-N-propylaniline

Your answer:

Solution:

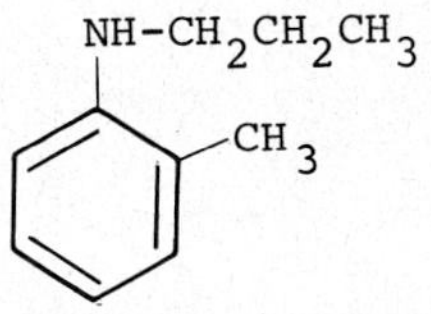

---

9. 2-amino-3-phenylpropanoic acid (*)

Your answer:

Solution:

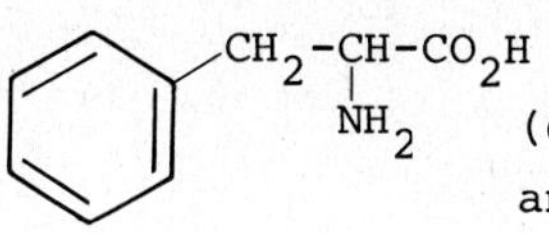

(α-amino-β-phenylpropionic acid or phenylalanine, an essential amino acid)

Questions 10-13 cover the common names of heterocyclic amines that are important bases in the biochemistry of life.

10.

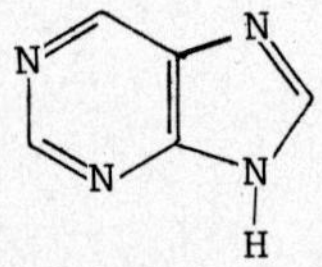

Your answer:

Solution: purine

11.

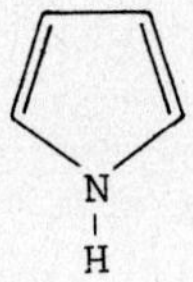

Your answer:

Solution: pyrrole

12.

Your answer:

Solution: pyrimidine

13.

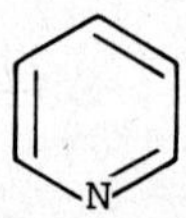

Your answer:

Solution: pyridine

---

17 Alcohols, Phenols, and Ethers

I. Nomenclature and Classification

1. Consider the following structural formulas.

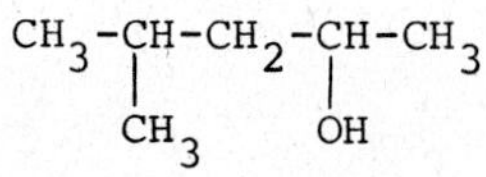
A

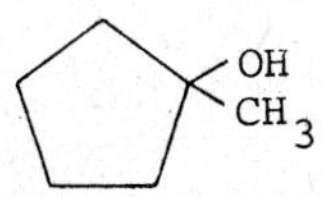
B

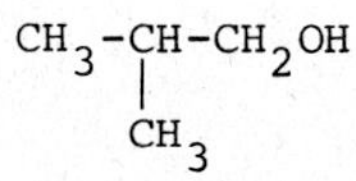
C

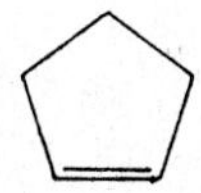
D

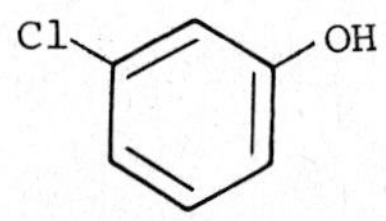
E

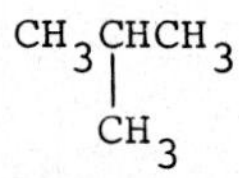
F

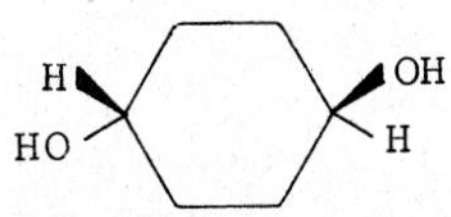
G

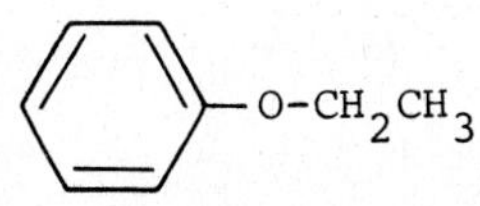
H

(a) Classify each compound by type. For each type of alcohol, indicate whether it is a primary (1°), secondary (2°), or tertiary (3°) alcohol.

(b) Name each compound. Use systematic names for all compounds except the aromatic compounds; for these compounds, use the common names.

Your answers:

	Classification	Name
A.		
B.		
C.		
D.		
E.		
F.		
G.		
H.		

Solution:

	Classification	Name
A.	2° alcohol	4-methyl-2-pentanol
B.	3° alcohol	1-methylcyclopentanol
C.	1° alcohol	2-methyl-1-propanol (isobutyl alcohol)
D.	alkene	cyclopentene
E.	a phenol (aromatic alcohol)	m-chlorophenol

F. alkane	methylpropane
G. 2° alcohol (a diol)	trans-1,4-cyclohexanediol
H. ether	ethyl phenyl ether

II. Reactions of Alcohols, Phenols, and Ethers

For each question use the structural formulas for the specified compounds given in Section I. Write a "type" reaction for the corresponding general type of compound, and describe the reaction in words. Write the equation(s) for the specific compound(s), and name the organic products (unless you have not covered them in lecture yet.)

1. Sodium metal (Na) reacts in a characteristic way with certain organic compounds. Illustrate this reaction with compounds A, B, E, and H.

Your answer:

Solution:

"Type" reaction: $R\text{-}OH + Na \longrightarrow RO^- \; Na^+ + \frac{1}{2} H_2(g)$ (all classes of alcohols, also phenols)

Although alcohols are less acidic than water, sodium is capable of extracting the proton (H^+). The evolution of H_2 gas provides positive identification of alcohols or phenols. However, ethers have no acidic hydrogen and therefore cannot react.

$R\text{-}O\text{-}R + Na \longrightarrow$ N.R.

A. $$CH_3\text{-}\underset{\displaystyle CH_3}{\underset{|}{CH}}\text{-}CH_2\text{-}\underset{\displaystyle OH}{\underset{|}{CH}}\text{-}CH_3 \xrightarrow{Na} CH_3\underset{\displaystyle CH_3}{\underset{|}{CH}}\text{-}CH_2\text{-}\underset{\displaystyle O^-\;Na^+}{\underset{|}{CH}}\text{-}CH_3 + \tfrac{1}{2} H_2(g)$$

sodium
4-methyl-2-pentoxide

B. 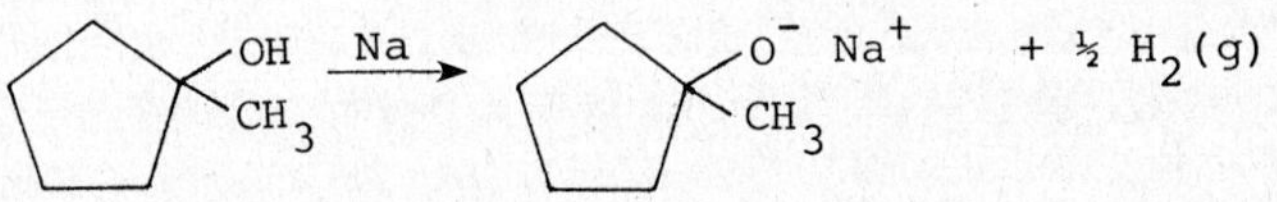

sodium
1-methylcyclopentoxide

E. 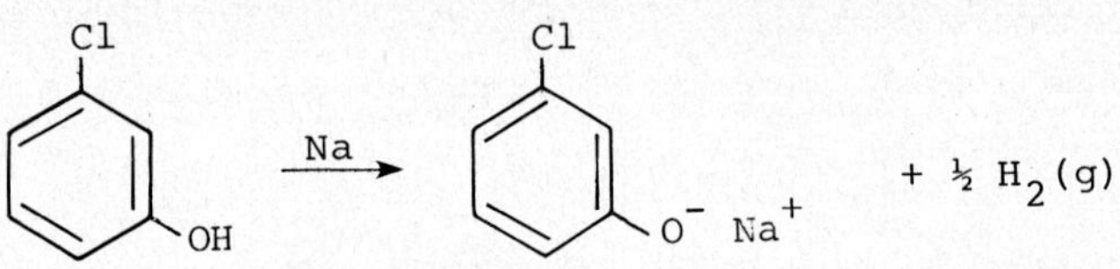

sodium
m-chlorophenoxide

H. 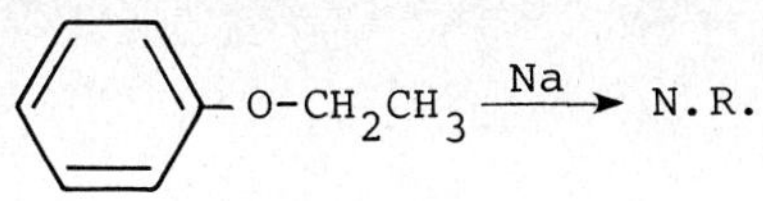

---

2. Write equations for the reactions of compounds C, E, and G with NaOH(aq), a familiar laboratory base.

Your answer:

Solution:

"Type" reactions: R-OH + NaOH (aq) ⟶ N.R. (alcohols are weaker acids than H_2O)

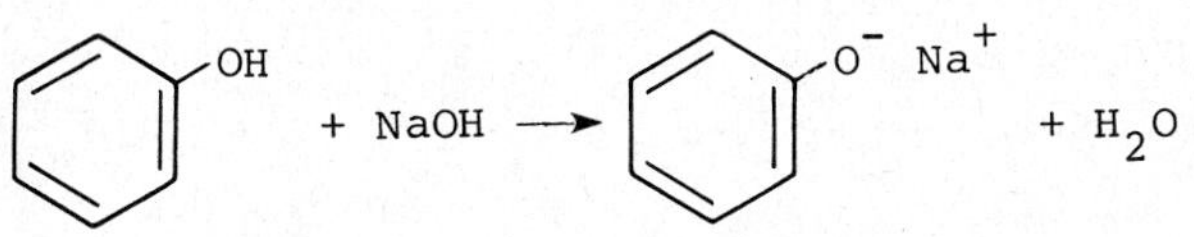

The acid strength for alcohols and related compounds is

R-O-H < H_2O < 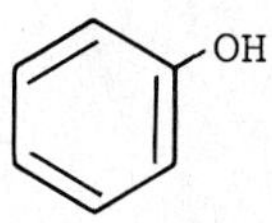

C. CH_3CHCH_2OH + NaOH(aq) ⟶ N.R.
(with a CH_3 branch on the second carbon)

E. 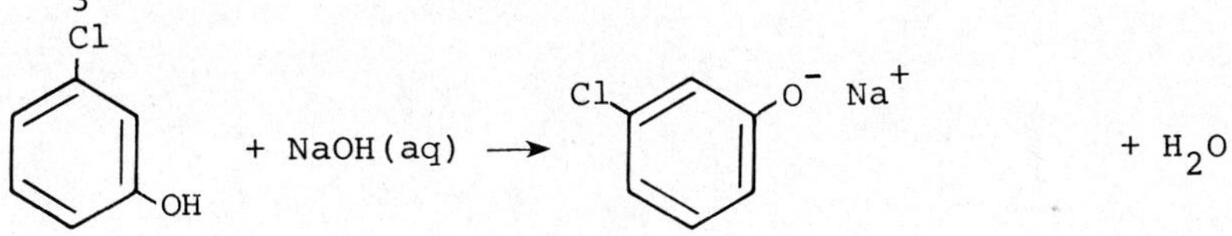

sodium m-chlorophenoxide

G. 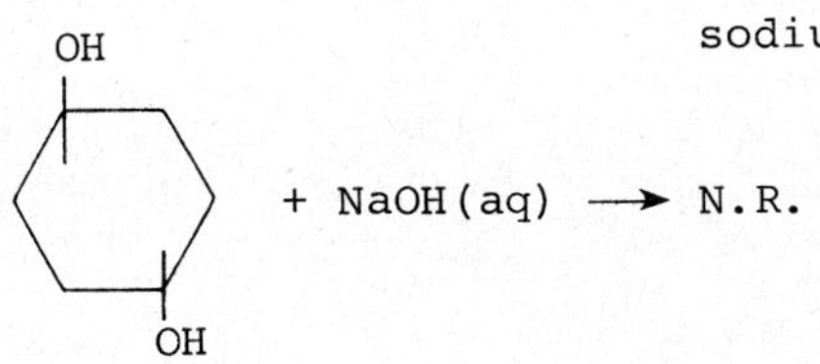

---

3. Potassium dichromate in an acidic solution acts as an oxidizing agent and is used as a test for alcohols. When an alcohol is oxidized (a positive test), the orange color of

dichromate changes to the green color of the chromic ion. Write a "type" reaction for each class of alcohol. Write specific reactions for compounds A, B, C, G, and H. (Notice that the dichromate reagent is written in a variety of ways.)

Your answer:

Solution:

"Type" reactions:

1° alcohol: 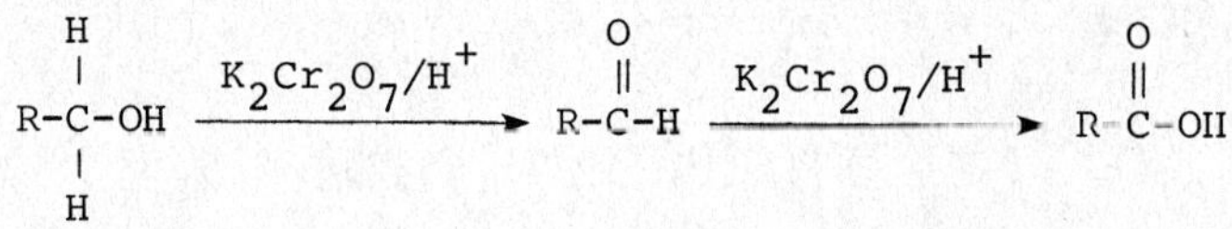

aldehyde carboxylic acid

Two hydrogens are removed during the oxidation of a primary alcohol to an aldehyde: one hydrogen from the carbon and the other from the oxygen. The aldehyde is unstable in the presence of the oxidizing agent and is further oxidized to a carboxylic acid by the extraction of another hydrogen and the addition of an oxygen.

2° alcohol: 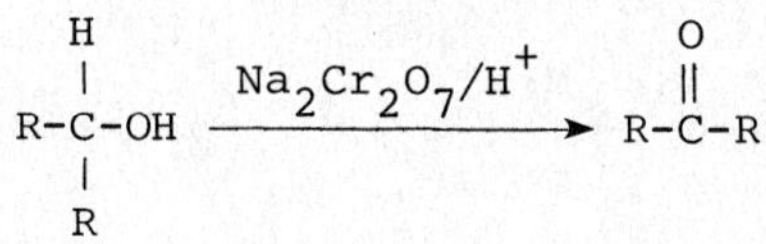

ketone

A secondary alcohol is oxidized to a ketone by the extraction of two hydrogens. However, the ketone is stable in the presence of oxidizing agents, and consequently is not oxidized further. (Notice that there is no hydrogen available for extraction on the oxidized carbon of the ketone.)

3° alcohol: 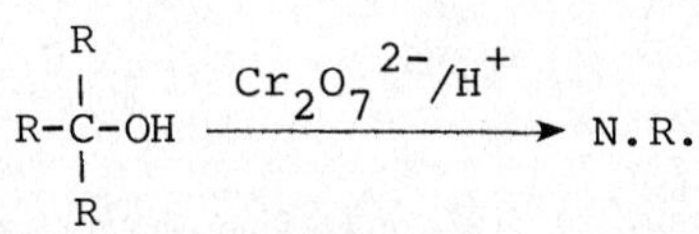

A tertiary alcohol cannot be oxidized because the tertiary carbon has no hydrogens that can be removed.

A. 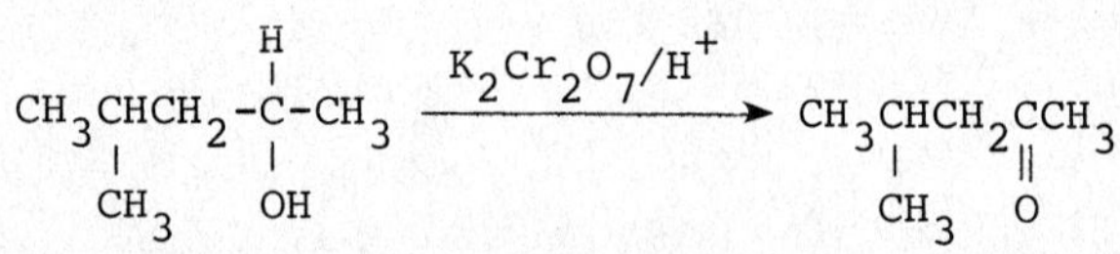

4-methyl-2-pentanone

B. 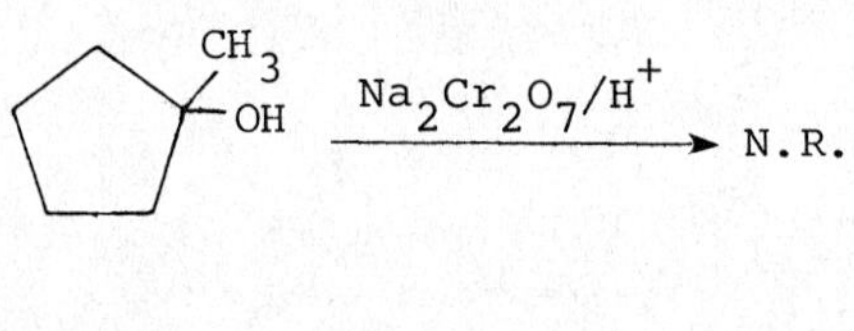

C. $CH_3CH(CH_3)CH_2OH \xrightarrow{K_2Cr_2O_7/H^+} CH_3CH(CH_3)CH{=}O \xrightarrow{K_2Cr_2O_7/H^+} CH_3\text{-}CH(CH_3)C(=O)\text{-}OH$

2-methylpropanal — 2-methylpropanoic acid

G. 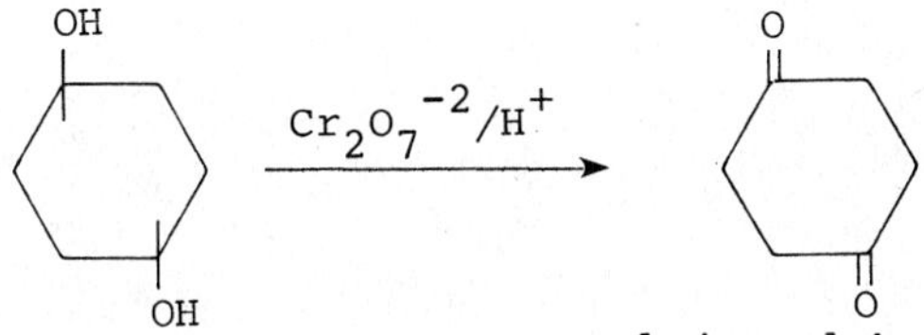

1,4-cyclohexanedione

H. 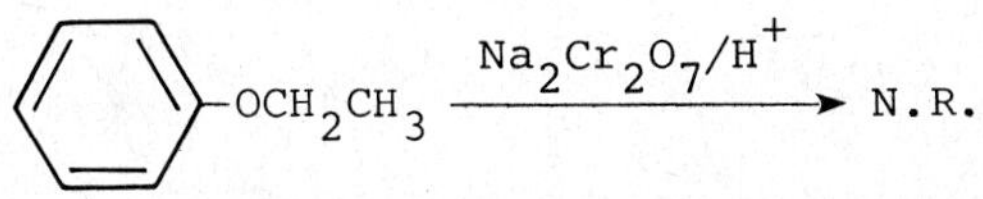

---

4. Consider the reaction of alcohols (and other compounds) with hot, concentrated sulfuric acid, H_2SO_4; these are conditions for dehydration and are often represented as H^+/heat. Write a "type" reaction and equations for the specific reactions of H_2SO_4 with compounds A, B, C, and F.

Your answer:

Solution:

"Type" reaction: $$R\text{-}CH(H)\text{-}C(H)(OH)\text{-}R \xrightarrow[-H_2O]{\text{hot } H_2SO_4 \text{ (conc)}} (H)(R)C{=}C(R)(H)$$

H_2SO_4 removes the elements of water (H and OH) from adjacent carbon atoms to form the alkene. If there is a choice of carbons (adjacent to the carbon containing the OH) from which the H is to be removed, the following order is utilized: a 3° hydrogen is removed more easily than a 2° hydrogen, which is, in turn, more easily removed than a 1° hydrogen (ease of extraction of hydrogens: $3^o > 2^o > 1^o$). However, a mixture of products usually results.

A. $$CH_3CH(CH_3)\text{-}CH_2\text{-}CH(OH)\text{-}CH_3 \xrightarrow{H_2SO_4,\ heat} CH_3CH(CH_3)CH{=}CHCH_3 + H_2O + CH_3CH(CH_3)CH_2CH{=}CH_2$$

cis- and trans-
4-methyl-2-pentene
(major product)

4-methyl-1-pentene
(minor product)

B.

1-methylcyclopentene (major product) — (minor product)

C. $CH_3-\underset{H}{\overset{CH_3}{|}}C-\underset{OH}{CH_2} \xrightarrow{H_2SO_4/-H_2O} (CH_3)_2C{=}CH_2 + H_2O$

methylpropene

F. $CH_3\underset{CH_3}{CH}CH_3 \xrightarrow{H_2SO_4/\Delta}$ N.R.

---

5. The Lucas test is used to differentiate primary, secondary, and tertiary alcohols. Write the "type" reaction for the Lucas reaction and write specific reactions for compounds B, C, E, and H. The Lucas reagent is HCl with $ZnCl_2$ catalyst.

Your answer:

Solution:

"Type" reaction: $R\text{-}OH + HCl \xrightarrow{ZnCl_2} [R^+] \xrightarrow{Cl^-} R\text{-}Cl + H_2O$

$ZnCl_2$ aids in the removal of the -OH to form an intermediate carbonium ion. The rate of reactivity of the alcohols is $3^o > 2^o > 1^o$. The reaction is extremely slow for primary alcohols. Ethers and phenols do not react at all. A positive reaction for the alcohols is recognized by the cloudy appearance of the solution, which eventually develops into two layers, an oily layer of alkyl halide (product) on top of an aqueous layer.

B.

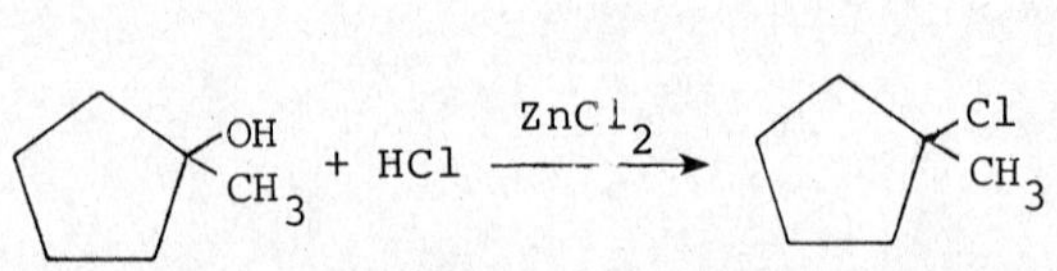

+ H_2O (fast reaction)

1-chloro-1-methylcyclopentane

C. $CH_3-\overset{CH_3}{\overset{|}{CH}}-CH_2-OH \xrightarrow{HCl/ZnCl_2} CH_3-\overset{CH_3}{\overset{|}{CH}}-CH_2-Cl + H_2O$ (very slow reaction)

1-chloro-2-methylpropane

E. 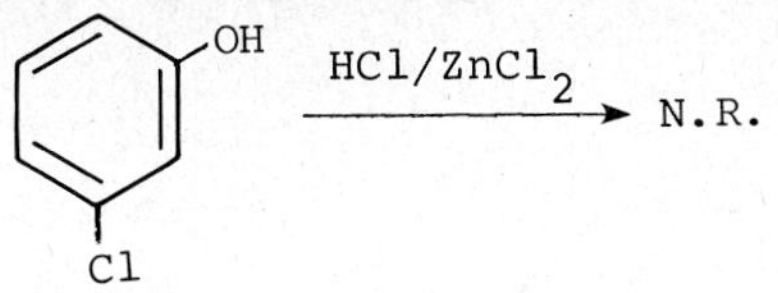

H. 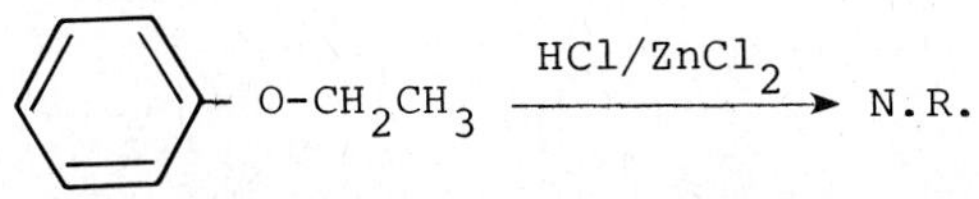

---

6. Replacement of the hydroxyl group of alcohols by a halogen is effectively accomplished using PBr_3, $SOCl_2$, or PCl_3. Write equations for the reactions of these reagents with compounds B, C, and E.

Your answer:

Solution:

"Type" reactions:

$3\ R\text{-}OH + PBr_3 \longrightarrow 3\ R\text{-}Br + P(OH)_3$

$R\text{-}OH + SOCl_2 \longrightarrow R\text{-}Cl + SO_2 + HCl$

These reagents react well with primary, secondary, or tertiary alcohols, but not with phenols.

B. 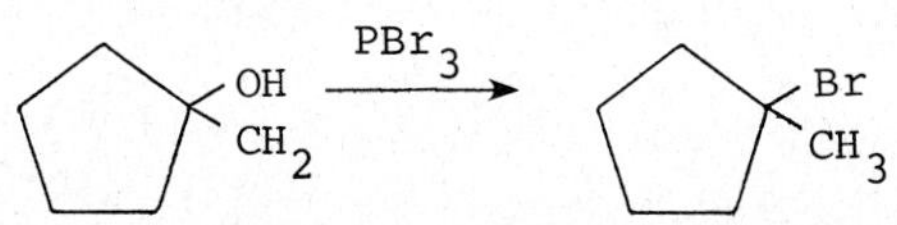 (unbalanced)

1-bromo- 1-methylcyclopentane

C. $CH_3CH(CH_3)CH_2OH + SOCl_2 \longrightarrow CH_3CH(CH_3)CH_2Cl$

1-chloro-2-methylpropane

E. 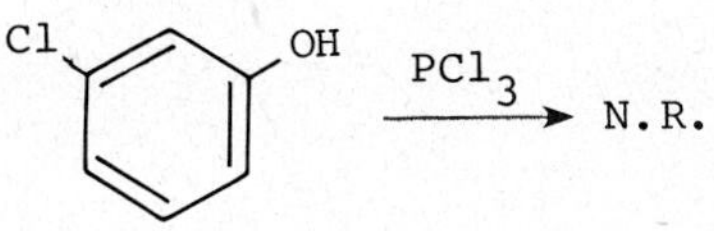

---

6. Esterification is a reaction in which a molecule of alcohol and a molecule of acid condense into an "ester" (with the removal of a molecule of water).

$$\text{R-OH} + \text{CH}_3\overset{\overset{\displaystyle\text{O}}{\|}}{\text{C}}\text{-OH} \rightleftharpoons \text{CH}_3\overset{\overset{\displaystyle\text{O}}{\|}}{\text{C}}\text{-OR} + \text{H}_2\text{O}$$

acetic acid

Write equations for the esterification of compound C with sulfuric acid, compound E with acetic acid, and compound G with nitric acid. (You will see more details for this reaction in a later chapter.)

Your answer:

<u>Solution:</u>

$$\text{C.}\quad \text{CH}_3\underset{\underset{\displaystyle\text{CH}_3}{|}}{\text{C}}\text{HCH}_2\text{OH} + \text{HO-}\overset{\overset{\displaystyle\text{O}}{\|}}{\underset{\underset{\displaystyle\text{O}}{\|}}{\text{S}}}\text{-OH} \overset{\text{H}^+}{\rightleftharpoons} \text{CH}_3\overset{\overset{\displaystyle\text{CH}_3}{|}}{\text{C}}\text{HCH}_2\text{-O-SO}_3\text{H} + \text{H}_2\text{O}$$

a sulfuric acid ester

(Some detergents are sulfuric acid esters of long-chain alcohols.)

E.

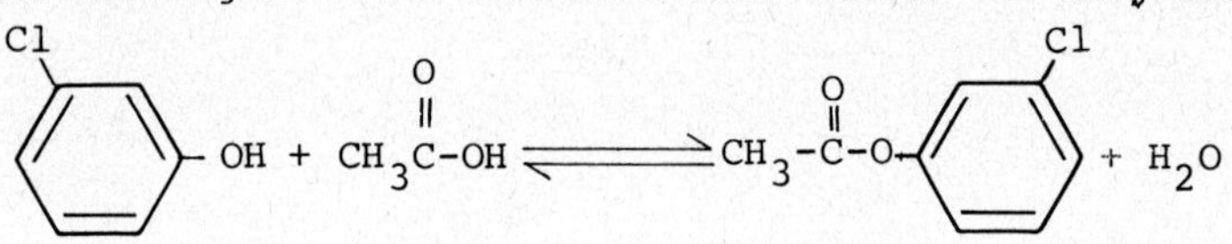

G.

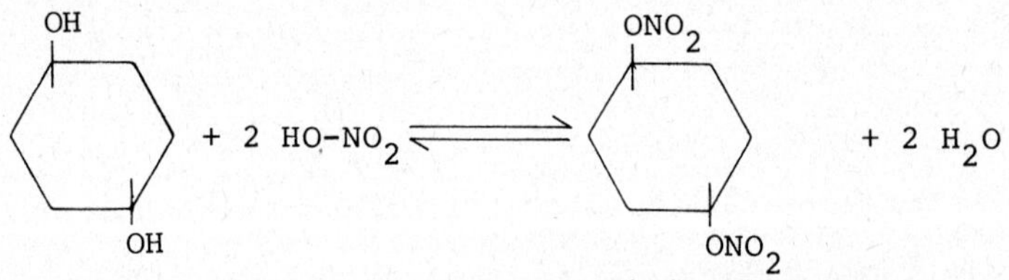

a dinitrate ester

III. Identification Problems

1. Compound X, $C_5H_{12}O$, reacts with Na to yield H_2 gas. It also reacts with the Lucas reagent, but does not react with $Na_2Cr_2O_7$ solution. What is the exact structure of compound X?

Your answer:

Solution:

Given information	Possible	Not possible
X is $C_5H_{12}O$	alcohol	cyclic compound
$C_nH_{2n+2}O$ (saturated compound)	ether	
$X \xrightarrow{Na} H_2$	alcohol	ether
positive Lucas test ($HCl/ZnCl_2$)	alcohol	
$X \xrightarrow{Na_2Cr_2O_7}$ N.R.	3° alcohol	1°, 2° alcohol

X is $CH_3\text{-}\overset{\displaystyle OH}{\underset{\displaystyle CH_3}{C}}\text{-}CH_2CH_3$, 2-methyl-2-butanol

---

2. Compound Y, $C_4H_{10}O$, does not react with Na metal, Br_2, or $Na_2Cr_2O_7$. Write structural formulas and names of the isomers that correspond to this description.

Your answer:

Solution:

Given information	Possible	Not possible
Y is $C_4H_{10}O$	alcohol	cyclic compound
$C_nH_{2n+2}O$ (saturated compound)	ether	
$Y \xrightarrow{Na}$ N.R.	ether	alcohol
$Y \xrightarrow{Na_2Cr_2O_7}$ N.R.	ether	alcohol

Y can be $CH_3\text{-}O\text{-}CH_2CH_2CH_3$, methyl propyl ether; $CH_3\text{-}O\text{-}\underset{\displaystyle CH_3}{C}HCH_3$, isopropyl methyl ether;

$CH_3CH_2\text{-}O\text{-}CH_2CH_3$, diethyl ether.

---

3. Suppose that you are asked to prepare 2-methyl-2-pentene by a dehydration reaction. The reagent shelf contains the following compounds: (i) 2-methyl-2-pentanol, (ii) 2-methyl-1-pentanol, (iii) 4-methyl-2-pentanol, and (iv) 2-methyl-3-pentanol. Choose the compound(s) that would be appropriate for the reaction and show the equations.

Your answer:

Solution: reactions (i) and (iv)

"Type" reaction: 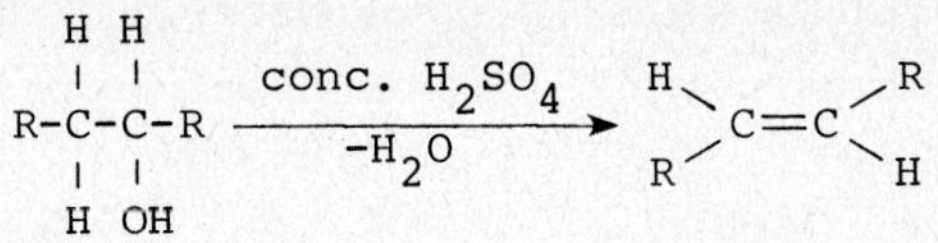

The removal of a 3° H from the carbon adjacent to the carbon containing the hydroxyl is better than the removal of a 2° H, which is better than the removal of a 1° H.

(i) 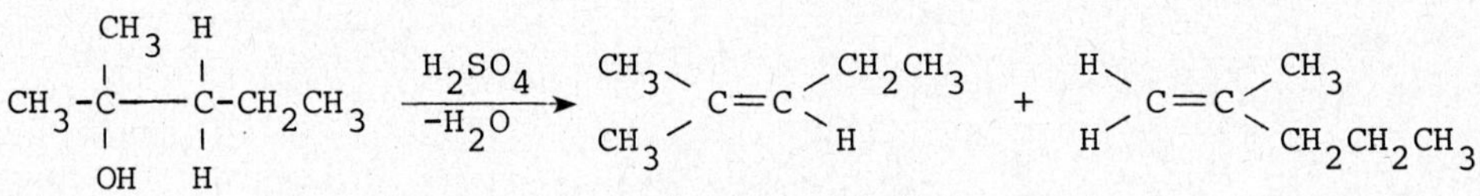

(major product)
2-methyl-2-pentene

(ii) 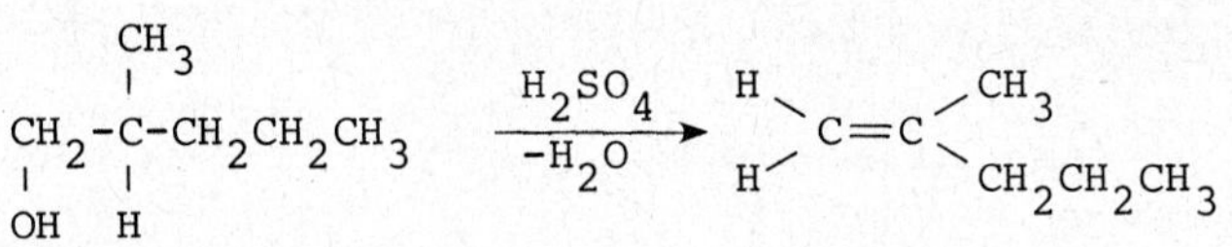

(iii) 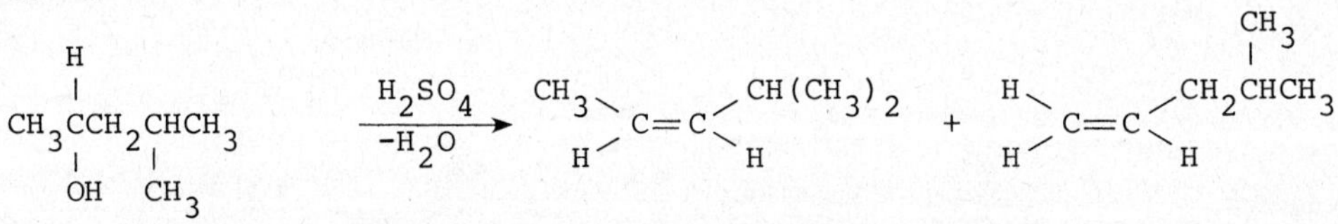

(iv) 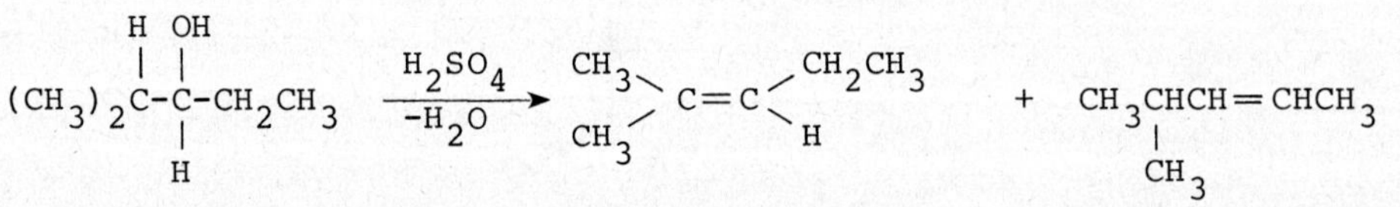

(major product) (minor product)
2-methyl-2-pentene

2-Methyl-2-pentene will be formed in higher proportion in reaction (iv) than in reaction (i) because it is formed by the removal of a 3° H in reaction (iv) but by the removal of a 2° H in reaction (i).

IV. Comparison of Some Physical Properties of Alcohols, Ethers, and Hydrocarbons

1. Compare (in the following table) the structural formulas, molecular weights, and types of intermolecular attractions of (a) hexane, (b) 2,3-dimethylbutane, (c) ethyl propyl

ether, (d) 1,2,3-propanetriol, (e) 2-methyl-2-butanol, and (f) 1-pentanol. Rank the compounds in order of increasing boiling points.

Your answer:

Compound	Structural formula	Molecular weight	Types of intermolecular attractions	Estimated rank of boiling point
(a)				
(b)				
(c)				
(d)				
(e)				
(f)				

Compound	Structural formula	Molecular weight	Types of intermolecular attractions	Rank of boiling point	Boiling point
(a)	$CH_3CH_2CH_2CH_2CH_2CH_3$	86	No hydrogen bonding; relatively nonpolar bonds, tetrahedral shape around each C, so no permanent dipole; only London forces between molecules	3	69°C
(b)	$CH_3\text{-}CH\text{-}CH\text{-}CH_3$ $\quad\;\; \vert \quad \vert$ $\quad\; CH_3CH_3$	86	Same as (a), only London forces, but these forces are greater in long straight-chain molecules than in branched molecules	1	58°C
(c)	$CH_3CH_2\text{-}O\text{-}CH_2CH_2CH_3$	88	No hydrogen bonding; a small amount of permanent dipole indicated by the structure; London forces	2	62°C(*)
(d)	$CH_2\text{-}CH\text{-}CH_2OH$ $\vert \quad\;\; \vert$ $OH \;\; OH$	92	Hydrogen bonding; much more hydrogen bonding than in regular alcohols; London forces	6	290°C
(e)	$\quad\;\; CH_3$ $\quad\;\;\; \vert$ $CH_3\text{-}C\text{-}CH_2CH_3$ $\quad\;\;\; \vert$ $\quad\;\; OH$	88	Hydrogen bonding; London forces (less than for straight-chain molecule)	4	129°C
(f)	$CH_3CH_2CH_2CH_2CH_2$ $\qquad\qquad\quad \vert$ $\qquad\qquad\; OH$	88	Hydrogen bonding; London forces	5	137°C

(*) The boiling point of the ether is a little lower than that predicted by its structure.

V. Practice and Review Reactions

Write the structural formulas of the reactant and product of each reaction. If no reaction occurs, explain why.

1. 2-propanol + Na ⟶

Your answer:

Solution: $CH_3CH(OH)CH_3 + Na \longrightarrow CH_3CH(O^- Na^+)CH_3 + \frac{1}{2} H_2(g)$

sodium 2-propoxide

---

2. ethyl phenyl ether + Na ⟶

Your answer:

Solution:

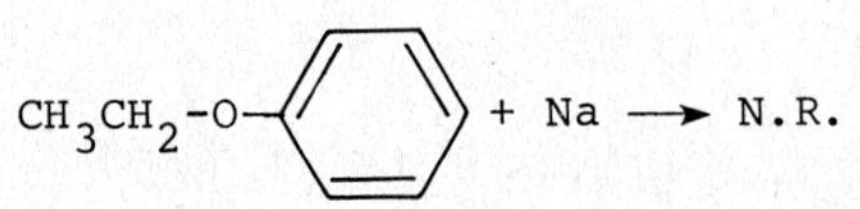

Ethers contain no acidic hydrogens and therefore cannot react.

---

3. m-methylphenol + NaOH(aq) ⟶

Your answer:

Solution:

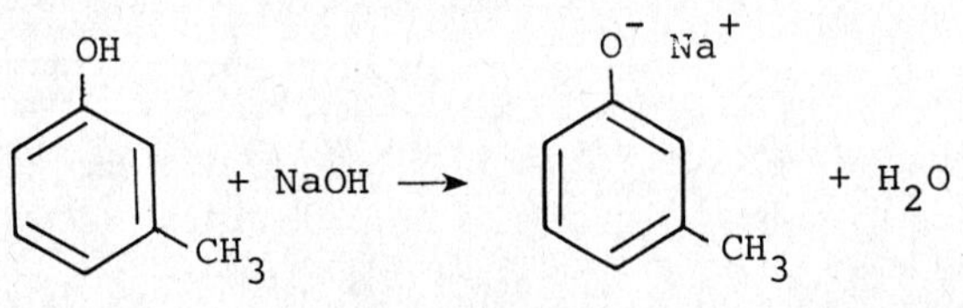

sodium m-methylphenoxide

---

4. 2-methyl-2-butanol $\xrightarrow{Na_2Cr_2O_7/H^+}$

Your answer:

Solution: 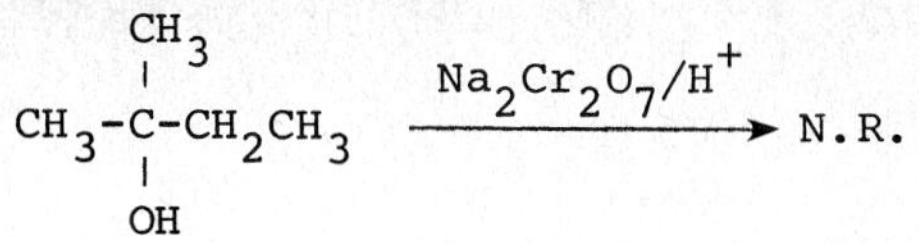

A tertiary alcohol cannot be oxidized by dichromate solution.

---

5. 2-propanol + acetic acid →

Your answer:

Solution: 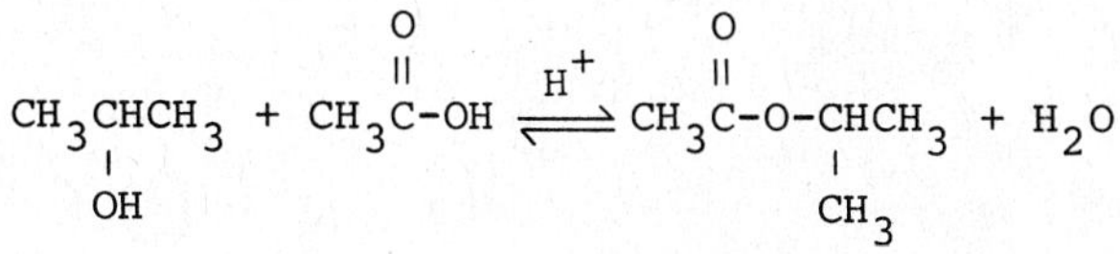

isopropyl ethanoate
(isopropyl acetate)

---

6. 3-methyl-2-butanol $\xrightarrow{Na_2Cr_2O_7/H^+}$

Your answer:

Solution: 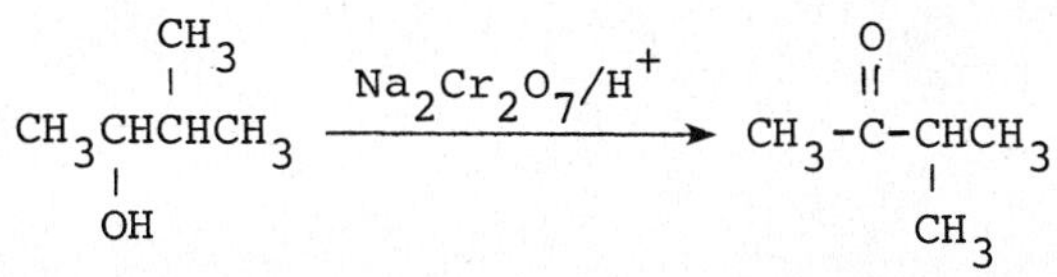

3-methyl-2-butanone

---

7. 3-methyl-2-butanol $\xrightarrow{H_2SO_4\ (conc.)/\Delta}$

Your answer:

Solution: 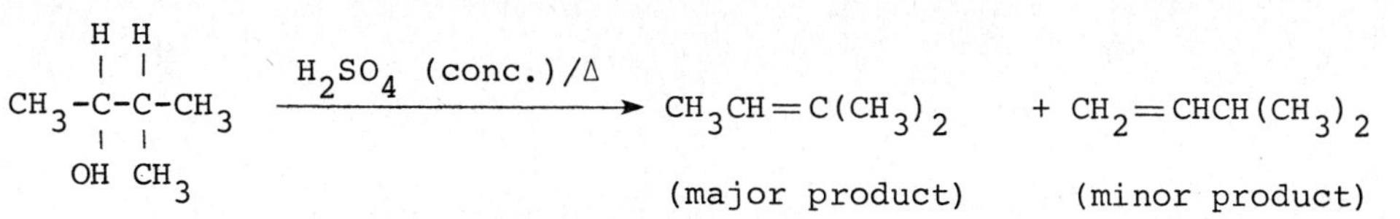

(major product) 2-methyl-2-butene

(minor product) 3-methyl-1-butene

---

8. 2,2-dimethyl-1-propanol $\xrightarrow{Na_2Cr_2O_7/H^+}$

Your answer:

Solution: 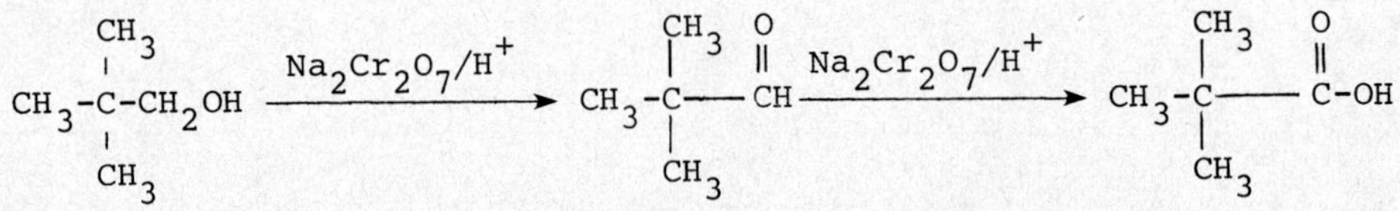

2,2-dimethylpropanal 2,2-dimethylpropanoic acid

---

9. cyclobutanol $\xrightarrow{Na_2Cr_2O_7/H^+}$

Your answer:

Solution: 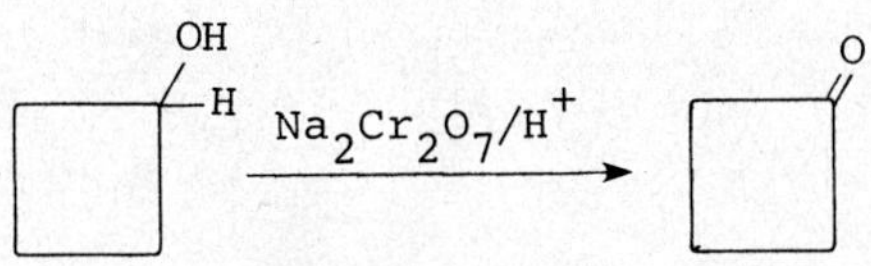

cyclobutanone

---

10. ethanol $\xrightarrow{HCl/ZnCl_2}$

Your answer:

Solution: $CH_3CH_2OH \xrightarrow{HCl/ZnCl_2} CH_3CH_2Cl$ (reaction occurs slowly)

chloroethane

---

11. 3-phenyl-2-pentanol $\xrightarrow[-H_2O]{H_2SO_4}$? $\xrightarrow{Br_2 \text{ (dark)}}$?

Your answer:

Solution:

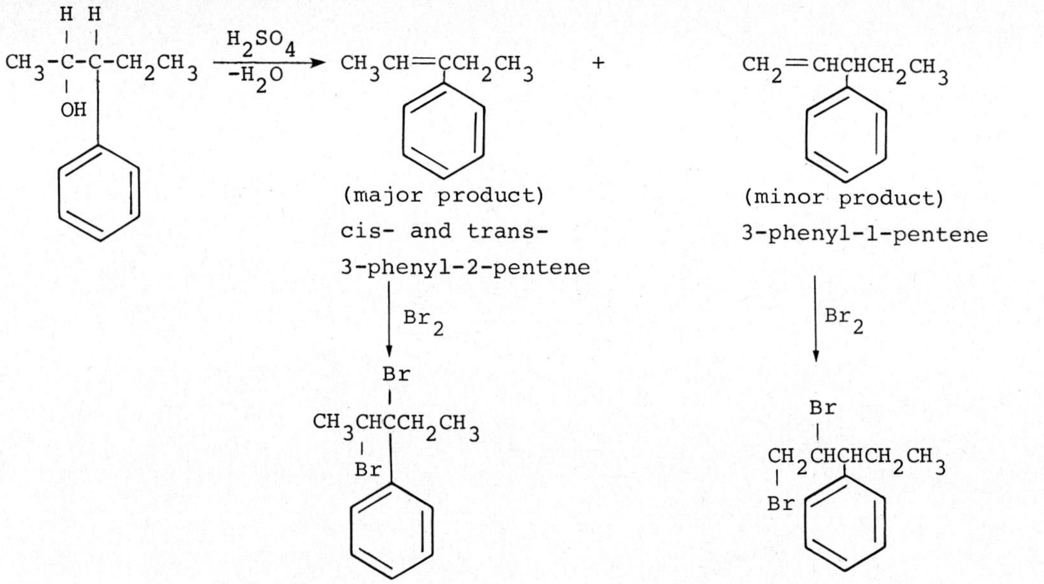

(major product) cis- and trans- 3-phenyl-2-pentene

(minor product) 3-phenyl-1-pentene

2,3-dibromo-3-phenylpentane

1,2-dibromo-3-phenylpentane

---

12. trans-3-methyl-1-phenyl-2-pentene $\xrightarrow{Br_2/FeBr_3}$

Your answer:

Solution:

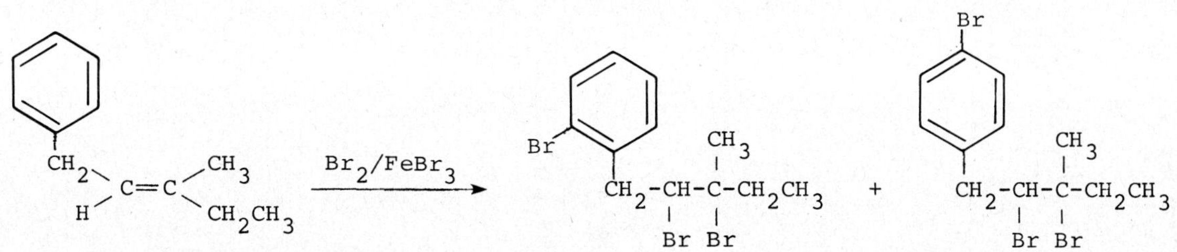

2,3-dibromo-3-methyl-1-(o-bromophenyl)pentane

2,3-dibromo-3-methyl-1-(p-bromophenyl)pentane

---

13. 3-methyl-2-butanol $\xrightarrow{H_2SO_4/\Delta}$? $\xrightarrow{?}$ 2-methylbutane

Your answer:

Solution:

$$CH_3-\underset{OH}{\overset{H}{C}}-\underset{H}{\overset{CH_3}{C}}-CH_3 \xrightarrow{H_2SO_4/\Delta} CH_3CH{=}C(CH_3)_2 \quad + \quad CH_2{=}CH\underset{CH_3}{CH}CH_3$$

(major product) 2-methyl-2-butene $\xrightarrow{H_2/Ni}$ $CH_3\overset{CH_3}{CH}CH_2CH_3$ 2-methylbutane

(minor product) 3-methyl-1-butene $\xrightarrow{H_2/Ni}$ $CH_3\underset{CH_3}{CH}CH_2CH_3$ 2-methylbutane

Both the major product and the minor product can be hydrogenated to 2-methylbutane.

---

14. glycerol + excess HNO_3 $\longrightarrow$

Your answer:

Solution:

$$\begin{matrix} CH_2-OH \\ | \\ CH-OH \\ | \\ CH_2-OH \end{matrix} + \text{excess } HNO_3 \longrightarrow \begin{matrix} CH_2-O-NO_2 \\ | \\ CH-O-NO_2 \\ | \\ CH_2-O-NO_2 \end{matrix}$$

trinitroglycerine

---

18 Aldehydes and Ketones

I. Structure, Nomenclature, and Isomerism

1. Write the structural formulas for all possible isomers of $C_5H_{10}O$ that contain a carbonyl group. Write the systematic name for each isomer.

Your answer:

Solution:

Structures	Systematic name
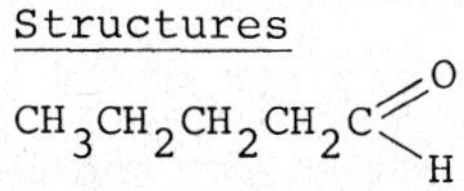	pentanal
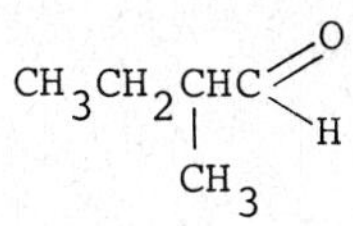	2-methylbutanal
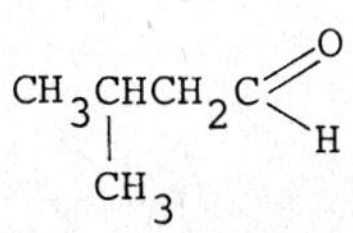	3-methylbutanal
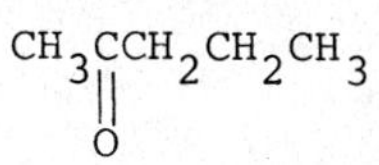	2-pentanone
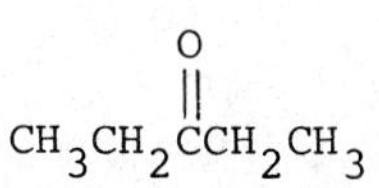	3-pentanone

Structures	Systematic name
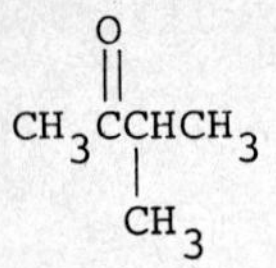	methylbutanone
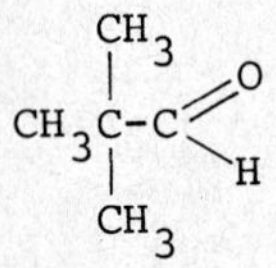	2,2-dimethylpropanal

2. What is the general molecular formula for the aldehydes and ketones which have noncyclic, saturated carbon chains?

Your answer:

Solution: $C_nH_{2n}O$

Subtract two hydrogens (to account for the pi bond in the carbonyl group) from the general molecular formula $C_nH_{2n+2}O$.

II. Reactions of Aldehydes and Ketones

For each reagent listed in the subsequent questions

(a) Write "type" reactions for an aldehyde and a ketone and describe each reaction in words.

(b) Write equations for the specified reactions; supply the missing structures and names. For small molecules, write common names in parentheses.

A. OXIDIZING AGENTS

1. potassium dichromate

   (b) Write the reactions for $K_2Cr_2O_7$ with the three isomers of C_4H_8O that contain a carbonyl group.

Your answer:

Solution:

(a) "Type" reaction for aldehydes: 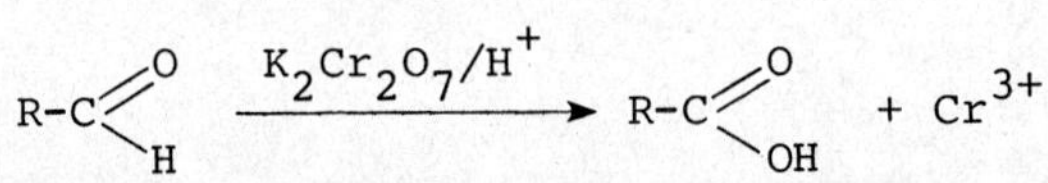

Aldehydes are oxidized more readily than any other functional group you have previously studied.

"Type" reaction for ketones: $R\text{-}C(=O)R' \xrightarrow{K_2Cr_2O_7/H^+}$ N.R.

More vigorous conditions are required to oxidize carbon-carbon bonds.

(b) $CH_3CH_2CH_2\overset{O}{\overset{\|}{C}}H \xrightarrow{K_2Cr_2O_7/H^+} CH_3CH_2CH_2\overset{O}{\overset{\|}{C}}\text{-}OH + Cr^{3+}$

butanal → butanoic acid

$CH_3\underset{CH_3}{\underset{|}{C}}H\overset{O}{\overset{\|}{C}}H \xrightarrow{K_2Cr_2O_7/H^+} CH_3\underset{CH_3}{\underset{|}{C}}H\overset{O}{\overset{\|}{C}}\text{-}OH + Cr^{3+}$

2-methylpropanal → 2-methylpropanoic acid

$CH_3\overset{O}{\overset{\|}{C}}CH_2CH_3 \xrightarrow{K_2Cr_2O_7/H^+}$ N.R.

butanone

---

2. Fehling's and Benedict's solution—very mild oxidizing agents (Cu^{2+} ions in a basic solution)

   (b) Write the reactions for Fehling's or Benedict's solution with cyclopentanone, propanone, and propanal.

Your answer:

Solution:

(a) "Type" reaction for aldehydes: $R\text{-}C(=O)H + Cu^{2+} \xrightarrow{OH^-} R\text{-}C(=O)O^- + Cu_2O$ (red precipitate)

The aldehyde is oxidized to a carboxylic acid ($R\text{-}C(=O)OH$) which is neutralized in the basic solution to the "salt" or anion ($R\text{-}C(=O)O^-$). A positive test is recognized by the disappearance of the blue solution of Cu^{2+} and the appearance of the red-brown Cu_2O precipitate. These mild oxidizing agents oxidize only aldehydes, not alcohols or ketones.

"Type" reaction for ketones: $R\text{-}C(=O)R' + Cu^{2+} \xrightarrow{OH^-}$ N.R. (a negative Benedict's or Fehling's test)

(b) 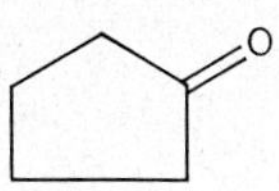 $\xrightarrow{Cu^{2+}/OH^-}$ N.R.

cyclopentanone

$$CH_3\overset{O}{\overset{\|}{C}}CH_3 \xrightarrow{Cu^{2+}/OH^-} N.R.$$

propanone
(acetone)

$$CH_3CH_2\overset{O}{\overset{\|}{C}}H \xrightarrow{Cu^{2+}/OH^-} CH_3CH_2\overset{O}{\overset{\|}{C}}\text{-}O^- + Cu_2O$$

propanal — propanoate ion

---

3. Tollens' reagent, $Ag(NH_3)_2^+$—a very mild oxidizing agent

   (b) Write reactions for Tollens' reagent and benzaldehyde, butanone, and 1-propanol.

Your answer:

Solution:

(a) "Type" reaction for aldehydes: $R\text{-}C(=O)H + Ag(NH_3)_2^+ \longrightarrow R\text{-}\overset{O}{\overset{\|}{C}}\text{-}O^- + Ag$

A positive test is recognized by the appearance of silver metal deposited on the sides of the test tube; for this reason the test is sometimes referred to as the "silver mirror" test. This mild oxidizing agent oxidizes only aldehydes, never alcohols or ketones.

"Type" reaction for ketones: $R\text{-}C(=O)R' + Ag(NH_3)_2^+ \longrightarrow N.R.$ (a negative Tollens' test)

(b) 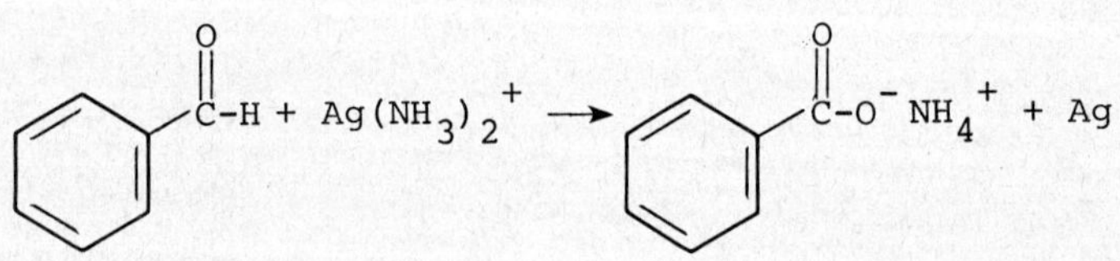

benzaldehyde — ammonium benzoate

$$CH_3\overset{O}{\overset{\|}{C}}CH_2CH_3 + Ag(NH_3)_2^+ \longrightarrow N.R.$$

butanone

$$CH_3CH_2CH_2OH + Ag(NH_3)_2^+ \longrightarrow N.R.$$

1-propanol

B. REDUCING AGENTS

1. lithium aluminum hydride, $LiAlH_4$

(b) Write reactions for $LiAlH_4$ with 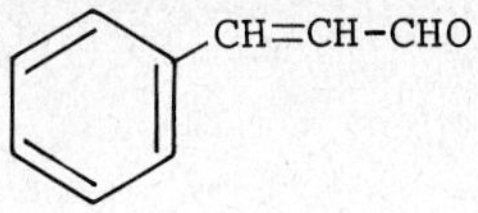 (cinnamaldehyde) and with 2-pentanone.

Your answer:

<u>Solution</u>:

(a) "Type" reaction for aldehydes or ketones: 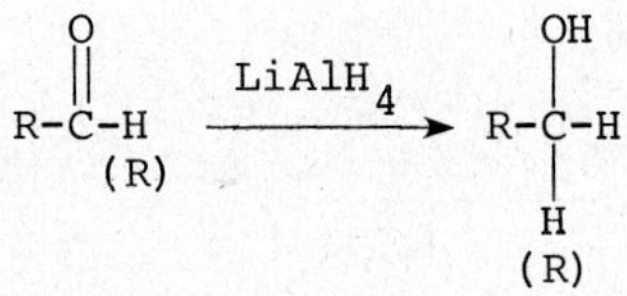

However: $>C{=}C< \xrightarrow{LiAlH_4}$ N.R.

This reagent adds two hydrogens across the carbonyl bond of aldehydes or ketones, but does not add to a double bond between two carbon atoms.

(b) 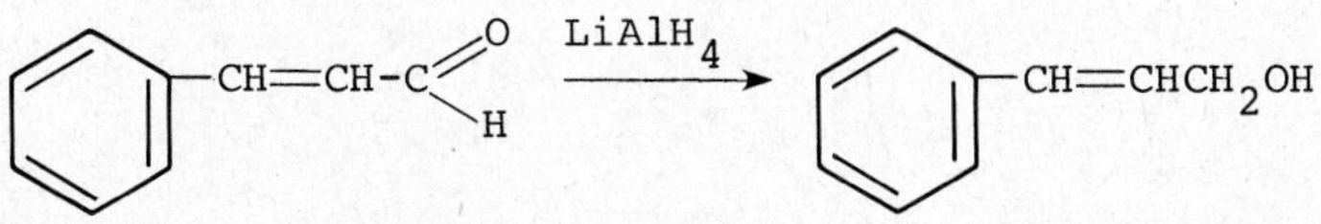

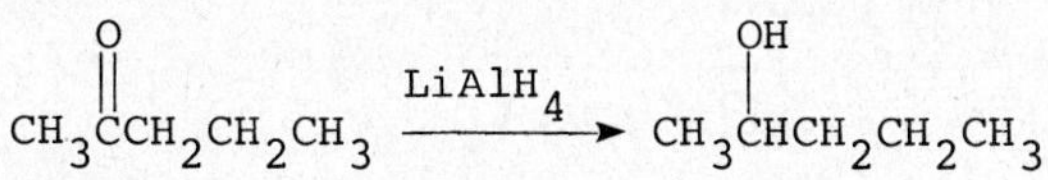

2-pentanone 2-pentanol

---

2. Hydrogenation or reduction with H_2 over a metal catalyst.

(b) Write the reaction of H_2 with cinnamaldehyde.

Your answer:

<u>Solution</u>:

"Type" reaction for aldehydes or ketones: 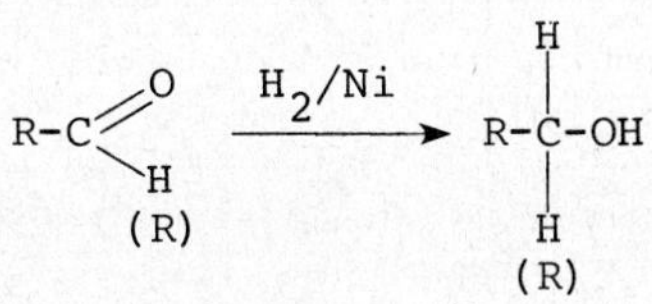

primary or secondary alcohol

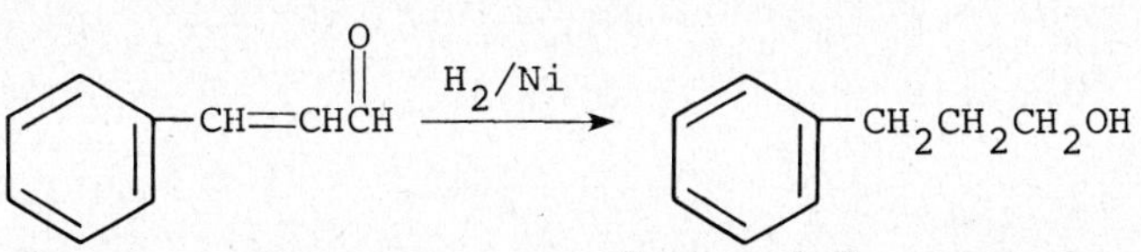

3-phenyl-1-propanol

C. ADDITION TO THE CARBONYL GROUP OF ALDEHYDES AND KETONES

Sometimes called nucleophilic addition because the incoming group contains a very electronegative atom that acts as a nucleophile.

1. addition of H_2O (hydration)
   (b) Write the reactions for the addition of H_2O to formaldehyde and benzaldehyde.

Your answer:

Solution:

(a) "Type" reaction for aldehydes and ketones: 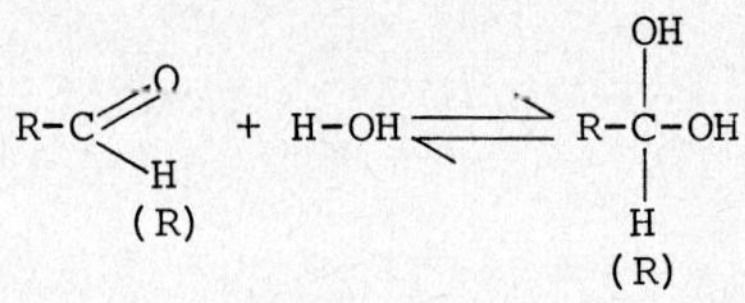

a hydrate

The carbonyl group is polar with the electron density drawn towards the oxygen atom, thus leaving the carbon atom relatively positive. The H_2O adds across the pi bond of the carbonyl: the nucleophilic group (OH^-) bonds to the (positive) carbon and the electrophilic group (H^+) bonds to the oxygen. The hydrates of most aldehydes and ketones are unstable, so the equilibrium lies to the left. However, the hydrate of formaldehyde is stable.

(b) 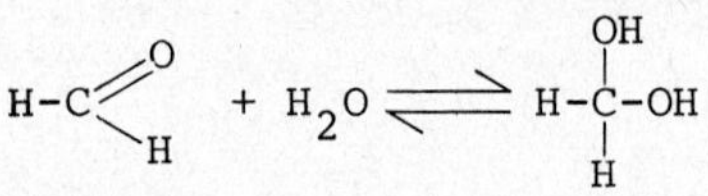

formaldehyde the hydrate of formaldehyde

A 37% (w/w %) solution of formaldehyde in water is known as formalin.

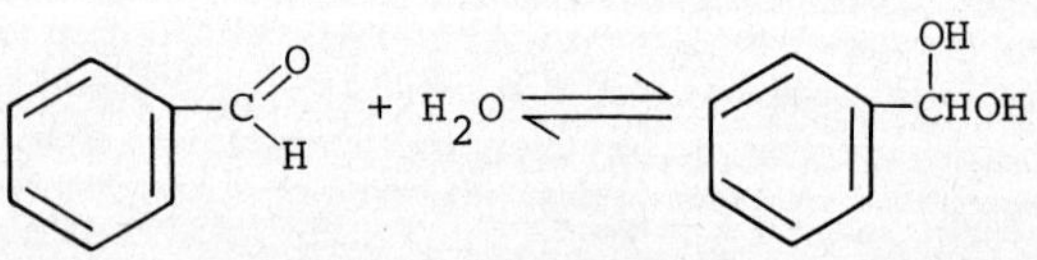

---

2. addition of alcohol in acid-catalyzed reaction
   (b) Write the reactions for the addition of isopropyl alcohol to ethanal and to butanone.

Your answer:

Solution:

(a) "Type" reaction for aldehydes and ketones:

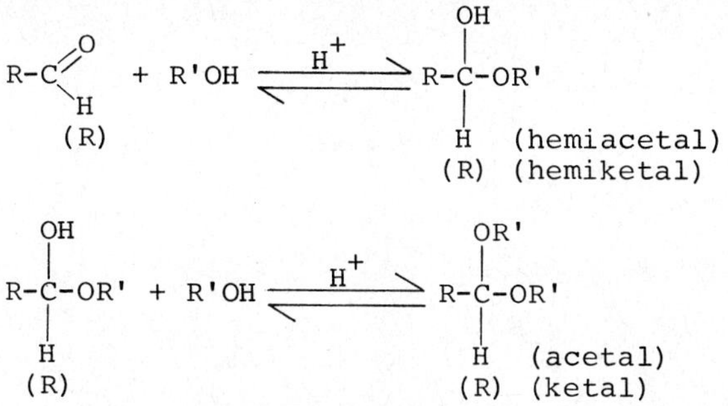

In the first reaction, the alcohol adds to the carbonyl bond to form the hemiacetal (or hemiketal). In the presence of excess alcohol, a molecule of water will split out from the hemiacetal (or hemiketal), and from another molecule of alcohol to form the acetal (or ketal). The hemiacetal (or hemiketal) and acetal (or ketal) will form an equilibrium mixture with the original carbonyl compound and alcohol.

(b)

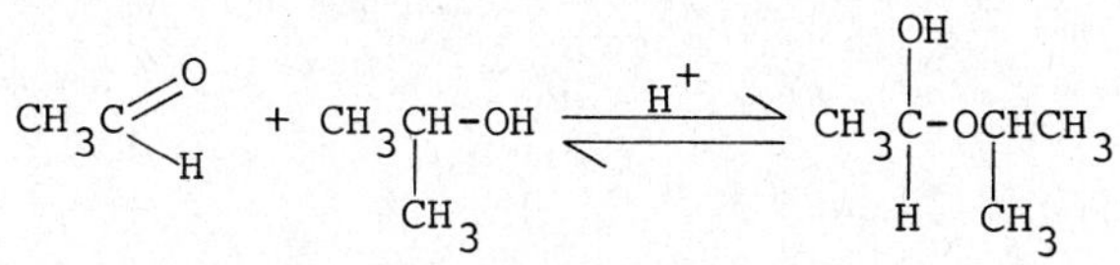

ethanal (acetaldehyde) a hemiacetal

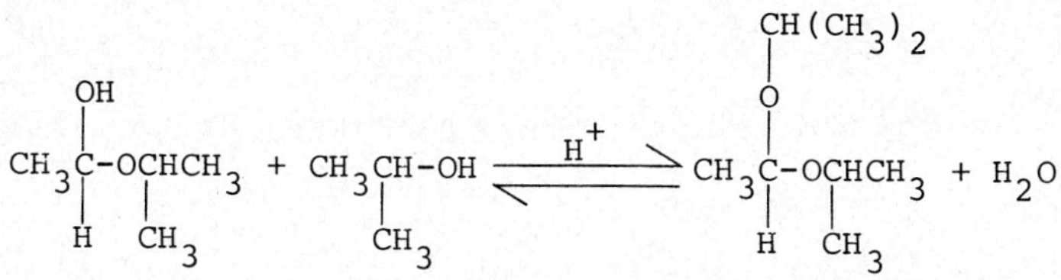

an acetal

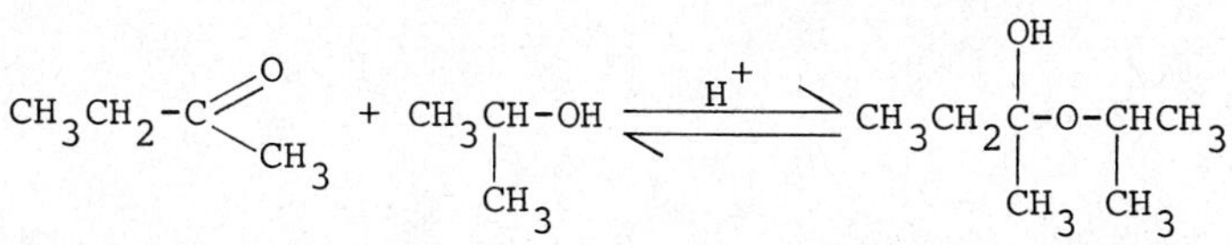

a hemiketal

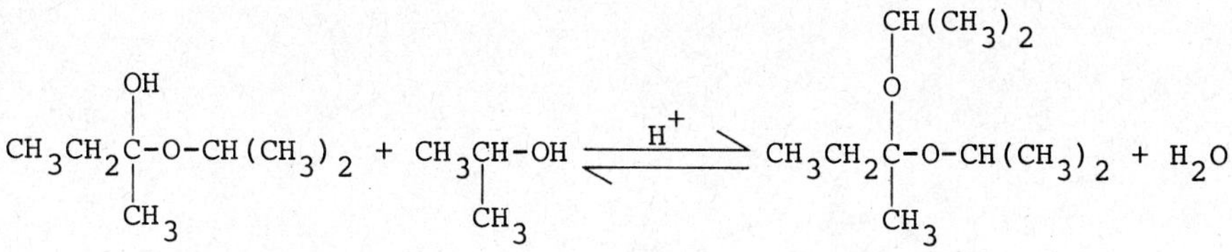

a ketal

---

3. addition of ammonia

   (b) Write the reactions for the addition of ammonia to the carbonyls of propanone and propanal.

Your answer:

Solution:

(a) "Type" reaction for aldehydes and ketones: 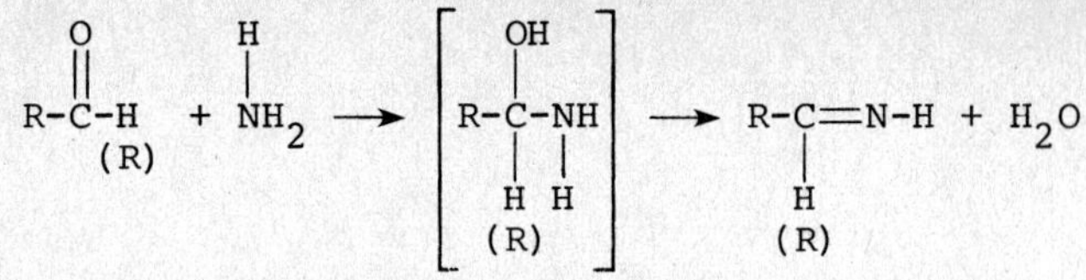

This "type" reaction applies to the addition of any NH_3-related compound (NH_2-NH_2, R-NH_2, etc.) to aldehydes or ketones. The temporary intermediate formed by nucleophilic addition to the carbonyl spontaneously loses water to form the type of compound shown on the right, which is called an imine or a Schiff's base.

(b) 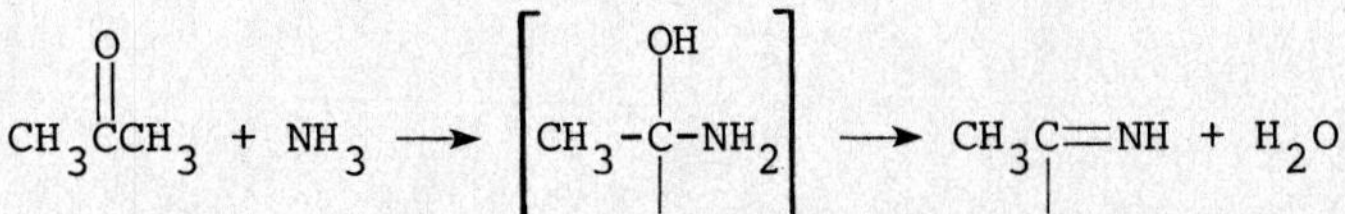

propanone
(acetone)

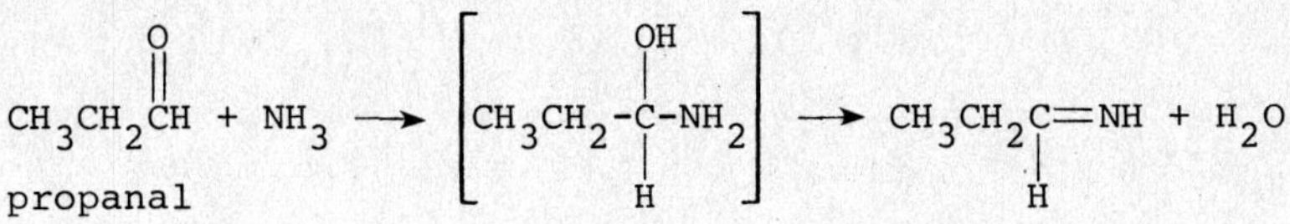

propanal

---

4. addition of hydrazine (NH_2-NH_2)
   (b) Write equations for the addition of hydrazine to cyclobutanone and to benzaldehyde.

Your answer:

Solution:

(a) "Type" reaction for aldehydes and ketones:

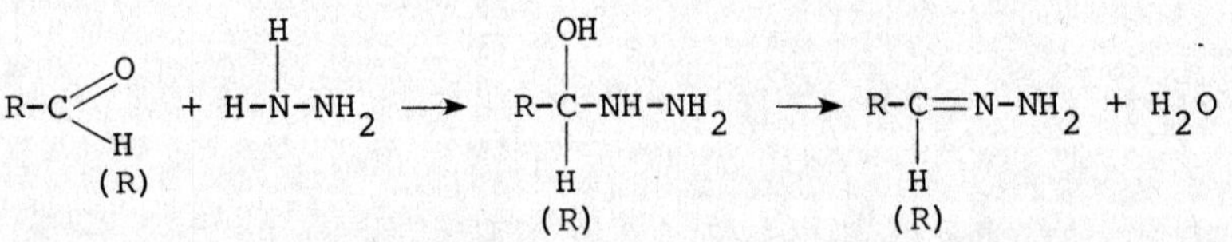

Notice that this is essentially the same reaction as the reaction shown for the addition of ammonia to aldehydes and ketones.

(b) 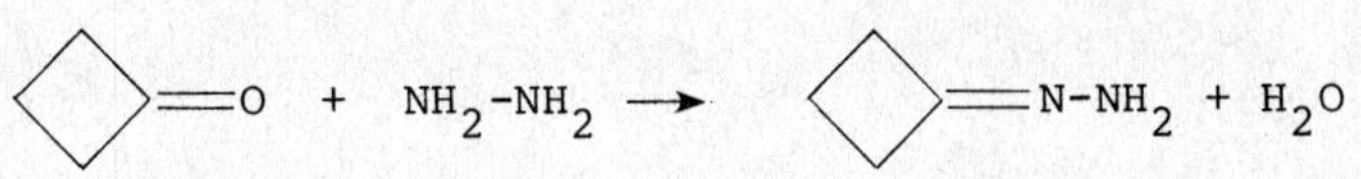

cyclobutanone

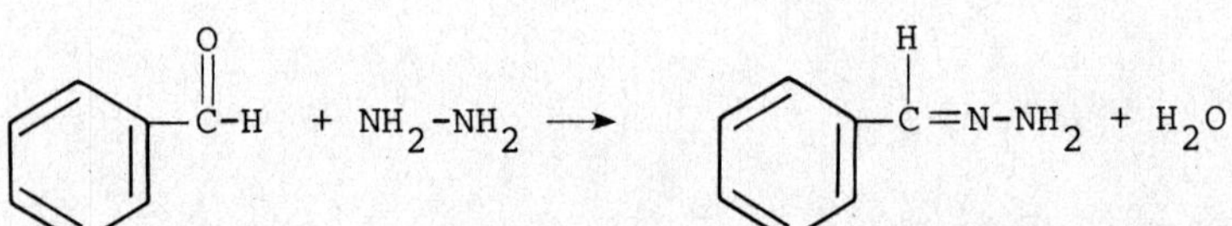

benzaldehyde

---

5. addition of 2,4-dinitrophenylhydrazine (DNP), a reagent used in the laboratory as a test for a carbonyl group. The reaction of DNP with aldehydes and ketones yields a yellow-to-orange precipitate; DNP does not react with alcohols.

   (b) Write the equations for the reactions of DNP with methylbutanone and with propanal.

Your answer:

Solution:

(a) "Type" reaction is the same as the "type" reaction given for the reaction of hydrazine with aldehydes and ketones (except for the fact that DNP has a nitrated phenyl group attached to the nitrogen).

(b) 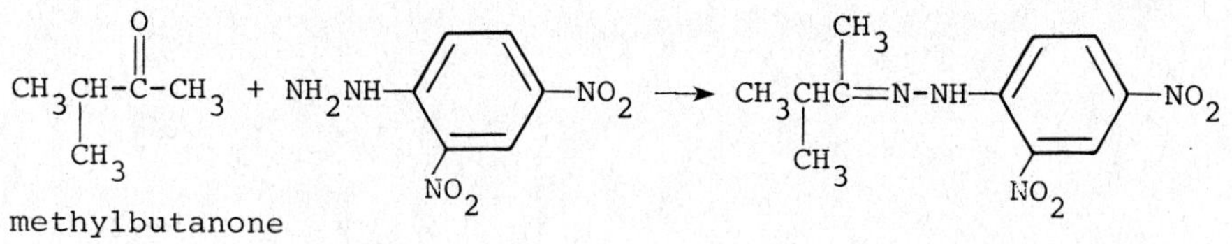

methylbutanone

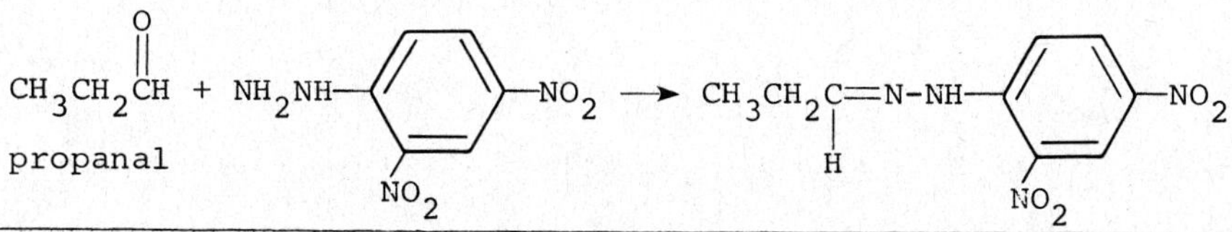

propanal

6. addition of amines to carbonyls

   (b) Write the equations for the reactions of acetone with $CH_3CH_2NH_2$ (ethylamine) and of formaldehyde with CH_3-NH_2 (methylamine).

Your answer:

Solution:

(a) "Type" reaction for aldehydes and ketones:

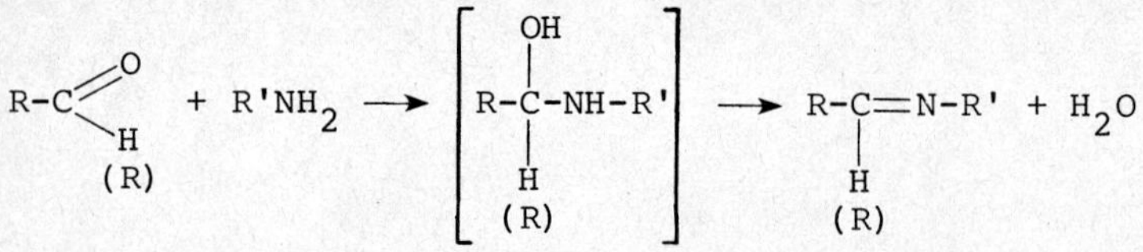

(b) 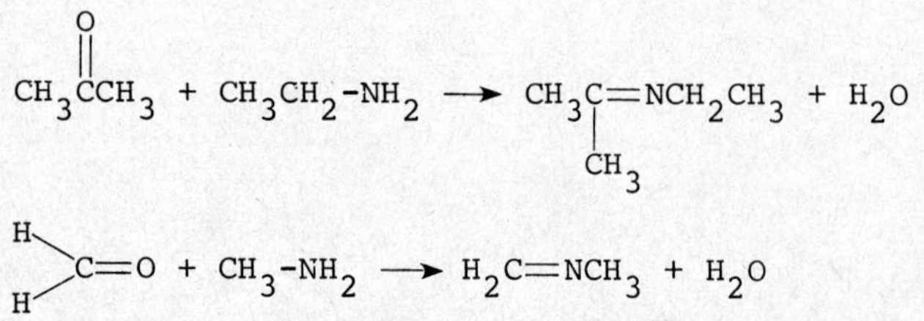

D. IODOFORM TEST

The iodoform test can be used to identify the special structure $CH_3\text{-}\overset{\overset{O}{\|}}{C}\text{-}$ or an alcohol that can be oxidized to this structure.

1. iodine in basic solution (I_2/OH^-, the iodoform reagent)
   (b) Write equations for the reactions with acetone, ethanol, and 3-pentanone.

Your answer:

Solution:

(a) "Type" reaction:

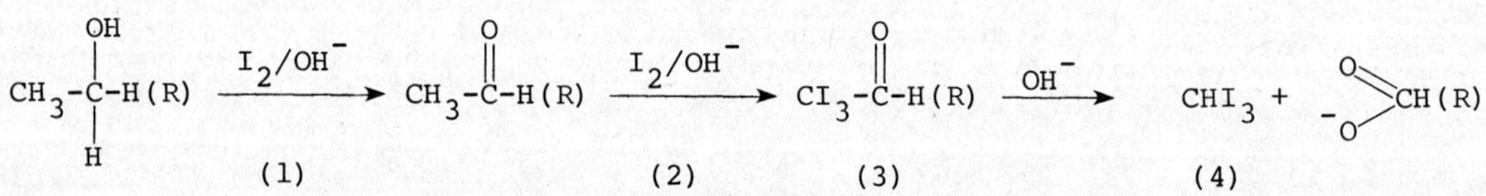

(1) Iodine is an oxidizing agent, so ethanol and some secondary alcohols are oxidized to a structure that can give a positive iodoform test.

(2) In basic solution, the hydrogens of a methyl group that is alpha (next) to a carbonyl are substituted by iodine atoms.

(3) Because there are electronegative groups on each side of the bond, the bond breaks.

(4) The light yellow precipitate of iodoform (CHI_3) is proof of a positive test. The carbon that was originally the carbonyl carbon is highly oxidized, but forms the anion (instead of forming the acid) because the solution is basic.

(b) 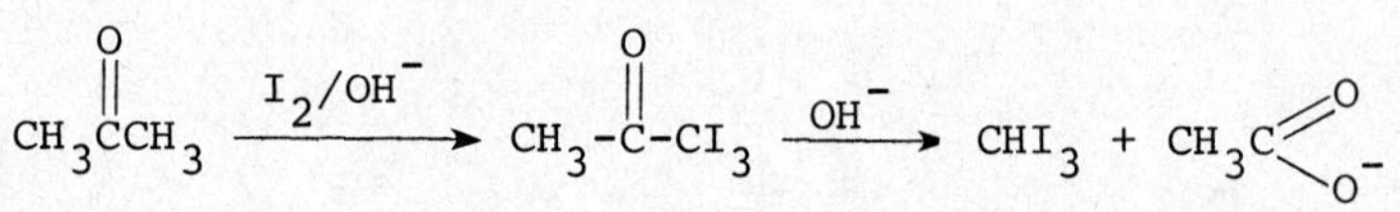

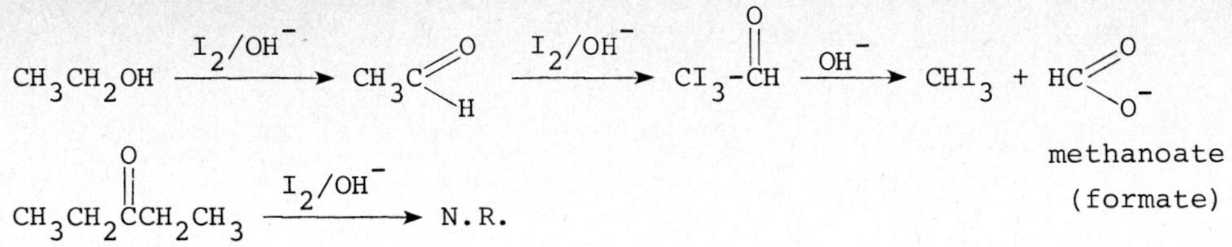

A reaction does not occur when there are less than three hydrogens on a carbon next to the carbonyl.

III. Chemical Tests to Differentiate Functional Groups or Compounds

For each of the following sets of compounds

(a) Choose one or more chemical tests which would differentiate the compounds; especially consider tests you have performed in the laboratory.

(b) Write an equation using structural formulas, that illustrates how each compound reacts in the presence of the reagent you list.

1. acetaldehyde and acetone

Your answer:

Solution:

(a) Fehling's or Benedict's test—these mild oxidizing agents will give a positive test for all aldehydes; also Tollens' test ($Ag(NH_3)_2^+$).

(b)

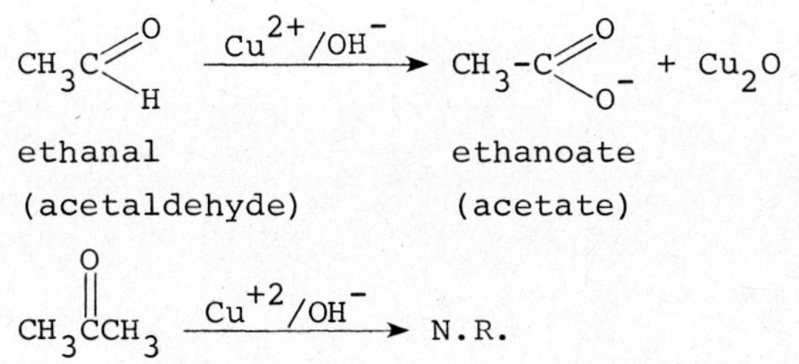

---

2. CH_3CH_2-O-CH_3 and $CH_3CH_2CH_2$-OH

Your answer:

Solution:

(a) Reaction with Na.

(b) $CH_3CH_2\text{-}O\text{-}CH_3 \xrightarrow{Na} N.R.$

ethyl methyl ether

$CH_3CH_2CH_2\text{-}OH + Na \longrightarrow CH_3CH_2CH_2\text{-}O^- \; Na^+ + \frac{1}{2} H_2(g)$

1-propanol   sodium 1-propoxide

---

3. CH_3CH_2CHO and CH_3COCH_3

Your answer:

Solution:

(a) Any mild oxidizing agent will oxidize propanal and not propanone (acetone): Fehling's or Benedict's solution (Cu^{2+}/OH^-), Tollens' solution ($Ag(NH_3)_2{}^+$); also potassium dichromate ($K_2Cr_2O_7/H^+$).

(b) $CH_3CH_2C(=O)H + Ag(NH_3)_2{}^+ \longrightarrow CH_3CH_2C(=O)O^- + Ag$

propanal   propanoate ion

$CH_3C(=O)CH_3 + Ag(NH_3)_2{}^+ \longrightarrow N.R.$

propanone
(acetone)

---

4. 2-pentanol, 2-methyl-1-butanol, and 2-methyl-2-butanol

Your answer:

Solution:

(a) Iodoform test together with oxidation by $K_2Cr_2O_7/H^+$ will differentiate the three compounds.

	Products Formed with	
	I_2/OH^-	$K_2Cr_2O_7/H^+$
$CH_3CH(OH)CH_2CH_2CH_3$ 2-pentanol	$^-O(O=)CCH_2CH_2CH_3 + CHI_3$	$CH_3C(=O)CH_2CH_2CH_3 + Cr^{3+}$
$CH_2(OH)-CH(CH_3)-CH_2CH_3$ 2-methyl-1-butanol	N.R.	$HO(O=)CCH(CH_3)CH_3 + Cr^{3+}$
$CH_3-C(CH_3)(OH)-CH_2CH_3$ 2-methyl-2-butanol	N.R.	N.R.

Only 2-pentanol gives a positive test in the presence of I_2/OH^- (Note: It is first oxidized to 2-pentanone.) 2-Pentanol and 2-methyl-1-butanol react with the dichromate. 2-Methyl-2-butanol reacts with neither reagent.

5. acetone and 2-propanol

Your answer:

Solution:

(a) DNP will add to carbonyls (acetone) and yield a characteristic precipitate, but will not react with alcohols (2-propanol).

(b)

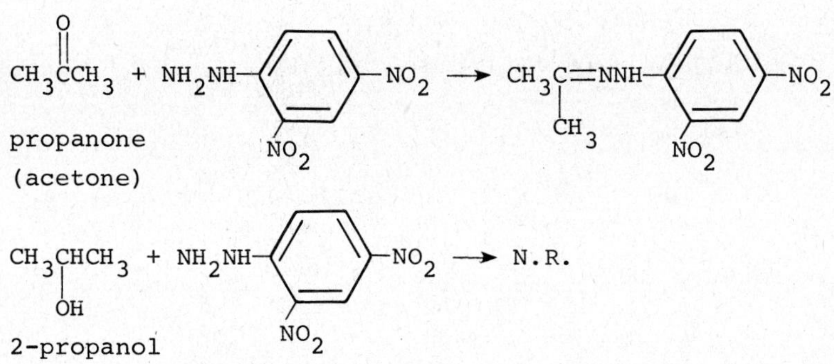

IV. Identification Problems

1. A compound X, $C_6H_{12}O$, does not react with Benedict's solution or with DNP (2,4-dinitrophenylhydrazine). It does not react with hydrogen in the presence of a metal catalyst. With an acidic dichromate solution, a sample of X oxidizes to Y, $C_6H_{10}O_2$. Give a possible structure for X and the corresponding structure for Y.

Your answer:

Solution:

Given information	Possible	Not possible
X is $C_6H_{12}O$ ($C_nH_{2n}O$) (a 2H deficiency)	aldehyde, ketone, cyclic or unsaturated alcohol or ether	
$X \xrightarrow{Cu^{2+}/OH^-} N.R.$	cyclic or unsaturated alcohol or ether	aldehyde, ketone
$X \xrightarrow{DNP} N.R.$		
$X \xrightarrow{H_2/Pt} N.R.$	cyclic alcohol or ether	unsaturated alcohol or ether
$X \xrightarrow{K_2Cr_2O_7/H^+} C_6H_{10}O_2$ (Y)	cyclic primary alcohol	secondary or tertiary alcohol or ether

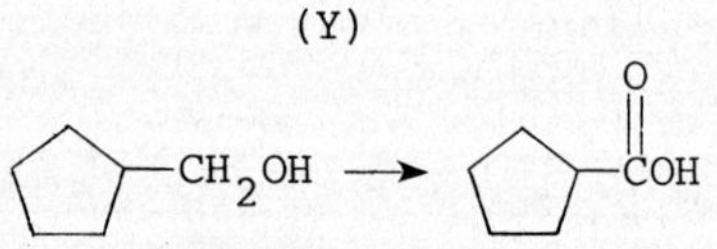

X Y

Smaller ring compounds are also possible, but less probable because the small rings will react with hydrogen in the presence of a metal catalyst.

---

2. Compound B, $C_6H_{12}O$, reacts with DNP to yield a yellow precipitate. The compound does not react with Tollens' reagent or with iodine in a basic solution (the iodoform test). Write two possible structures for compound B.

Your answer:

Solution:

Given information	Possible	Not possible
B is $C_6H_{12}O$ ($C_nH_{2n}O$) (a 2H deficiency)	aldehyde, ketone, unsaturated or cyclic alcohol or ether	
B $\xrightarrow{DNP}$ yellow precipitate	aldehyde, ketone	alcohol, ether
B $\xrightarrow{Ag(NH_3)_2^+}$ N.R.	ketone	aldehyde
B $\xrightarrow{I_2/OH^-}$ N.R.	$CH_3CH_2\overset{O}{\overset{\|}{C}}CH_2CH_2CH_3$ 3-hexanone or $CH_3CH_2\overset{O}{\overset{\|}{C}}CH(CH_3)_2$ 2-methyl-3-pentanone	$CH_3\overset{O}{\overset{\|}{C}}-$

3. Compound X, C_3H_6O, reacts with 2,4-dinitrophenylhydrazine and with Benedict's solution. A sample of X reacts with $LiAlH_4$ to yield compound Y. In the presence of an acid catalyst, compound X will react with compound Y (in excess). Write equations illustrating the equilibria that exist between X and Y and their products.

Your answer:

Solution:

Given information	Possible	Not possible
X is C_3H_6O ($C_nH_{2n}O$) (a 2H deficiency)	aldehyde, ketone, cyclic or unsaturated alcohol or ether	
X reacts with DNP	aldehyde, ketone	alcohol, ether
X reacts with Benedict's	aldehyde	ketone

$$X \xrightarrow{LiAlH_4} Y\ (C_3H_8O)$$

Y is primary alcohol: $CH_3CH_2CH_2OH$.

(X is therefore CH_3CH_2CHO.)

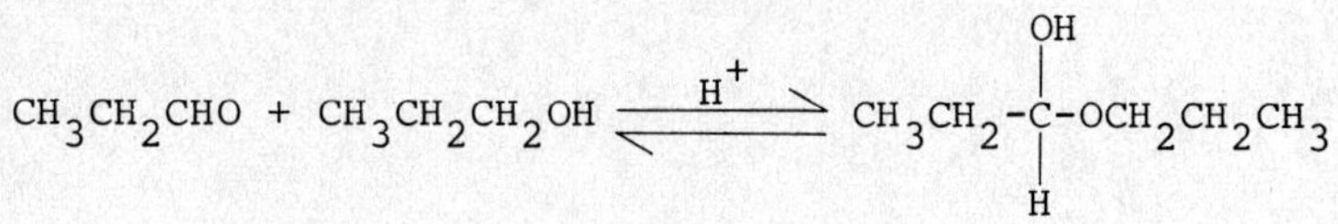

a hemiacetal

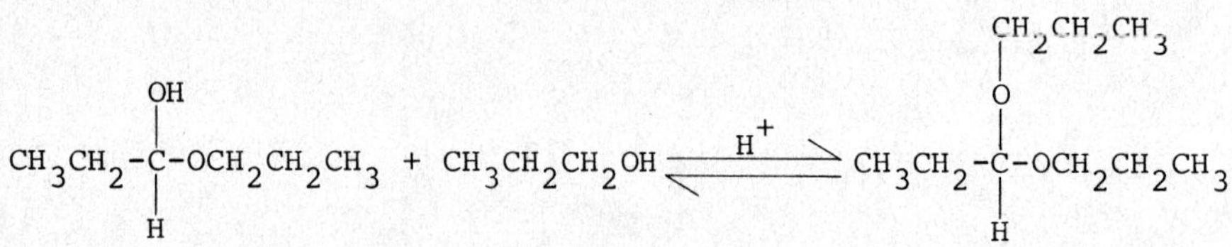

an acetal

V. Practice and Review Reactions

Complete the equations by writing the structural formula(s) and the name(s) of the product(s). Mentally recall the "type" reaction as you work.

1. acetaldehyde + Benedict's solution (or Fehling's solution) $\longrightarrow$

Your answer:

Solution: 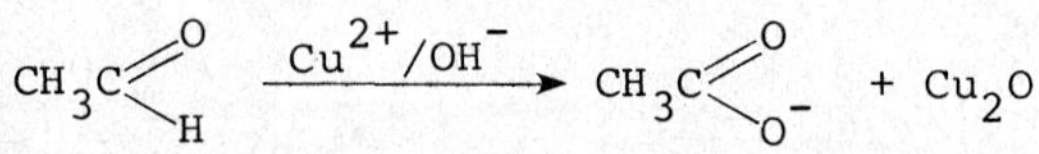

ethanoate
(acetate)

---

2. 3-methyl-2-butanol $\xrightarrow{K_2Cr_2O_7/H^+}$? $\xrightleftharpoons{\text{excess ethanol}/H^+}$

Your answer:

Solution:

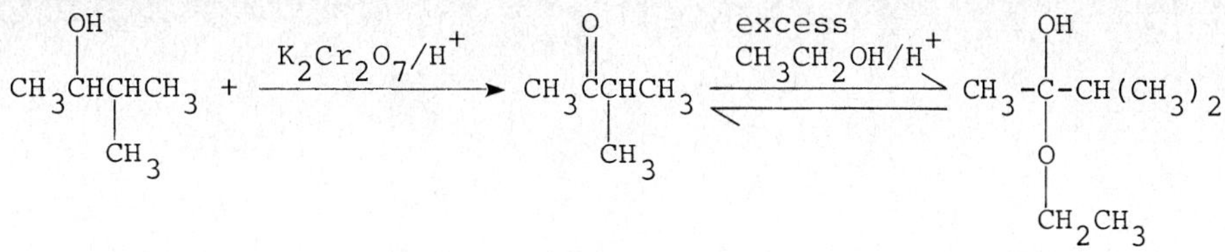

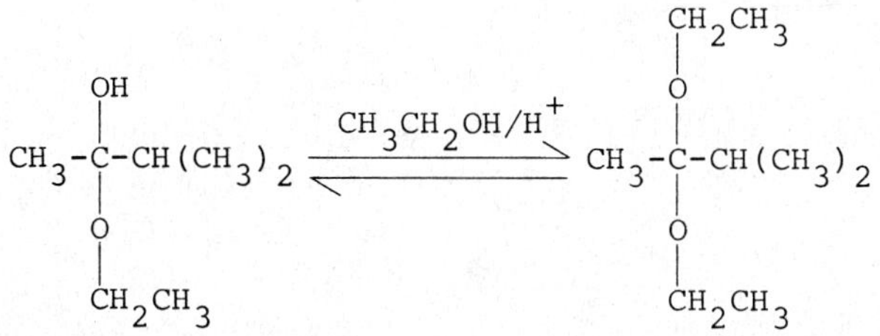

a hemiketal a ketal

---

3. cyclohexanone + $Ag(NH_3)_2^+$ →

Your answer:

Solution:

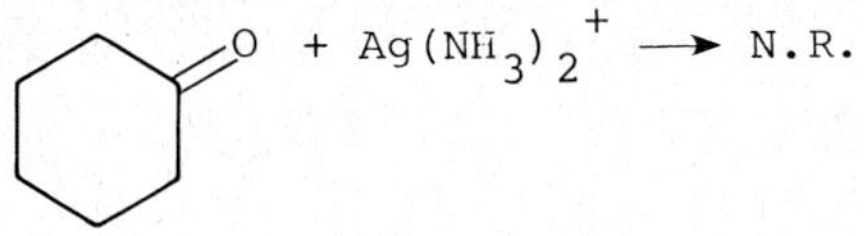

---

4. m-hydroxybenzaldehyde + NaOH →

Your answer:

Solution:

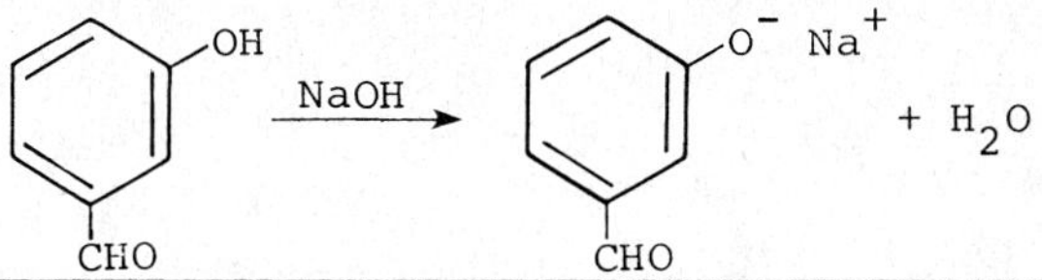

---

5. acetone $\xrightarrow{LiAlH_4}$? $\xrightarrow{HCl/ZnCl_2}$?

Your answer:

Solution:

$$\underset{\text{propanone}}{CH_3\overset{\overset{\displaystyle O}{\|}}{C}CH_3} \xrightarrow{LiAlH_4} \underset{\text{2-propanol}}{CH_3\overset{\overset{\displaystyle OH}{|}}{C}HCH_3} \xrightarrow[\text{(Lucas reagent)}]{HCl/ZnCl_2} \underset{\text{2-chloropropane}}{CH_3\overset{\overset{\displaystyle Cl}{|}}{C}HCH_3}$$

---

6. 2-methyl-1-propanol $\xrightarrow{K_2Cr_2O_7/H^+}$

Your answer:

Solution: $$\underset{\text{(isobutyl alcohol)}}{CH_3\overset{\overset{\displaystyle CH_3}{|}}{C}HCH_2OH} \xrightarrow{K_2Cr_2O_7/H^+} \underset{\text{2-methylpropanoic acid}}{CH_3\overset{\overset{\displaystyle CH_3}{|}}{C}HC{\lower{0.5ex}\overset{\displaystyle \nearrow O}{\searrow OH}}}$$

---

7. 4-penten-2-one $\xrightarrow{LiAlH_4}$? $\xrightarrow{Br_2}$

Your answer:

Solution: $$CH_2{=}CHCH_2\overset{\overset{\displaystyle O}{\|}}{C}CH_3 \xrightarrow{LiAlH_4} \underset{\text{4-penten-2-ol}}{CH_2{=}CHCH_2\overset{\overset{\displaystyle OH}{|}}{C}HCH_3} \xrightarrow{Br_2} \underset{\text{4,5-dibromo-2-pentanol}}{\underset{\displaystyle Br}{\underset{|}{CH_2}}\underset{\displaystyle Br}{\underset{|}{C}}HCH_2\overset{\overset{\displaystyle OH}{|}}{C}HCH_3}$$

---

8. cyclopentanone + 2,4-dinitrophenylhydrazine

Your answer:

Solution:

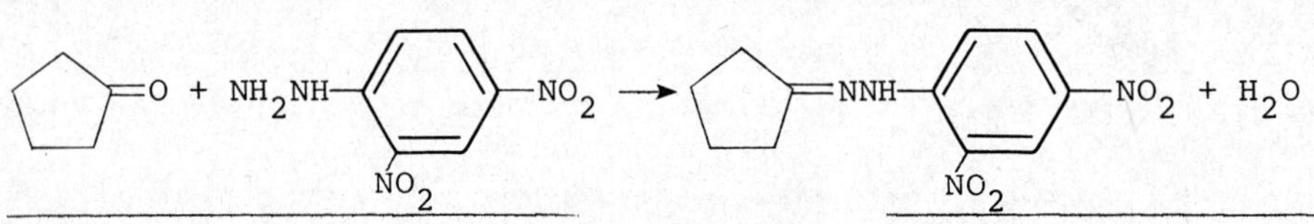

9. 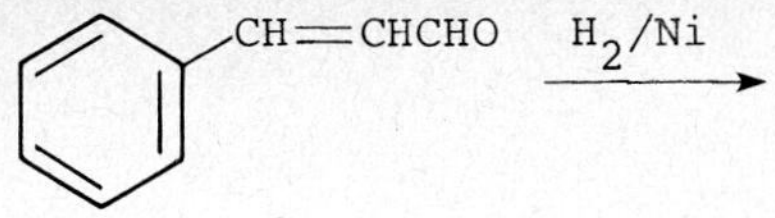

Your answer:

Solution:

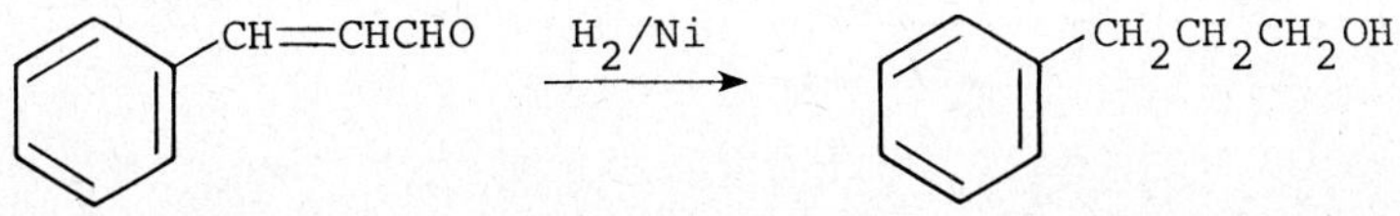

(cinnamaldehyde) 3-phenyl-1-propanol

---

10. 1-methylcyclopentanol $\xrightarrow{I_2/OH^-}$

Your answer:

Solution:

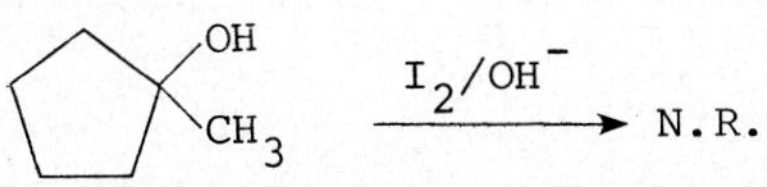

A tertiary alcohol cannot be oxidized so the iodoform test is negative.

---

11. cis-2-methylcyclopentanol $\xrightarrow{I_2/OH^-}$

Your answer:

Solution:

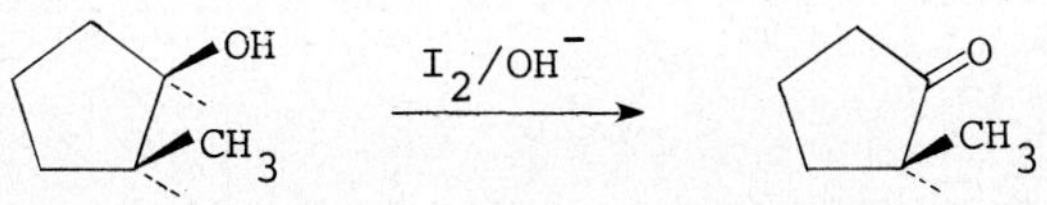

2-methylcyclopentanone

The iodine (in a basic solution) will oxidize the alcohol group, but CHI_3 (iodoform) will not form because there is no methyl group next to the carbonyl.

---

12. p-ethylbenzaldehyde + $KMnO_4$ (hot) $\longrightarrow$

Your answer:

Solution:

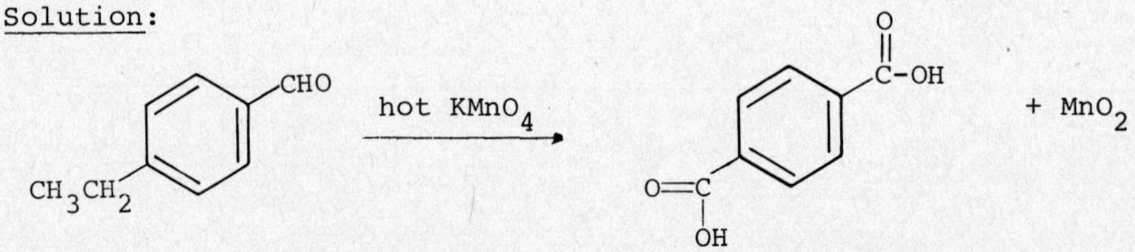

13. cyclopentanone + 1,2-ethanediol (ethylene glycol) $\xrightleftharpoons{H^+}$

Your answer:

Solution:

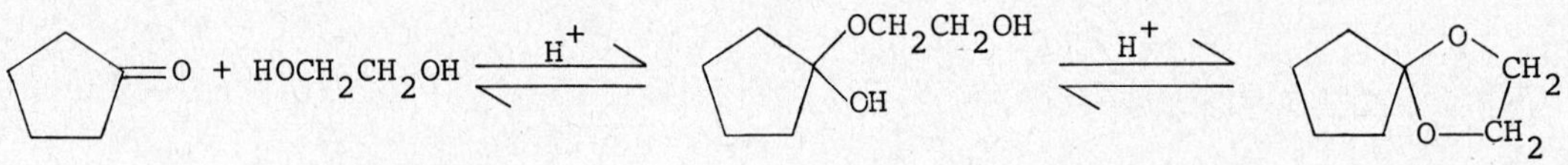

a hemiketal — an intramolecular ketal

14. butanal + $CH_3CH_2NH_2$ $\longrightarrow$

Your answer:

Solution: $CH_3CH_2CH_2C(=O)H + CH_3CH_2NH_2 \longrightarrow CH_3CH_2CH_2CH{=}N{-}CH_2CH_3$

ethylamine

19 Carboxylic Acids and Derivatives

The functional groups involved in this set of questions are carboxylic acid, salts of carboxylic acids, acyl (acid) halides, acid anhydrides, esters and amides. The questions presuppose that you are adept at recognizing these groups in structural formulas, but if you are not, work on the problems in Section III of Chapter 16, "Nomenclature of Organic Compounds Containing Oxygen and Nitrogen."

I. Dissociation (Ionization) of Carboxylic Acids in Water

1. Write equations to illustrate the dissociations of (a) acetic acid and (b) benzoic acid in water. In which direction does each equilibrium lie?

Your answer:

Solution:

(a) $CH_3\overset{O}{\overset{\|}{C}}OH + H_2O \rightleftharpoons CH_3\overset{O}{\overset{\|}{C}}O^- + H_3O^+$

ethanoate (acetate) ion

(b) 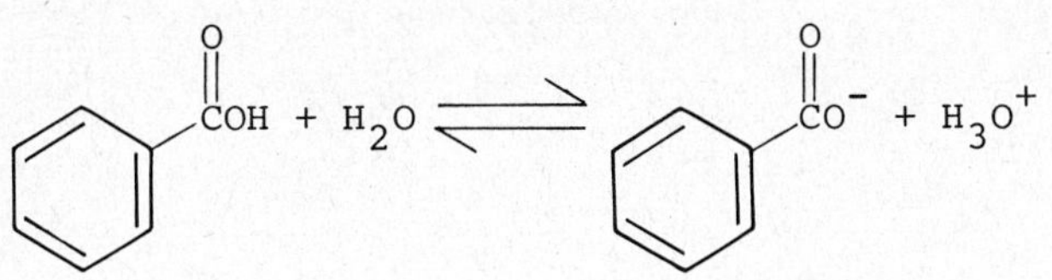

benzoate ion

Both acids are weak acids and therefore exist primarily in the molecular (undissociated) form: the equilibrium lies to the left.

---

2. Write the equilibrium expression in terms of K_a (also called K_d) for (a) formic acid and (b) 2-chloropropanoic acid.

Your answer:

Solution:

(a) $K_a = \dfrac{[H_3O^+][HCO_2^-]}{[HCO_2H]}$ (Note: Sometimes $[H^+]$ is written instead of $[H_3O^+]$.)

(b) $K_a = \dfrac{[H_3O^+][CH_3\overset{\displaystyle Cl \atop |}{C}HCO_2^-]}{[CH_3\underset{\displaystyle | \atop Cl}{C}HCO_2H]}$

---

3. Use the given K_a values to rank the four acids in order of increasing acid strength (from weakest to strongest acid). If the pH of 0.1 M solutions of each of the acids is measured, which solution would have the lowest pH (and would therefore be the most acidic)?

acetic acid, CH_3CO_2H	$K_a = 1.8 \times 10^{-5}$
pyruvic acid, CH_3COCO_2H	$K_a = 3.2 \times 10^{-3}$
lactic acid, $CH_3CHOHCO_2H$	$K_a = 1.4 \times 10^{-4}$
trichloroacetic acid, CCl_3CO_2H	$K_a = 3.0 \times 10^{-1}$

Your answer:

Solution:

(weakest acid, least dissociated) acetic acid $<$ lactic acid $<$ pyruvic acid $<$ trichloroacetic acid (strongest acid, most dissociated)

Trichloroacetic acid solution (0.1 M) has the lowest pH and is the most acidic solution.

---

4. Rank the following compounds in order of increasing acid strength: acetic acid, water, methanol, hydrochloric acid, and phenol.

Your answer:

Solution: methanol < water < phenol < acetic acid < hydrochloric acid
HCl is a strong acid (implying that it is virtually 100% dissociated in water solutions). The acidity of alcohols and phenols relative to water is discussed in Chapter 17.

II. Reactions of Carboxylic Acids

Write a "type" reaction for each new reagent presented. Write the missing names or structures of the reactants and products.

1. Neutralization of a carboxylic acid: Write equations for (a) the reaction of formic acid and aqueous sodium hydroxide and (b) the reaction of benzoic acid and ammonia.

Your answer:

Solution: "Type" reaction: $R\text{-}\overset{O}{\overset{\|}{C}}OH + \text{base } (OH^-) \longrightarrow R\text{-}\overset{O}{\overset{\|}{C}}\text{-}O^- + \text{H-base } (H_2O)$

(a) $HCO_2H + NaOH(aq) \longrightarrow HCO_2^-\ Na^+ + H_2O$

methanoic acid (formic acid) → sodium methanoate (sodium formate)

"Neutralization" refers to the reaction of an acid and a base; a salt and water are often formed.

(b) 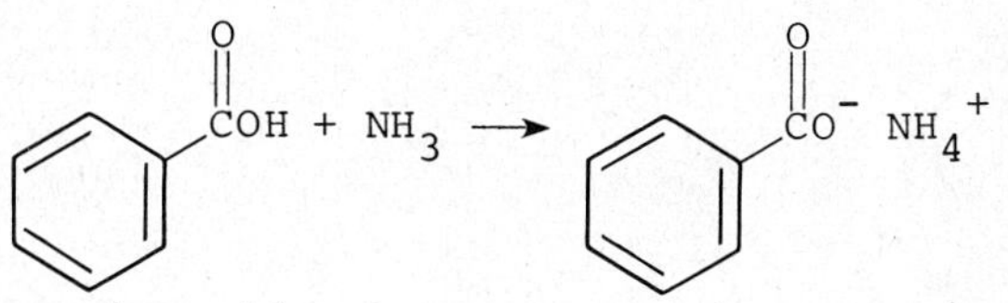

benzoic acid → ammonium benzoate

If it is heated, an ammonium salt is dehydrated to form an amide. (An amine, $R\text{-}NH_2$, will also neutralize a carboxylic acid, producing an ammonium salt that can be dehydrated to produce an amide. See Chapter 20.)

---

2. Preparation of an acyl (acid) halide: Write equations for (a) the reaction of propanoic acid with PBr_3 and (b) the reaction of butanoic acid with $SOCl_2$.

Your answer:

<u>Solution</u>: "Type" reaction: $R\text{-}\overset{O}{\overset{\|}{C}}OH + PBr_3$ (or PCl_3 or $SOCl_2$) $\longrightarrow R\text{-}\overset{O}{\overset{\|}{C}}\text{-}Br + P(OH)_3$

This reaction can be visualized as a substitution of the halide (-Br or -Cl) for the -OH of the carboxyl group.

(a) $CH_3CH_2\overset{O}{\overset{\|}{C}}OH + PBr_3 \longrightarrow CH_3CH_2\overset{O}{\overset{\|}{C}}Br + P(OH)_3$

propanoyl bromide

You can observe that this equation is not chemically balanced; when an organic reaction is written, it is acceptable to show only the alterations in the organic structure.

(b) $CH_3CH_2CH_2\overset{O}{\overset{\|}{C}}OH + SOCl_2 \longrightarrow CH_3CH_2CH_2\overset{O}{\overset{\|}{C}}Cl + SO_2$

butanoyl chloride

---

3. Esterification: Write equations for (a) the reaction of formic acid and 1-propanol, (b) the reaction of benzoic acid and ethyl alcohol and (c) the reaction of propanoic acid and phenol. Each reaction requires an acid catalyst.

Your answer:

<u>Solution</u>: "Type" reaction: $R\text{-}\overset{O}{\overset{\|}{C}}OH + R'OH \overset{H^+}{\rightleftharpoons} R\text{-}\overset{O}{\overset{\|}{C}}\text{-}OR' + H_2O$

The ester forms as an H from the alcohol and an OH from the acid are removed. An equilibrium mixture forms because the -OH and -OR have approximately the same tendency to replace each other on the carbonyl carbon.

(a) $HCO_2H + CH_3CH_2CH_2\text{-}OH \overset{H^+}{\rightleftharpoons} H\overset{O}{\overset{\|}{C}}\text{-}OCH_2CH_2CH_3 + H_2O$

propyl methanoate
(propyl formate)

(b) 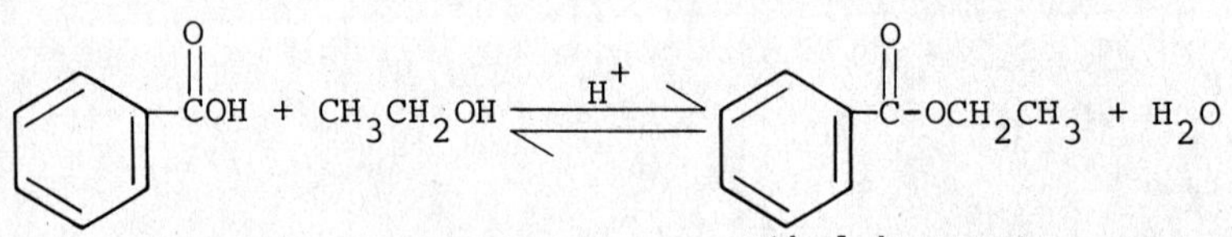

ethyl benzoate

(c) 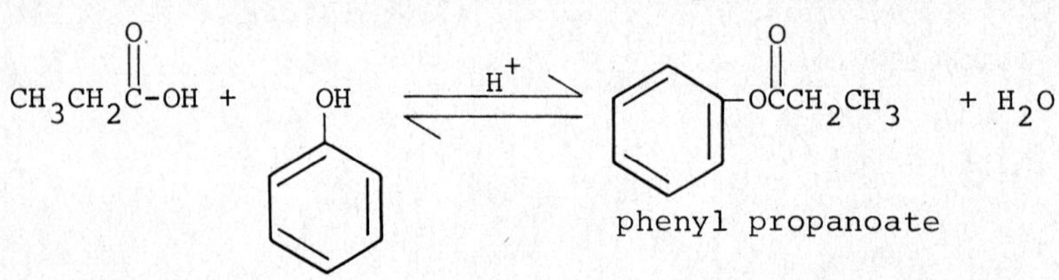

phenyl propanoate

III. Reactions of Salts of Carboxylic Acids

Write a "type" reaction and the missing names or structures of the reactants and products.

1. Regeneration of the acid: Write equations for (a) the reaction of sodium benzoate and hydrochloric acid and (b) the reaction of ammonium propanoate and hydrochloric acid.

Your answer:

Solution: "Type" reaction: $R\text{-}\overset{O}{\overset{\|}{C}}O^-\ Na^+ + \underset{\text{(any strong acid)}}{HCl} \longrightarrow R\text{-}\overset{O}{\overset{\|}{C}}\text{-}OH + NaCl$

The weak acid, $R\text{-}CO_2H$, forms because its conjugate base (in the salt) readily accepts a proton from HCl.

(a) 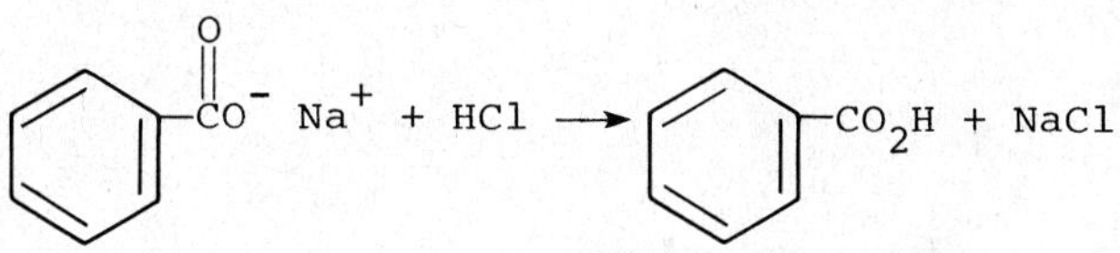

benzoic acid

(b) $CH_3CH_2\overset{O}{\overset{\|}{C}}O^-\ NH_4^+ + HCl \longrightarrow \underset{\text{propanoic acid}}{CH_3CH_2\overset{O}{\overset{\|}{C}}OH} + NH_4Cl$

---

2. Reaction of ammonium salts of carboxylic acids when heated: Write equations to illustrate the reactions that occur when (a) ammonium acetate is heated and (b) when N-methylammonium benzoate is heated.

Your answer:

Solution: Type reaction: $R\text{-}\overset{O}{\overset{\|}{C}}O^-\ NH_4^+ \xrightarrow{\Delta} \underset{\text{amide}}{R\text{-}\overset{O}{\overset{\|}{C}}\text{-}NH_2} + H_2O$

Ammonium salts are dehydrated to form amides by heating.

(a) $CH_3\overset{O}{\overset{\|}{C}}-O^- \ NH_4^+ \xrightarrow{\Delta} CH_3\overset{O}{\overset{\|}{C}}-NH_2 + H_2O$

ammonium ethanoate → ethanamide (acetamide)

(b) 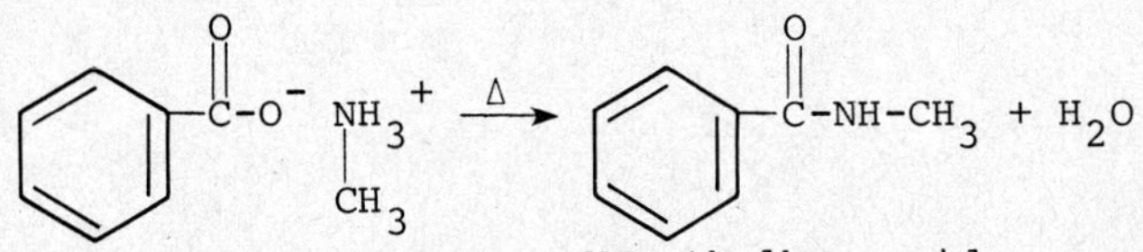

N-methylbenzamide

IV. Reactions of Esters

Write a "type" reaction and the missing names or structures of the reactants and products.

1. Acidic hydrolysis: Write equations to illustrate the hydrolysis of (a) ethyl benzoate and (b) methyl acetate in slightly acidic solution.

Your answer:

Solution: "Type" reaction: $R-\overset{O}{\overset{\|}{C}}-OR' + H_2O \underset{}{\overset{H^+}{\rightleftharpoons}} R-\overset{O}{\overset{\|}{C}}-OH + R'OH$

An equilibrium mixture forms in this reaction.

(b) 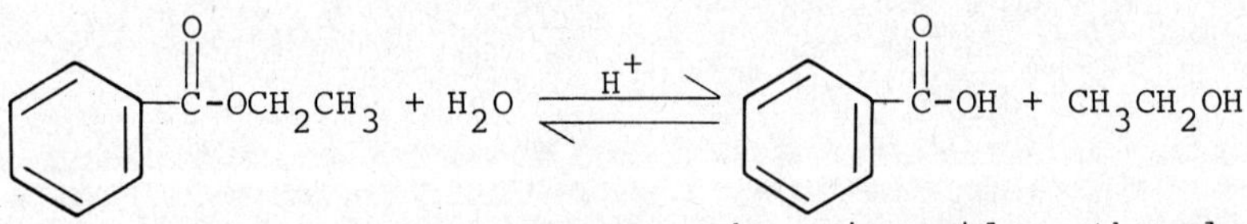

benzoic acid ethanol

(c) $CH_3\overset{O}{\overset{\|}{C}}-OCH_3 + H_2O \overset{H^+}{\rightleftharpoons} CH_3CO_2H + CH_3OH$

ethanoic acid (acetic acid) methanol

2. Basic hydrolysis: Write equations for the saponification reactions of methyl acetate and methyl butanoate.

Your answer:

Solution: "Type" reaction: $R-\overset{O}{\overset{\|}{C}}-OR' + NaOH\ (aq) \xrightarrow{\Delta} R-\overset{O}{\overset{\|}{C}}-O^- \ Na^+ + R'OH$

Compare this reaction with the equilibrium reaction for the acidic hydrolysis of esters; in saponification, the strong base shifts the equilibrium away from the ester by forming a salt (which cannot react with the alcohol to reverse the reaction). The term saponification means "soap making"; grandma's soap was made by heating a mixture of animal fat (containing esters of carboxylic acids) with lye (sodium hydroxide).

(a) $CH_3\overset{O}{\overset{\|}{C}}-OCH_3 + NaOH\ (aq) \xrightarrow{\Delta} CH_3\overset{O}{\overset{\|}{C}}-O^- \ Na^+ + CH_3OH$

(b) $CH_3CH_2CH_2\overset{O}{\overset{\|}{C}}-OCH_3 + NaOH\ (aq) \xrightarrow{\Delta} CH_3CH_2CH_2\overset{O}{\overset{\|}{C}}-O^- \ Na^+ + CH_3OH$

V. Reactions of Acid Derivatives

"Type" reaction—can be visualized as the substitution of one nucleophilic group for another

W = any nucleophilic group (e.g., Cl^-, OH^-)

Z = another nucleophilic group

$R-\overset{O}{\overset{\|}{C}}-W$ = acyl halide, acid anhydride, carboxylic acid, ester, or amide

$$R-\overset{O}{\overset{\|}{C}}-W + H-Z \longrightarrow R-\overset{O}{\overset{\|}{C}}-Z + H-W$$

This reaction occurs if W "leaves" more easily than does Z. However, if Z "leaves" more readily than W does, the forward reaction does not occur to any appreciable extent, and we can write N.R. for "no reaction."

The "leaving group" order is

$$-Cl > -O-\overset{O}{\overset{\|}{C}}-R > -OH \cong -OR > -NH_2$$

Notice that in this type of reaction, the carbonyl group remains intact in the product (in contrast to many of the reactions of aldehydes and ketones in which the pi bond is broken).

1. Use this "leaving group" theory to write equations for the hydrolysis of each of the following.
   (a) propanoyl chloride
   (b) acetic anhydride
   (c) methyl formate
   (d) benzamide

Your answer:

Solution: Type reaction: $R\text{-}\overset{\overset{O}{\|}}{C}\text{-}W + H_2O \longrightarrow R\text{-}\overset{\overset{O}{\|}}{C}\text{-}OH + H\text{-}W$

This reaction will occur if W is a better "leaving group" than -OH.

(a) $CH_3CH_2\overset{\overset{O}{\|}}{C}\text{-}Cl + H_2O \longrightarrow CH_3CH_2CO_2H + HCl$

This reaction occurs because -Cl is a better "leaving group" than -OH.

(b) $CH_3\text{-}\overset{\overset{O}{\|}}{C}\text{-}O\text{-}\overset{\overset{O}{\|}}{C}\text{-}CH_3 + H_2O \longrightarrow 2\ CH_3CO_2H$

(c) $H\overset{\overset{O}{\|}}{C}OCH_3 + H_2O \underset{}{\overset{H^+}{\rightleftharpoons}} HCO_2H + CH_3OH$

An equilibrium mixture forms because -OH and -OR have the same relative ability "to leave."

(d) 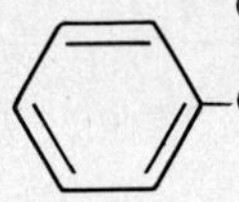 $\text{-}\overset{\overset{O}{\|}}{C}NH_2 + H_2O \longrightarrow$ N.R.

No reaction is predicted to occur unless the reactants are subjected to heat and 6 M acid or base or unless a specific enzyme can catalyze the reaction.

---

2. Write equations for the reactions of (a) acetyl chloride and (b) propanamide with ethyl alcohol.

Your answer:

Solution: Type reaction: $R\text{-}\overset{\overset{O}{\|}}{C}\text{-}W + R'OH \longrightarrow R\text{-}\overset{\overset{O}{\|}}{C}\text{-}OR' + H\text{-}W$

This reaction will occur if W is a better "leaving group: than -OR.

(a) $CH_3\overset{\overset{O}{\|}}{C}\text{-}Cl + CH_3CH_2OH \longrightarrow CH_3\overset{\overset{O}{\|}}{C}\text{-}OCH_2CH_3 + HCl$

ethyl ethanoate
(ethyl acetate)

(b) $CH_3CH_2\overset{\overset{O}{\|}}{C}\text{-}NH_2 + CH_3CH_2OH \longrightarrow$ N.R.

No reaction occurs because $-NH_2$ is a very poor "leaving group."

VI. Practice and Review Reactions of Acids and Acid Derivatives

1. Complete the following equations and name the organic products. For each organic reactant, mentally identify the type of functional group(s) present in the molecule and recall the corresponding "type" reaction.

(a) propanoyl chloride + methylamine $\longrightarrow$

Your answer:

<u>Solution</u>: $CH_3CH_2\overset{O}{\overset{\|}{C}}\text{-}Cl + CH_3NH_2 \longrightarrow CH_3CH_2\overset{O}{\overset{\|}{C}}\text{-}NHCH_3 + HCl$

N-methylpropanamide

(b) $CH_3CH_2\underset{O}{\underset{\|}{C}}\text{-}O\text{-}\underset{O}{\underset{\|}{C}}CH_2CH_3 + CH_3CH_2NH_2 \longrightarrow$

Your answer:

<u>Solution</u>: $CH_3CH_2\overset{O}{\overset{\|}{C}}\text{-}O\text{-}\overset{O}{\overset{\|}{C}}CH_2CH_3 + CH_3CH_2NH_2 \longrightarrow CH_3CH_2\overset{O}{\overset{\|}{C}}\text{-}NHCH_2CH_3 + CH_3CH_2CO_2H$

propanoic anhydride — N-ethylpropanamide — propanoic acid

(c) 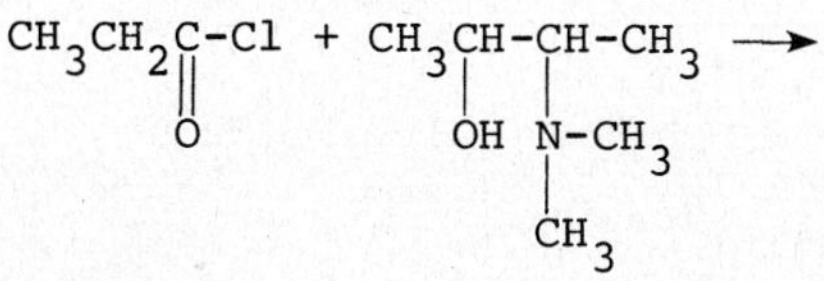

Your answer:

<u>Solution</u>: 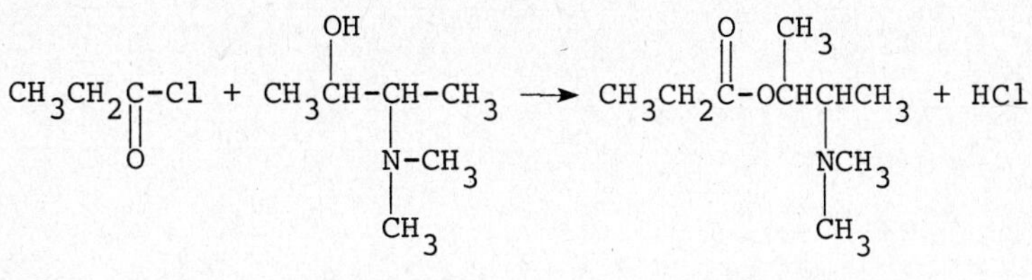

an ester

An acyl halide reacts with an alcohol or a 1° or 2° amine, but not with a 3° amine.

(d) trans-4-hexen-2-one + $LiAlH_4 \longrightarrow$

Your answer:

Solution:

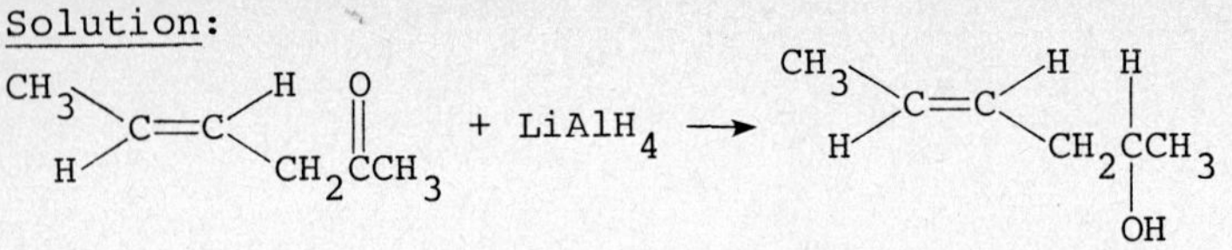

trans-4-hexen-2-ol

---

2. Write an equation illustrating the saponification of methyl butanoate.

Your answer:

Solution: $CH_3CH_2CH_2\overset{O}{\overset{\|}{C}}\text{-}OCH_3$ + NaOH/heat ⟶ $CH_3CH_2CH_2\overset{O}{\overset{\|}{C}}O^- \; Na^+$ + CH_3OH

methyl butanoate — sodium butanoate — methanol

---

3. Suggest an appropriate reagent to convert sodium o-chlorobenzoate to its corresponding acid. Write the equation for the reaction.

Your answer:

Solution:

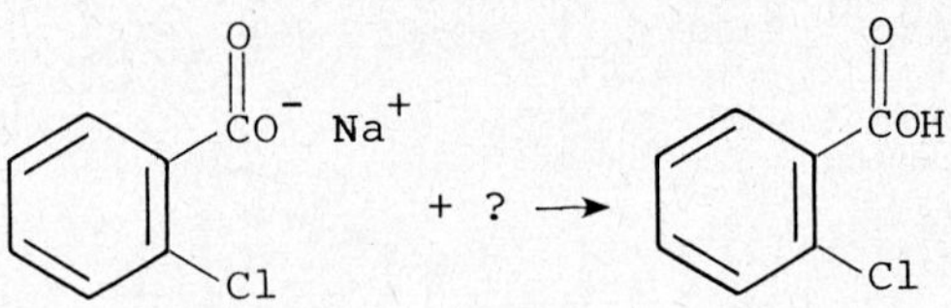

A source of protons is required for this reaction to react with the benzoate ion (the conjugate base of a weak acid); a strong acid such as HCl would enable the o-chlorobenzoate to form o-chlorobenzoic acid.

---

4. Isopropyl propanoate can be converted to propanoic acid and 2-propanol by a series of two reactions. Write equations illustrating these reactions, and specify the reagents and conditions needed for the reactions.

Your answer:

Solution:

(1) $CH_3CH_2\overset{O}{\overset{\|}{C}}\text{-O-}CH(CH_3)_2 + \text{NaOH/heat} \longrightarrow CH_3CH_2CO_2^- \; Na^+ + (CH_3)_2CH\text{-OH}$

(2) Separate the two products and react the sodium propanoate with HCl.

$CH_3CH_2COO^- \; Na^+ + HCl \longrightarrow CH_3CH_2CO_2H$

---

5. What two reactants are needed to form propanoyl chloride? Write the equation for the reaction.

Your answer:

Solution: $CH_3CH_2\overset{O}{\overset{\|}{C}}OH + PCl_3 \longrightarrow CH_3CH_2\overset{O}{\overset{\|}{C}}Cl$

$? + ? \longrightarrow CH_3CH_2\overset{O}{\overset{\|}{C}}Cl$

An acyl halide is prepared from the carboxylic acid and any one of these reagents: PCl_3, PCl_5, or $SOCl_2$.

---

6. Write the equation for the reaction of aniline and propanoic acid.

Your answer:

Solution:

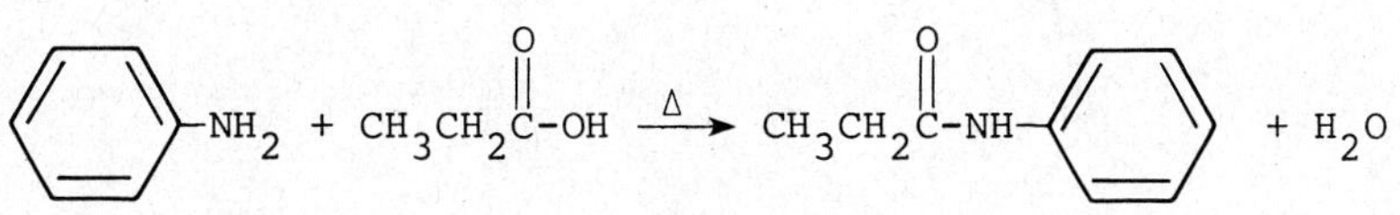

N-phenylpropanamide

The amine and acid react to form a salt that forms an amide on heating.

---

7. Methyl butanoate can be prepared by reacting any of three compounds with the same alcohol. Write equations for the three reactions.

Your answer:

Solution: Consider the components of methyl butanoate.

$$\underbrace{CH_3CH_2CH_2}\overset{O}{\overset{\|}{C}}-\underbrace{OCH_3}$$

This portion is from butanoic acid or its acyl halide or acid anhydride.

This portion is from methanol.

$$\underset{\text{butanoic acid}}{CH_3CH_2CH_2CO_2H} + CH_3OH \overset{H^+}{\rightleftharpoons} CH_3CH_2CH_2\overset{O}{\overset{\|}{C}}OCH_3 + H_2O$$

This equilibrium can be shifted to the right (to increase ester yield) by removing the water from the reaction vessel.

$$\underset{\text{butanoyl chloride}}{CH_3CH_2CH_2\overset{O}{\overset{\|}{C}}-Cl} + CH_3OH \longrightarrow CH_3CH_2CH_2\overset{O}{\overset{\|}{C}}-OCH_3 + HCl$$

$$\underset{\text{butanoic acid anhydride}}{CH_3CH_2CH_2\overset{O}{\overset{\|}{C}}-O-\overset{O}{\overset{\|}{C}}CH_2CH_2CH_3} + CH_3OH \longrightarrow CH_3CH_2CH_2\overset{O}{\overset{\|}{C}}-OCH_3 + \underset{\text{butanoic acid}}{CH_3CH_2CH_2CO_2H}$$

---

8. Write an equation for the reaction of 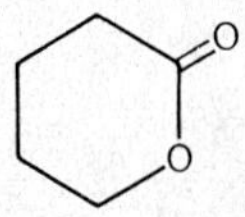 with NaOH. What is this process called?

Your answer:

Solution:

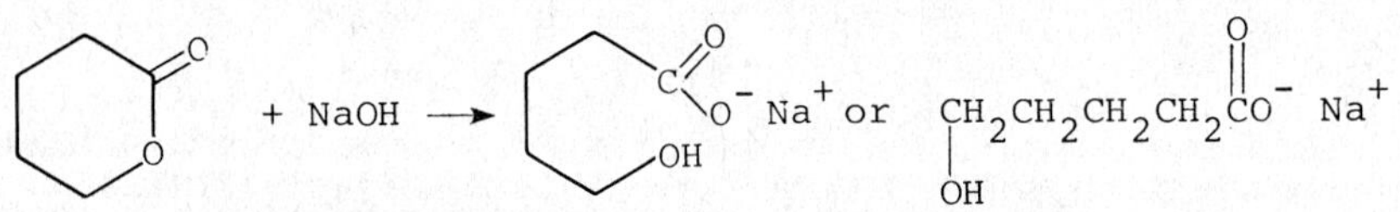

(an intramolecular ester)

sodium 5-hydroxypentanoate

The process is saponification.

---

9. Write equations for (a) the reaction of acetaldehyde with ethanol and (b) the reaction of acetone with ethyl amine.

Your answer:

Solution:

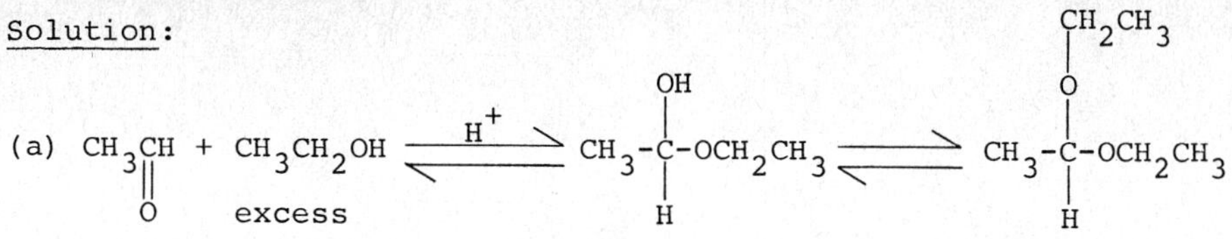

ethanal ethanol hemiacetal acetal

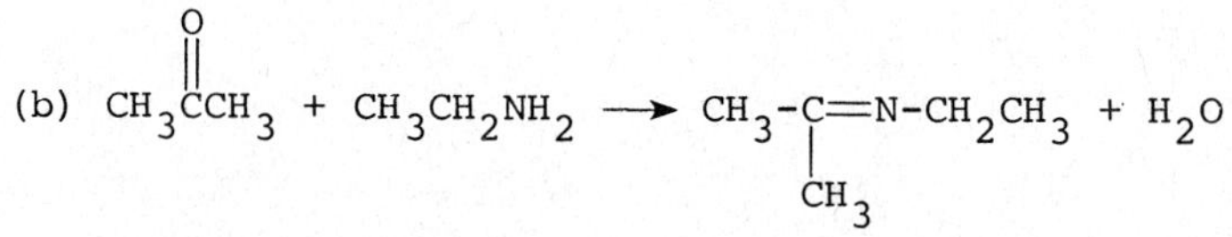

acetone ethylamine an imine or Schiff's base

VII. Identification Problems

1. A compound, $C_5H_8O_2$, reacts with NaOH to yield a salt and water. The compound does not react with Br_2, but does react with PCl_3 or $SOCl_2$ to produce C_5H_7OCl. Write all structures for $C_5H_8O_2$ that are consistent with the experimental data. Choose one of these structures and write equations for each reaction.

Your answer:

Solution:

Given information	Possible	Not possible
$C_5H_8O_2$ $C_nH_{2n-2}O$ (a 4 H deficiency)	RC(=O)OH or RC(=O)OR' with ring or C=C; two carbonyl groups	
$C_5H_8O_2 \xrightarrow{NaOH}$ salt + H_2O	carboxylic acid with ring or with double bond	two carbonyl groups, ester + NaOH ⟶ salt + alcohol
$C_5H_8O_2 + Br_2 \longrightarrow$ N.R.	carboxylic acid with ring	C=C
$C_5H_8O_2 + PCl_3 \longrightarrow C_5H_7OCl$	R-C(=O)-OH with ring	

Possible structures are

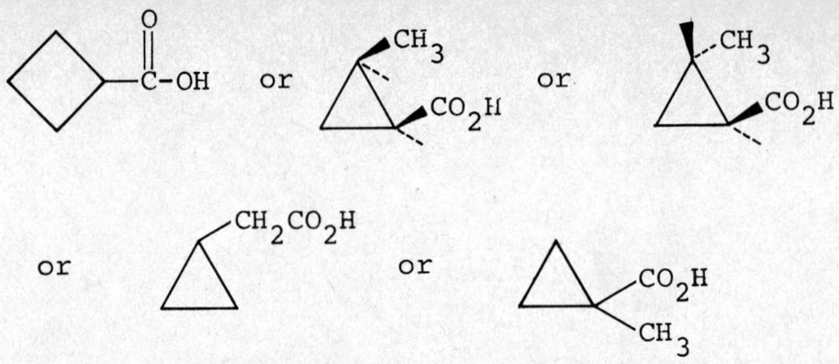

Reactions are

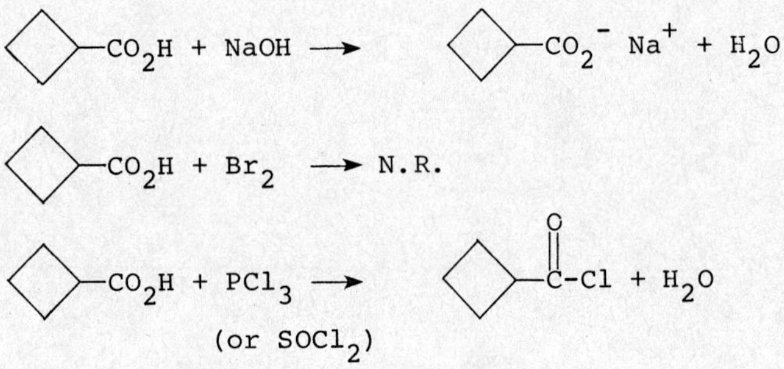

---

2. A compound X, $C_5H_{10}O_2$, reacts with NaOH to yield two organic compounds, Y and Z. Z reacts with Na and also very rapidly with the Lucas reagent. Write structural formulas for X, Y, and Z.

Your answer:

Solution

Given information	Possible	Not possible
X is $C_5H_{10}O_2$ $C_nH_{2n}O_2$ (a 2 H deficiency)	R-C(=O)OH; R-C(=O)OR'; aldehyde or ketone with alcohol; cyclic diether; cyclic diol	
$C_5H_{10}O_2$ + NaOH $\longrightarrow$ Y + Z	ester (R-C(=O)OR')	acid (RCO_2H); carbonyl with alcohol; cyclic diether or diol

The compound cannot be an acid because the products of the reaction with NaOH and an acid are salt and water (and water is not an organic product as required by the problem); when an ester is saponified, the products are an alcohol and an organic salt.

Z $\xrightarrow{Na}$ H_2	alcohol	

Given information	Possible
Z $\xrightarrow{HCl/ZnCl_2}$ fast reaction	tertiary alcohol

Smallest 3° alcohol is $CH_3-\underset{CH_3}{\overset{CH_3}{\overset{|}{\underset{|}{C}}}}-OH$ (compound Z). If Z is t-butyl alcohol, then the acid that is esterified to the alcohol is formic acid (methanoic acid), HCO_2H.

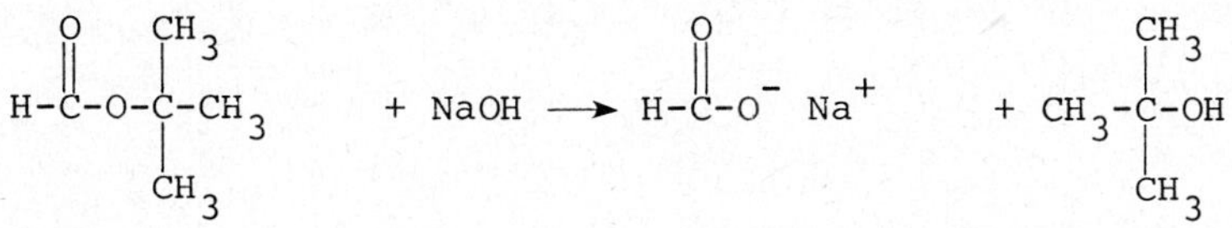

X t-butyl formate Y sodium formate Z t-butyl alcohol

---

3. A compound, Z, with molecular formula $C_4H_6O_2$, does not react with NaOH, Br_2, Benedict's (or Fehling's) solution, Na, or $K_2Cr_2O_7$. When a sample of Z reacts with 2,4-dinitrophenylhydrazine, a yellow precipitate is produced. Write the structural formula for the $C_4H_6O_2$ compound. Use the information in the problem to systematically eliminate functional groups; exclude ethers also.

Your answer:

Solution:

Given information	Possible	Not possible
Z is $C_4H_6O_2$ $C_nH_{2n-2}O_2$ (a 4H deficiency)	acid or ester with C=C or ring; some combination of carbonyls	
Z + NaOH → N.R.		acid, ester
Z + Br_2 → N.R.		C=C
Z $\xrightarrow{Cu^{2+}/OH^-}$ N.R.		aldehyde
Z $\xrightarrow{Na}$ N.R.		alcohol
Z $\xrightarrow{K_2Cr_2O_7/H^+}$ N.R.		aldehyde
Z + 2,4-DNP → yellow precipitate	ketones (two, because two oxygens in molecular formula)	

The compound is butanedione, $CH_3\overset{O}{\overset{\|}{C}}-\overset{O}{\overset{\|}{C}}CH_3$

Verify your answer by rechecking this structure against the evidence in the problem: butanedione will not react with NaOH, Benedict's (or Fehling's) solution, Na, or $K_2Cr_2O_7$, but it will react with 2,4-dinitrophenylhydrazine.

---

20 Amines

The exercises in this chapter emphasize amines but include a few other compounds for contrast. If you do not feel competent in the recognition of functional groups containing nitrogen, you should work the problems in section IV of Chapter 16, "Nomenclature of Organic Compounds Containing Oxygen and Nitrogen."

I. Reactions of Amines

For each reaction, write a "type" reaction and specify the classes of amines that can undergo the reaction. Then describe the reaction.

A. AMINES AS BASES

1. Write equations showing the equilibrium reactions of (a) methylamine with water and (b) aniline with water. Are amines weak or strong bases? Explain in terms of the equations you have written.

Your answer:

Solution:

"Type" reaction: $RNH_2 + H_2O \rightleftharpoons RNH_3^+ + OH^-$

1°, 2°, and 3° amines can undergo this reaction.

The equilibrium lies far to the left. Because there are relatively few OH^- ions available in the solution, the amines are weak bases.

(a) $CH_3NH_2 + H_2O \rightleftharpoons CH_3NH_3^+ + OH^-$

(b) 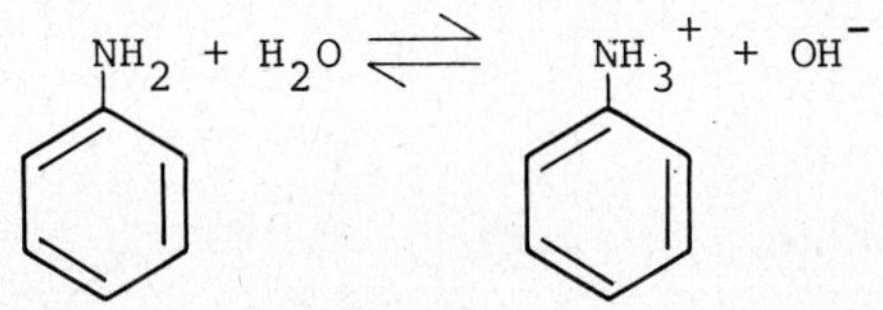

---

2. Write the equilibrium expression for K_b (also called K_d) for an aqueous solution of trimethylamine.

Your answer:

Solution: $K_b = \dfrac{[(CH_3)_3NH^+][OH^-]}{[(CH_3)_3N]}$

The equation for the reaction of trimethylamine with water is

$$(CH_3)_3N + H_2O \rightleftharpoons (CH_3)_3NH^+ + OH^-$$

---

3. Rank the following bases in order of increasing base strength.

	K_b
aniline	4×10^{-10}
ethylamine	5×10^{-4}
diethylamine	1×10^{-3}
ammonia	2×10^{-5}
pyridine	2×10^{-9}
NaOH	

Your answer:

Solution: aniline < pyridine < ammonia < ethylamine < diethylamine < NaOH

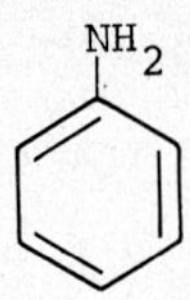

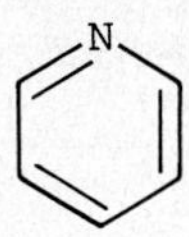

NH_3 $\quad CH_3CH_2NH_2 \quad (CH_3CH_2)_2NH$

---

4. For neutralization by acids, the "type" reaction is

$$R\text{-}NH_2 \text{ (or } R_2NH \text{ or } R_3N) + HA \text{ (any acid)} \longrightarrow RNH_3^+\ A^- \text{ (or } R_2NH_2^+\ A^- \text{ or } R_3NH^+\ A^-)$$

Complete each of the following equations.

(a) $(CH_3CH_2)_2NH + HCl \longrightarrow$

Your answer:

Solution: $(CH_3CH_2)_2NH + HCl \longrightarrow (CH_3CH_2)_2NH_2^+\ Cl^-$

diethylamine $\qquad$ diethylammonium chloride

(b) $CH_3CH_2\underset{\displaystyle NH_2}{\underset{|}{C}H}\text{-}CH_3 + CH_3C\begin{smallmatrix}\nearrow O \\ \searrow OH\end{smallmatrix} \longrightarrow$

Your answer:

Solution: 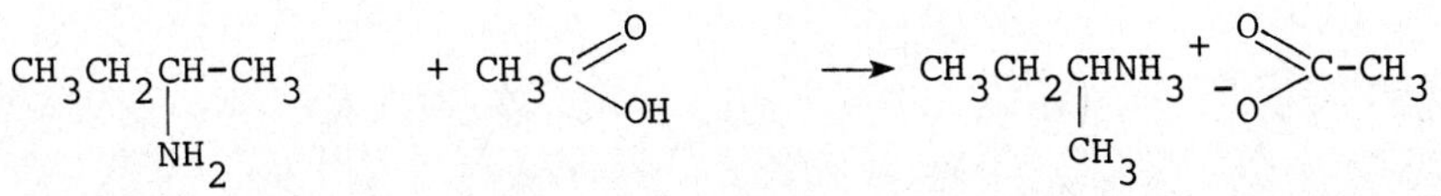

2-aminobutane (sec-butylamine) | ethanoic acid (acetic acid) | (sec-butylammonium acetate)

If the product of this reaction is heated, an amide will form. In general, if the ammonium salt of a carboxylic acid is heated, the salt is converted to an amide.

(c) $H_2N\text{-}CH(CH_3)\text{-}CO_2H$ + benzoic acid $\longrightarrow$

Your answer:

Solution: 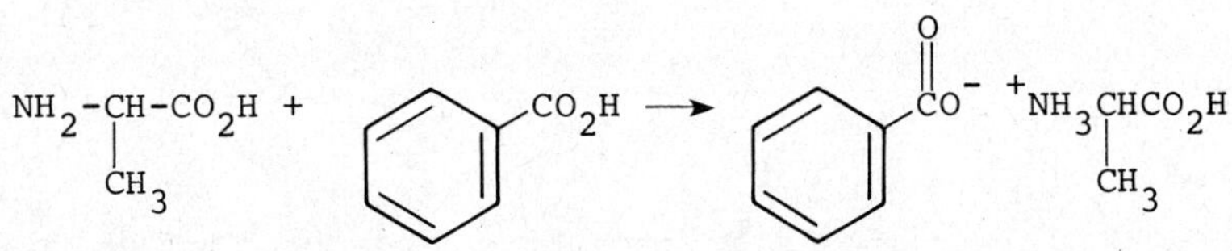

B. NITROUS ACID REACTION (A TEST THAT CLASSIFIES AMINES)

1. Write structural formulas and names for all the 1°, 2°, and 3° amines that have a molecular formula of C_3H_9N. Name them.

Your answer:

Solution:

$CH_3CH_2CH_2NH_2$	propylamine	1°
$CH_3CH(CH_3)$-NH_2	isopropylamine	1°
CH_3CH_2NH-CH_3	ethylmethylamine	2°
CH_3-$N(CH_3)$-CH_3	trimethylamine	3°

---

2. Write "type" reactions for 1°, 2°, and 3° amines, each with nitrous acid, HNO_2.

Your answer:

Solution:

HNO_2, nitrous acid, is unstable, so it is prepared immediately before a reaction is run:
$NaNO_2 + HCl \longrightarrow HNO_2 + NaCl$.

"Type" reactions:

1° amine $\quad R\text{-}NH_2 + HONO \longrightarrow N_2\ (g) + ROH + H_2O$

The bubbles of nitrogen gas constitute proof of the presence of a primary amine. The volume of the nitrogen gas produced can be used to quantitatively determine the number of primary amine groups per mole of compound.

2° amine $\quad R\text{-}NH\text{-}R + HONO \longrightarrow R\text{-}N(R)\text{-}NO + H_2O$

The compounds produced are nitrosamines, which form a yellow, oily liquid layer in the test tube or beaker.

3° amine $\quad R\text{-}N(R)\text{-}R + HONO \longrightarrow R\text{-}\overset{+}{N}(H)(R)\text{-}R\ \ NO_2^-$

There is no visible sign of reaction because the salt formed is soluble even if the original amine is not.

---

3. Write equations for each of the amine isomers of C_3H_9N with nitrous acid.

Your answer:

Solution:

(i) $CH_3CH_2CH_2NH_2 + HNO_2 \longrightarrow CH_3CH_2CH_2OH + N_2\ (g) + H_2O$

(ii) $CH_3\underset{\displaystyle CH_3}{\underset{|}{C}}H\text{-}NH_2 + HNO_2 \longrightarrow CH_3\text{-}\underset{\displaystyle CH_3}{\underset{|}{C}}H\text{-}OH + N_2\ (g) + H_2O$

(iii) $CH_3CH_2\text{-}NH\text{-}CH_3 + HNO_2 \longrightarrow CH_3CH_2\text{-}\underset{\displaystyle CH_3}{\underset{|}{N}}\text{-}NO + H_2O$

a nitrosamine

(iv) $CH_3\text{-}\overset{\displaystyle CH_3}{\underset{\displaystyle CH_3}{\overset{|}{\underset{|}{N}}}} + HNO_2 \longrightarrow (CH_3)_3\text{-}N^+\ NO_2^-$

trimethylammonium nitrite

C. REACTIONS OF AMINES WITH DERIVATIVES OF CARBOXYLIC ACIDS

1. Write a "type" reaction to illustrate the reactions of an amine with each of the following: an acyl halide, an acid anhydride, and an ester. Specify which classes of amines will react and which, if any, will not.

Your answer:

Solution:

(i) $R\text{-}\overset{\displaystyle O}{\overset{\|}{C}}\text{-}Cl + R'NH_2 \longrightarrow R\text{-}\overset{\displaystyle O}{\overset{\|}{C}}\text{-}NH\text{-}R' + HCl$

acyl halide amide

1° and 2° amines react, but 3^{c} amines do not (because 3° amines lack the H that would combine with the Cl)

(ii) $R\text{-}\overset{\displaystyle O}{\overset{\|}{C}}\text{-}O\text{-}\overset{\displaystyle O}{\overset{\|}{C}}\text{-}R + R'NH_2 \longrightarrow R\text{-}\overset{\displaystyle O}{\overset{\|}{C}}\text{-}NHR' + R\overset{\displaystyle O}{\overset{\|}{C}}\text{-}OH$

acid anhydride amide acid

1° and 2° amines react, but 3° amines do not.

(iii) $R\text{-}\overset{\displaystyle O}{\overset{\|}{C}}\text{-}OR'' + R'NH_2 \longrightarrow R\text{-}\overset{\displaystyle O}{\overset{\|}{C}}\text{-}NH\text{-}R' + R'OH$

ester amide alcohol

1° and 2° amines react, but not 3° amines.

---

2. Complete each of the following reactions.

(a) aniline + propanoyl chloride ⟶

Your answer:

Solution: 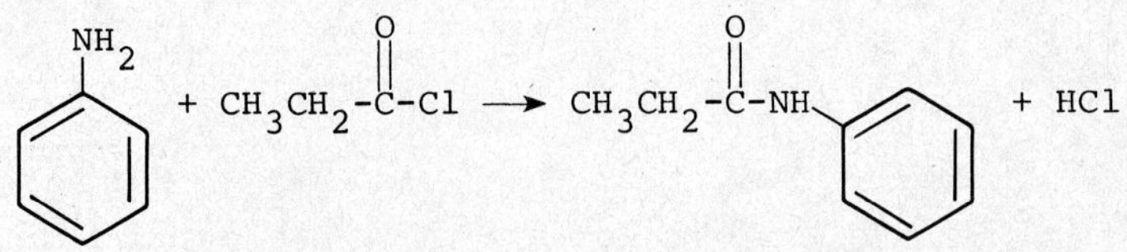

N-phenylpropanamide

(b) ethyl benzoate + triethylamine ⟶

Your answer:

Solution: 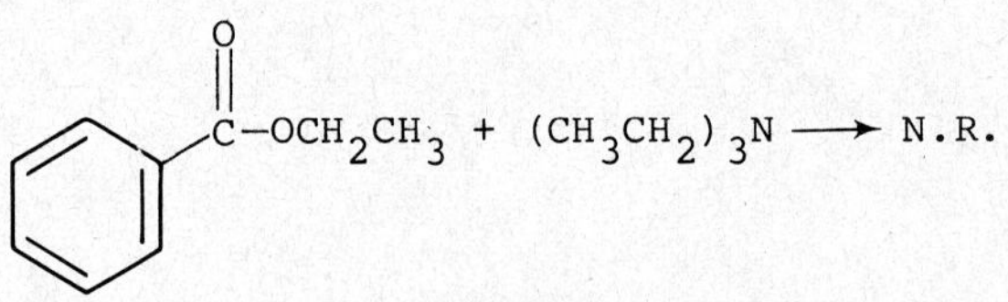 (3° amines do not react)

(c) acetic anhydride + ethylmethylamine ⟶

Your answer:

Solution: $CH_3\text{-}\overset{O}{\overset{\|}{C}}\text{-}O\text{-}\overset{O}{\overset{\|}{C}}\text{-}CH_3 + CH_3CH_2NH\text{-}CH_3 \longrightarrow CH_3\overset{O}{\overset{\|}{C}}\text{-}\underset{\underset{CH_3}{|}}{N}\text{-}CH_2CH_3 + CH_3CO_2H$

N-ethyl-N-methylethanamide
(N-ethyl-N-methylacetamide)

D. REACTIONS OF AMINES WITH ALDEHYDES AND KETONES (REVIEW)

1. Write a "type" reaction to illustrate the reaction of amines with aldehydes and ketones.

Your answer:

Solution: 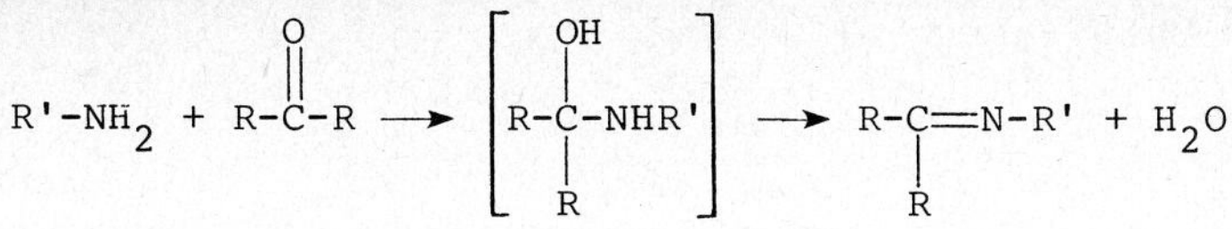

Two hydrogens are eliminated from the amine to form the imine (Schiff's base); only 1° amines can form Schiff's bases.

---

2. Complete each of the following:

(a) isobutylamine + acetone ⟶

Your answer:

Solution: 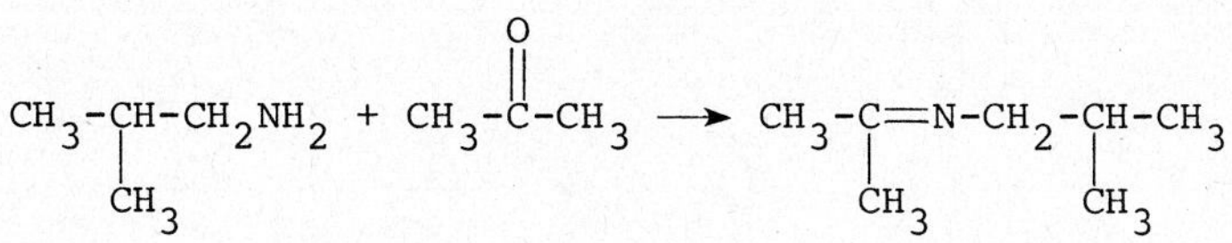

an imine or Schiff's base

(b) $HO\text{-}CH_2CH_2\text{-}NH\text{-}CH_3$ + methanal (formaldehyde) ⟶

Your answer:

Solution: 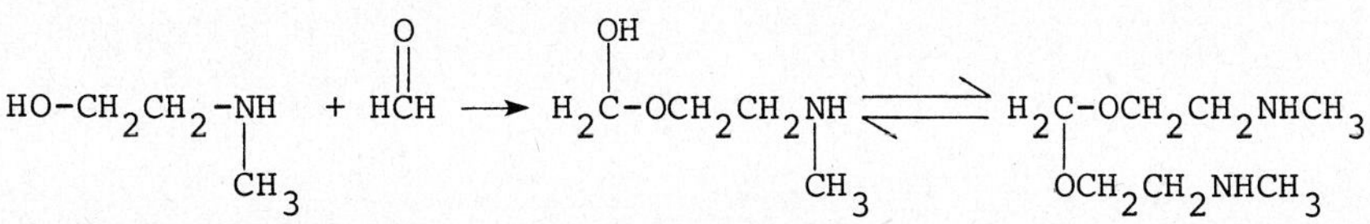

A 2° amine will not add to carbonyls, but the alcohol does add and yields an equilibrium mixture of the hemiacetal and acetal compounds.

II. Chemical Tests to Differentiate Amines and Other Compounds

Choose the reagent(s) that would adequately identify each set of compounds. Write equations for the reactions, and indicate how you would visually recognize the reaction.

1. octylamine, butylpentylamine, diethylisobutylamine

Your answer:

Solution: Nitrous acid.

$CH_3(CH_2)_7NH_2 + HNO_2 \longrightarrow N_2\ (g) + CH_3(CH_2)_7OH + H_2O$

octylamine 1-octanol

The primary amine is identified by the bubbles of nitrogen gas.

$CH_3(CH_2)_3NH$ + HNO_2 $\longrightarrow$ $CH_3(CH_2)_3$-N-NO

$\quad\quad (CH_2)_4CH_3 \quad\quad\quad\quad\quad\quad (CH_2)_4CH_3$

butylpentylamine a nitrosamine

The yellow, oily nitrosamine will identify the secondary amine.

CH_3-CH-CH_2-$N(CH_2CH_3)_2$ + HNO_2 $\longrightarrow$ CH_3CH-CH_2NH^+ NO_2^-

$\quad CH_3 \quad\quad\quad\quad\quad\quad\quad\quad\quad\quad CH_3 \quad (CH_2CH_3)_2$

diethylisobutylamine an ammonium salt

No visible reaction occurs because the ammonium salt dissolves in the aqueous solution.

---

2. $CH_3CH_2-NH_2$ and CH_3CO_2H

Your answer:

Solution:

(i) Simplest test: in a water solution, ethylamine, a base, will turn red litmus paper blue, and the acetic acid will turn blue litmus paper red.

(ii) React each compound with HCl.

$CH_3CH_2NH_2 + HCl \longrightarrow CH_3CH_2NH_3^+\ Cl^-$

$CH_3CO_2H + HCl \longrightarrow$ N.R.

Although HCl will not visibly differentiate these two compounds (the ammonium salt is water soluble, so both solutions will appear to be transparent), the HCl will differentiate larger-molecular-weight amines from acids. For example, HCl will differentiate aniline from benzoic acid.

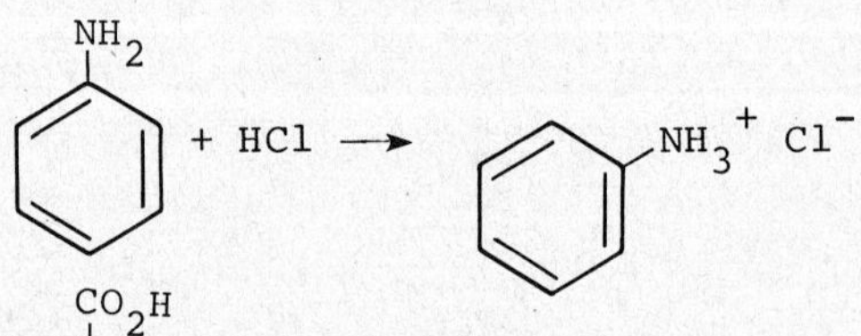

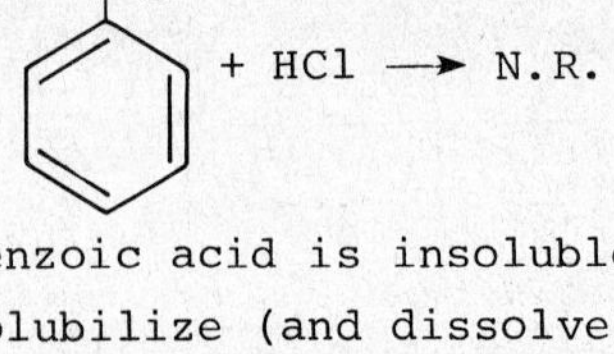

Benzoic acid is insoluble in an acid solution, but the ammonium salt of aniline will solubilize (and dissolve).

(iii) React each compound with NaOH.

$CH_3CH_2NH_2 + NaOH \longrightarrow$ N.R.

$$CH_3CO_2H + NaOH \longrightarrow CH_3CO_2^-Na^+ + H_2O$$

NaOH will not visibly differentiate these two compounds because the ethylamine and the sodium acetate are both soluble in water solution. However, the NaOH will differentiate larger-molecular-weight amines from acids. Consider the reactions of NaOH with aniline and with benzoic acid.

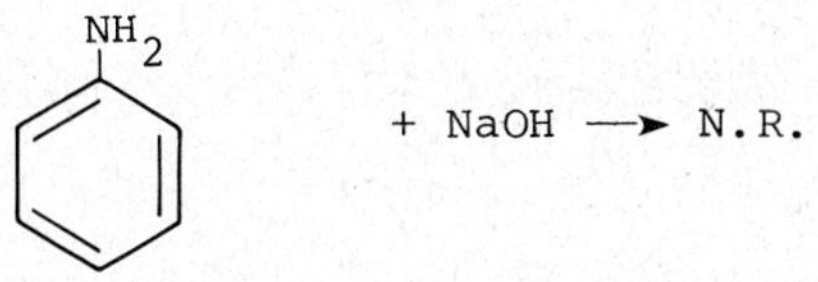

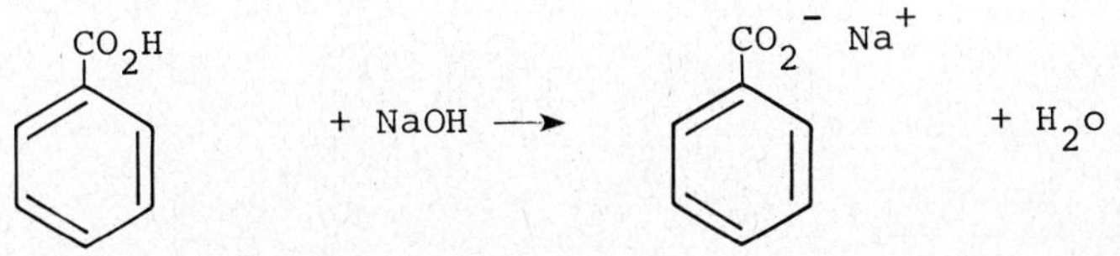

Aniline is insoluble in a basic solution, but the sodium benzoate will dissolve because it is ionic.

---

3. $CH_3\overset{O}{\overset{\|}{C}}\text{-}NH_2$ and $CH_3CH_2\text{-}NH_2$

Your answer:

Solution: nitrous acid

Ethylamine will react with nitrous acid, HNO_2, and will produce ethanol and bubbles of nitrogen gas (positive proof of the presence of the primary amine).

$$CH_3CH_2NH_2 + HNO_2 \longrightarrow CH_3CH_2OH + N_2(g) + H_2O$$

Ethanamide does not release nitrogen bubbles in the presence of HNO_2.

---

4. CH_3CH_2OH, $CH_3CH_2NH_2$, $CH_3CH_2\overset{O}{\overset{\|}{C}}H$

Your answer:

Solution:

To differentiate these three compounds, think of reagents that yield a visible reaction with only one of the functional groups represented, for example, Na for alcohols, nitrous acid for primary amines, and Tollens' or Benedict's (or Fehling's) reagent for aldehydes.

$$CH_3CH_2OH + Na \longrightarrow CH_3CH_2O^- \; Na^+ + \tfrac{1}{2} H_2(g)$$

(Ethylamine and propanal will not react with Na.)

$$CH_3CH_2NH_2 + HNO_2 \longrightarrow CH_3CH_2OH + N_2(g) + H_2O$$

(Ethanol and propanal will not react with HNO_2.)

$$CH_3CH_2\overset{O}{\overset{\|}{C}}H + Ag(NH_3)_2^+ \text{ (Tollens')} \longrightarrow CH_3CH_2\overset{O}{\overset{\|}{C}}O^- + Ag^o$$

$$CH_3CH_2\overset{O}{\overset{\|}{C}}H + Cu^{2+}/OH^- \text{ (Benedict's or Fehling's)} \longrightarrow CH_3CH_2CO_2^- + Cu_2O$$

(Ethanol and ethylamine will not react with any of these mild oxidizing agents.)

---

5. 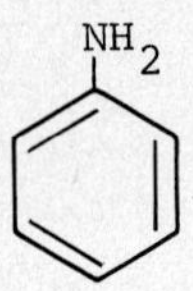 and 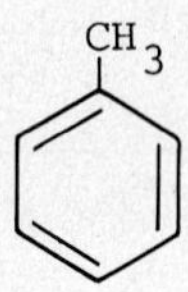

Your answer:

Solution:

Aniline is an aromatic primary amine; toluene is an aromatic hydrocarbon. Both compounds undergo typical substitution reactions on the aromatic ring, but such reactions would not visually distinguish the two compounds. Neither compound is water soluble. Some possible reactions that will differentiate the two include

(i)

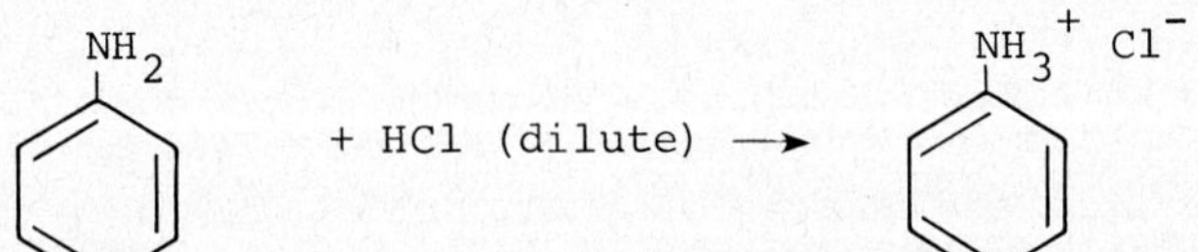

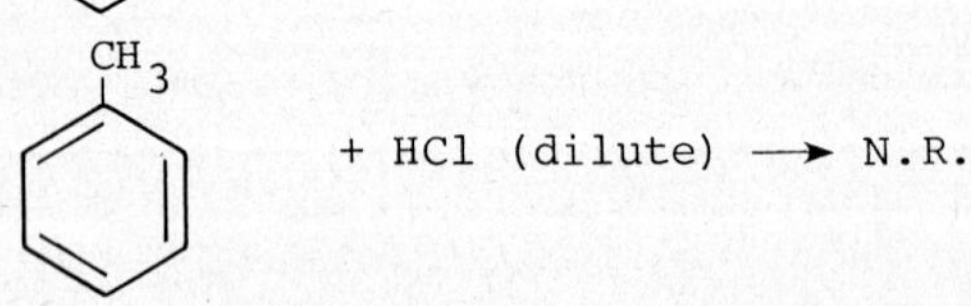

The ammonium salt formed from aniline is water soluble; toluene is not and will form a separate layer in the test tube.

(ii)

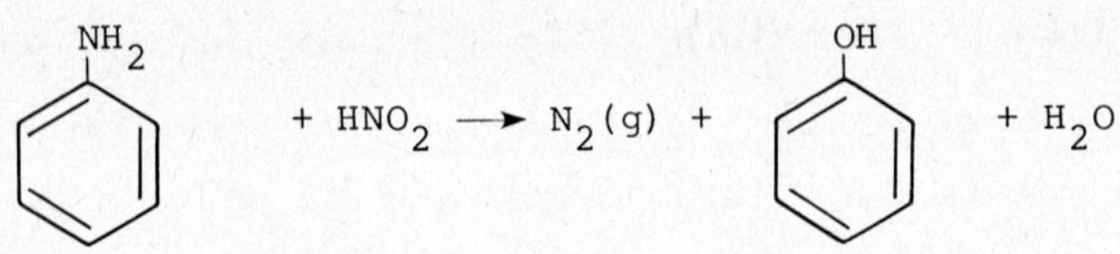

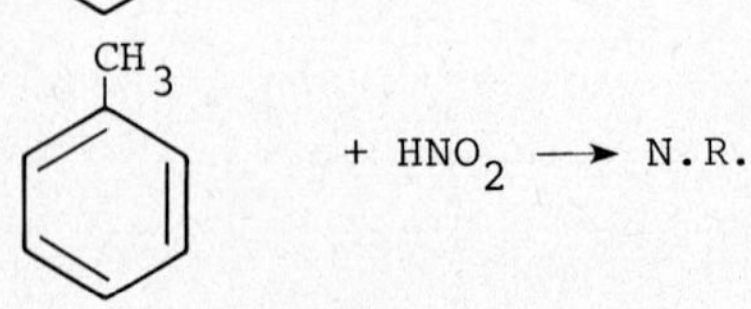

(iii)

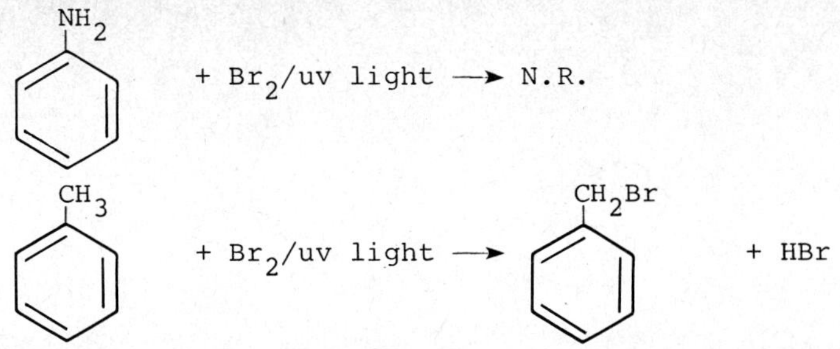

---

6. $CH_3\overset{\overset{O}{\|}}{C}CH_3$ and $CH_3\text{-}O\text{-}CH_2CH_3$

Your answer:

Solution:

Ethyl methyl ether is very unreactive, but acetone, a ketone, will react with I_2/OH^- to form iodoform and will react with 2,4-dinitrophenylhydrazine.

$$CH_3\overset{\overset{O}{\|}}{C}CH_3 + I_2/OH^- \longrightarrow CHI_3 + {}^-O\text{-}\overset{\overset{O}{\|}}{C}\text{-}CH_3$$

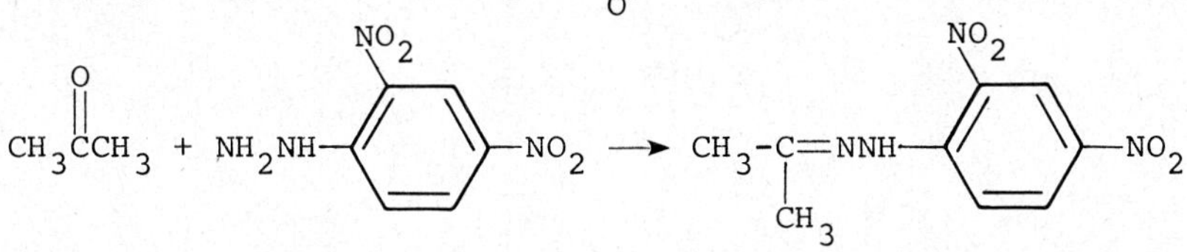

a yellow precipitate

Ethyl methyl ether will give negative results in either test.

---

21 Review of Organic Reactions

In this chapter special emphasis is placed on "instant recall" of reagents, catalysts, and conditions necessary for a reaction as well as on the functional groups of the reactants and products. The functional groups are to be recognized from the names of classes of compounds or from structural formulas of biologically important multifunctional molecules.

I. "Type" Formulas, Reactions, and Terminology of Organic Reactions

The terms in the following list describe or identify reactions. One or more of these terms apply to each reaction in the summary diagram (Figure 21.1).

substitution for H in an
    alkyl (uv required)
electrophilic addition
aromatic substitution
nucleophilic substitution
nucleophilic addition
    (to a carbonyl)
reactions involving a
    carbocation intermediate
reduction

oxidation
    combustion (complete oxidation)
    dichromate, permanganate
        (Baeyer's)
    Tollens' (mild)
    Benedict's or Fehling's (mild)
    ozonolysis
dehydration
hydration
dehydrogenation

hydrogenation
hydrolysis
    acidic
    basic
catalyzed reaction
condensation
saponification
esterification
neutralization
saturation (of a
    carbon-carbon
    double bond)

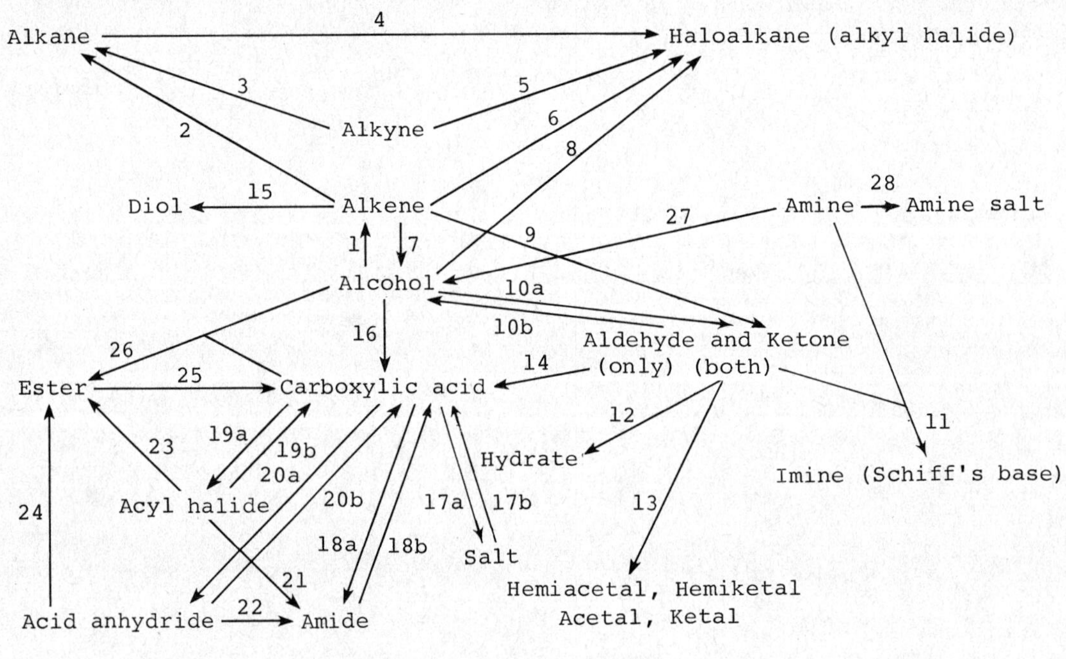

Figure 21.1 Summary Diagram of Interrelated Reactions

Neither ethers nor aromatics are represented in this diagram; however, you should review their chemical properties at this time also.

The numeral on each arrow identifies the reaction in the question.

(1) Write the "type" formulas of the organic reactant and product, and indicate all needed reagents or catalysts.

(2) Write a specific reaction. For example, use 1-butanol as the reactant for reaction 1 and trace its carbon chain through the interrelated reactions on the summary diagram. For reaction 10a, start with 2-butanol as the reactant and trace the interrelated compounds through reactions 10b and 11-13.

(3) Identify the reaction; list all terms from the reference list that are appropriate.

If you are "rusty" on any reaction, return to the appropriate chapter and review the specifics of the reaction.

1. alcohol ⟶ alkene (use 1-butanol)

Your answer:

Solution:

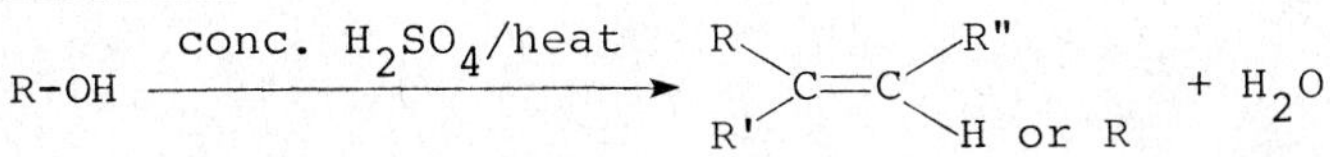

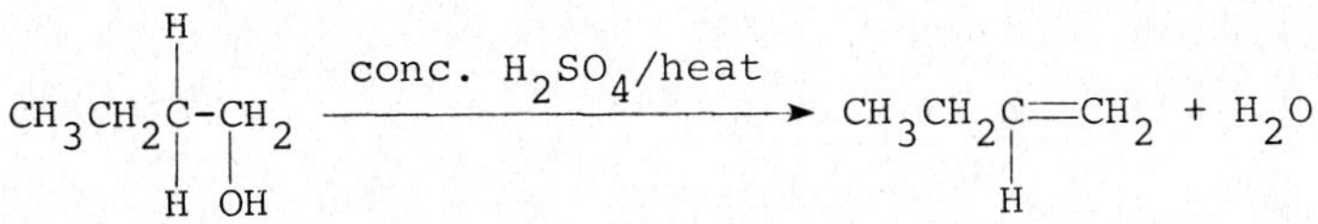

terms: dehydration, catalyzed reaction

---

2. alkene ⟶ alkane

Your answer:

Solution:

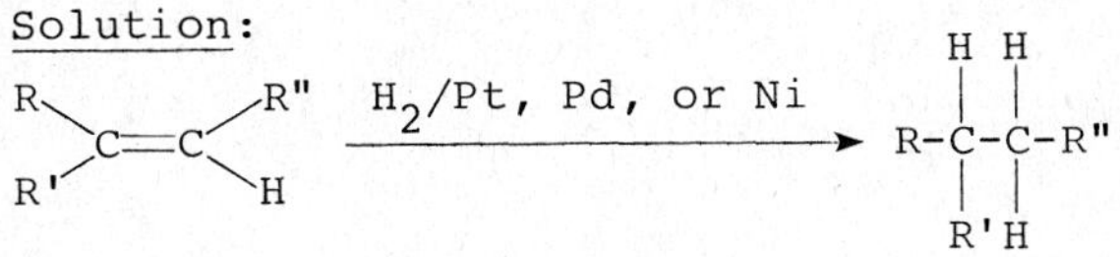

$$CH_3CH_2CH{=}CH_2 \xrightarrow{H_2/Ni} CH_3CH_2CH_2CH_3$$

terms: reduction, hydrogenation, catalyzed reaction, saturation

---

3. alkyne ⟶ alkane (use 2-butyne)

Your answer:

Solution:

$$R\text{-}C{\equiv}C\text{-}H\text{(or R)} \xrightarrow{\text{excess } H_2/\text{Pt, Pd, or Ni}} R\text{-}CH_2CH_3 \quad \text{(or R-H)}$$

$$CH_3C{\equiv}C\text{-}CH_3 \xrightarrow{\text{excess } H_2/\text{Pt}} CH_3CH_2CH_2CH_3$$

terms: reduction, hydrogenation, catalyzed reaction, saturation,

---

4. alkane ⟶ haloalkane (alkyl halide)   (use butane)

Your answer:

Solution:

$$R\text{-}H \xrightarrow{Cl_2\text{(or } Br_2\text{)/uv light or sunlight}} R\text{-}Cl \text{ (or R-Br)} + HCl \text{ (or HBr)}$$

$$CH_3CH_2CH_2CH_3 \xrightarrow{Cl_2/\text{sunlight}} C_4H_9Cl + HCl$$

The specific isomer is not designated because principles concerning its prediction are not usually covered.

terms: substitution for H in an alkyl

---

5. alkyne ⟶ haloalkane (alkyl halide)

Your answer:

Solution:

$$R\text{-}C{\equiv}C\text{-}H\text{(or R)} \xrightarrow{Cl_2 \text{ (or } Br_2\text{)}} R\text{-}CCl_2\text{-}CHCl_2 \text{ (or } R\text{-}CBr_2\text{-}CHBr_2\text{)}$$

$$\xrightarrow{HCl \text{ (or HBr)}} R\text{-}CCl_2\text{-}CH_3 \text{ (or } R\text{-}CBr_2\text{-}CH_3\text{)}$$

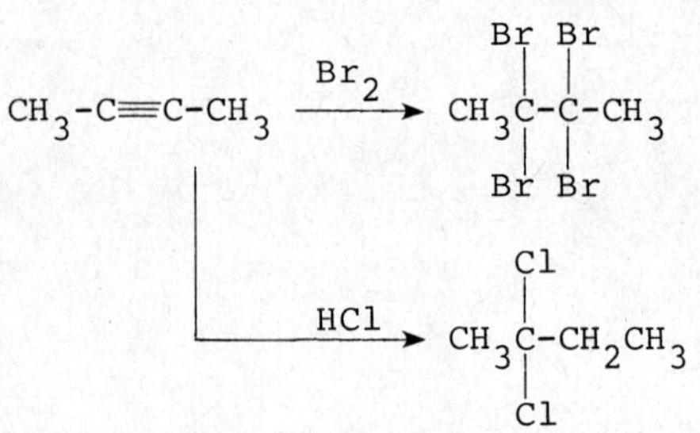

terms: electrophilic addition, carbocation intermediate

6. alkene → haloalkane (alkyl halide)

Your answer:

<u>Solution:</u>

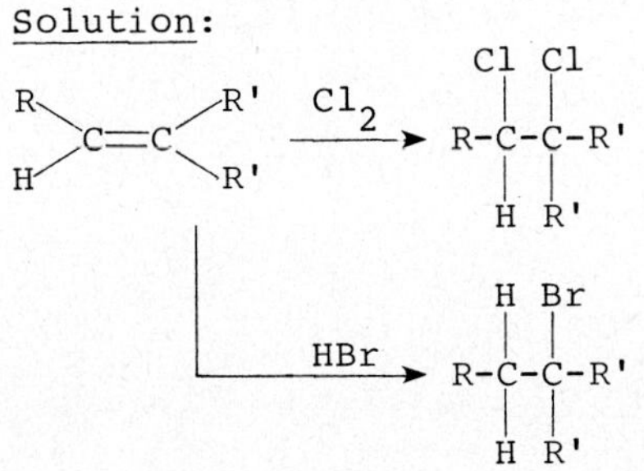

terms: electrophilic addition, carbocation intermediate

---

7. alkene → alcohol (use 1-butene)

Your answer:

<u>Solution:</u>

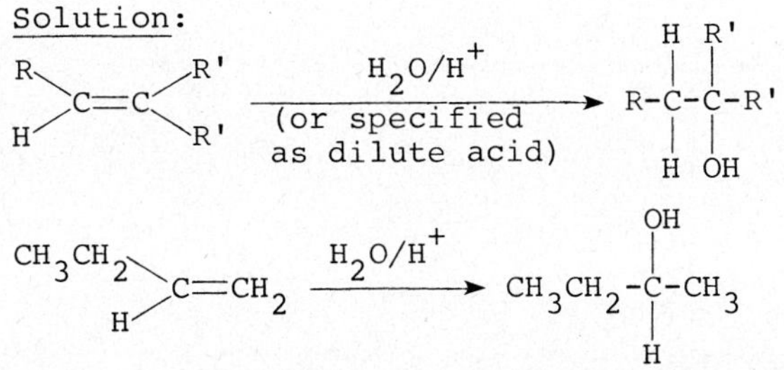

terms: electrophilic addition, carbocation intermediate, hydration, catalyzed reaction

---

8. alcohol → alkyl halide

Your answer:

<u>Solution:</u>

R-OH $\xrightarrow{PCl_3,\ PBr_5,\ SOCl_2}$ R-Br (or R-Cl)

R-OH $\xrightarrow{HCl/ZnCl_2\text{(Lucas reagent)}}$ R-Cl (3° fast, 2° slow, 1° very slow)

terms: nucleophilic substitution, carbocation intermediate (in Lucas reaction)

---

9. alkene ⟶ aldehyde and/or ketone (use 1-butene)

Your answer:

Solution:

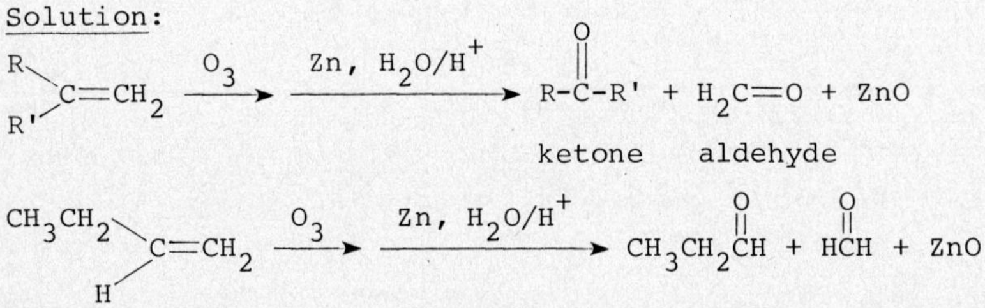

terms: oxidation, ozonolysis

---

10. (a) alcohol ⟶ ketone (use 2-butanol)

Your answer:

Solution:

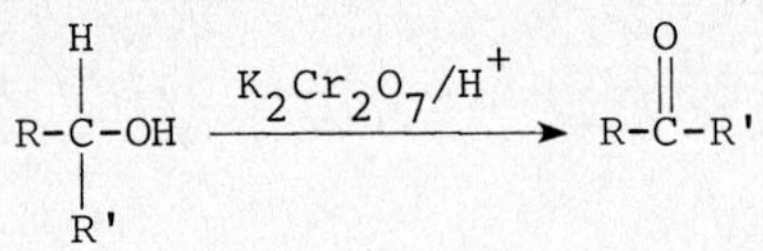

2° alcohol (only) gives a ketone.

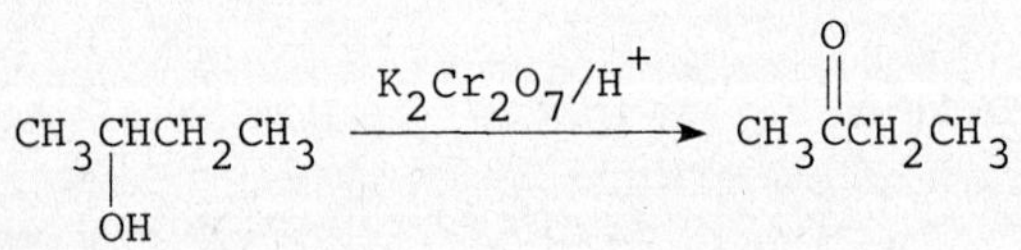

terms: oxidation (dichromate)

(b) ketone (or aldehyde) ⟶ alcohol

Your answer:

Solution:

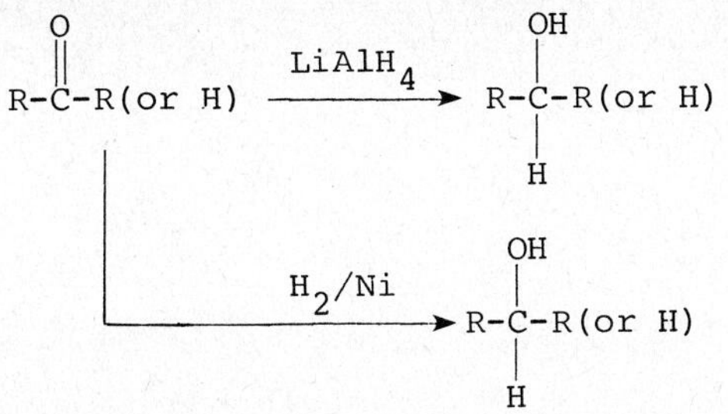

2° alcohol from ketone
1° alcohol from aldehyde

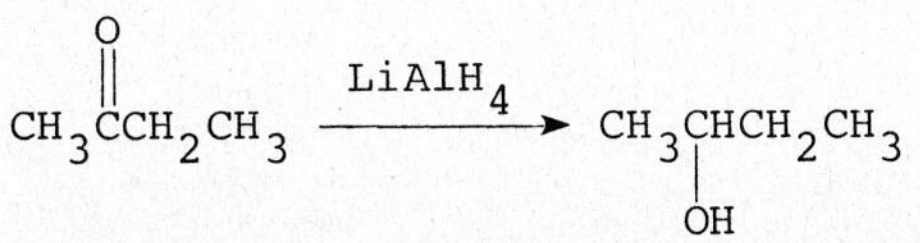

terms: reduction, hydrogenation, catalyzed reaction (H_2/Ni)

---

11. ketone (or aldehyde) $\xrightarrow{\text{amine}}$ imine (Schiff's base) (use methylamine and butanone)

Your answer:

Solution:

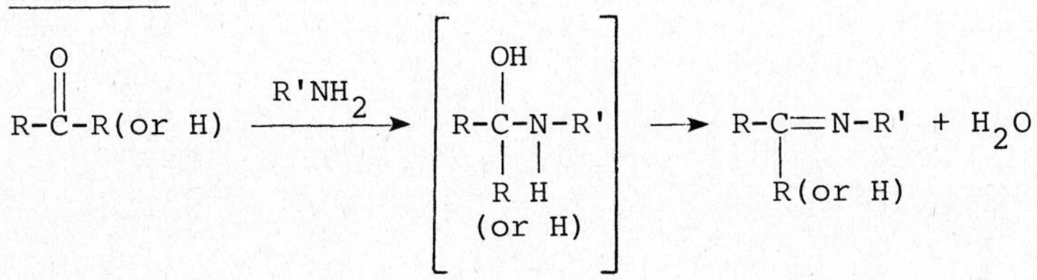

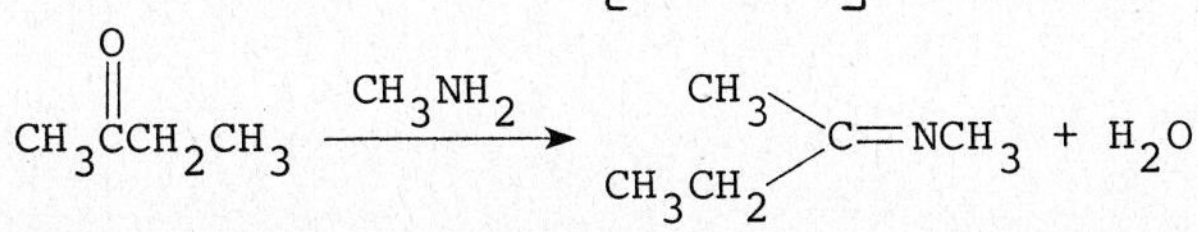

terms: nucleophilic addition to a carbonyl, followed by dehydration

---

12. ketone (or aldehyde) $\rightleftharpoons$ ketone (or aldehyde) hydrate

Your answer:

Solution:

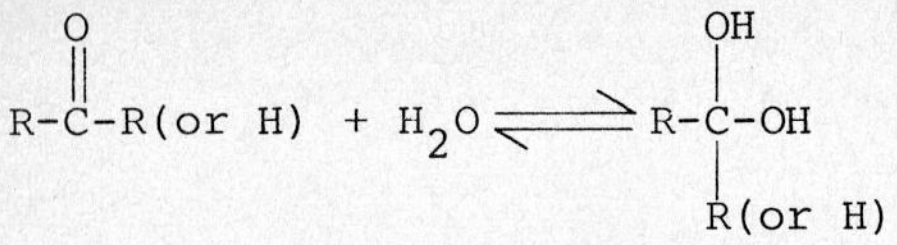

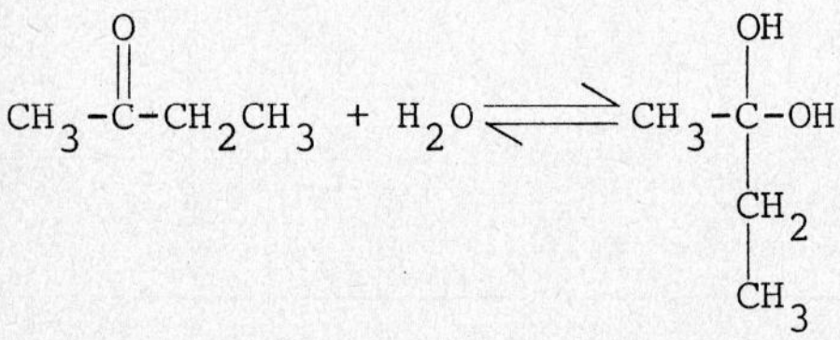

Hydrates, with the exception of the hydrate of formaldehyde, are quite unstable.

terms: nucleophilic addition, hydration

---

13. ketone (or aldehyde) $\rightleftharpoons$ hemiketal (or hemiacetal) or ketal (or acetal)

Your answer:

Solution:

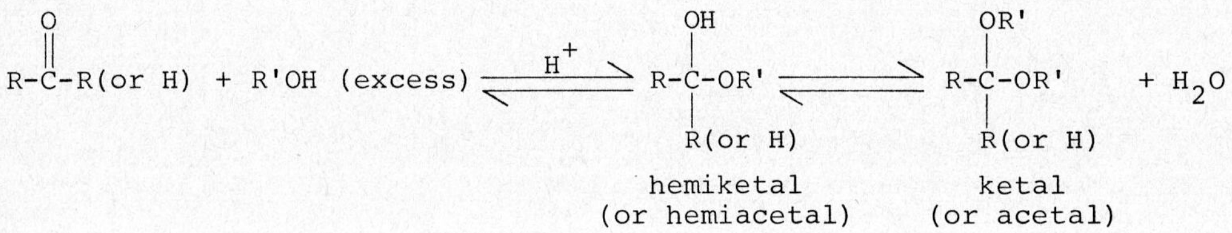

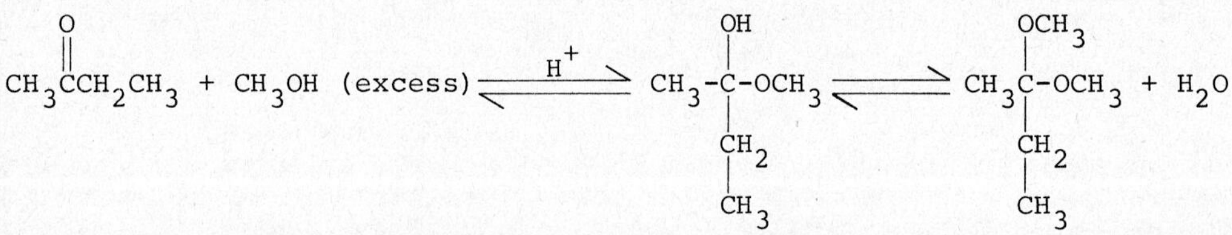

terms: nucleophilic addition, dehydration, catalyzed reaction

---

14. aldehyde $\longrightarrow$ carboxylic acid (use butanal)

Your answer:

Solution:

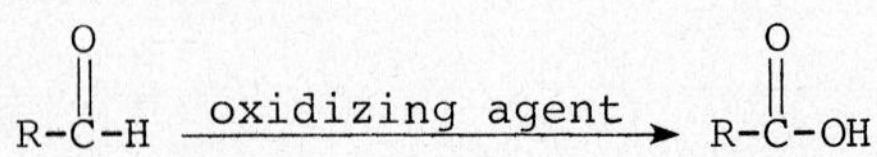

Note: Oxidizing agents are $K_2Cr_2O_7/H^+$, which produces RCO_2H from aldehyde; $Ag(NH_3)_2^+$

(Tollens'), which produces RCO_2^- + Ag^0; and Cu^{2+}/OH^- (Benedict's or Fehling's), which produces RCO_2^- + Cu_2O.

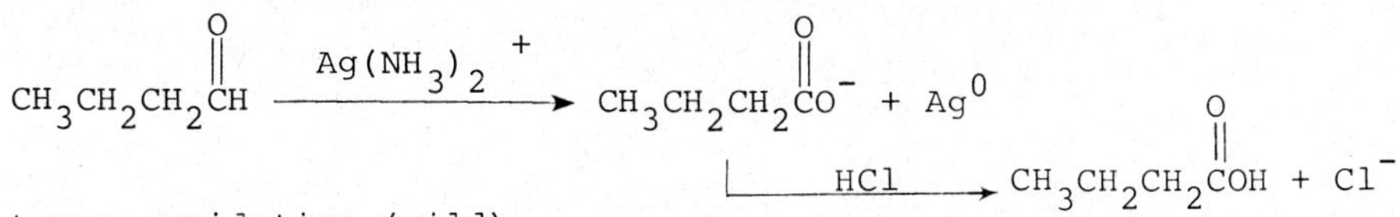

terms: oxidation (mild)

---

15. alkene ⟶ diol (glycol) (use 1-butene)

Your answer:

Solution:

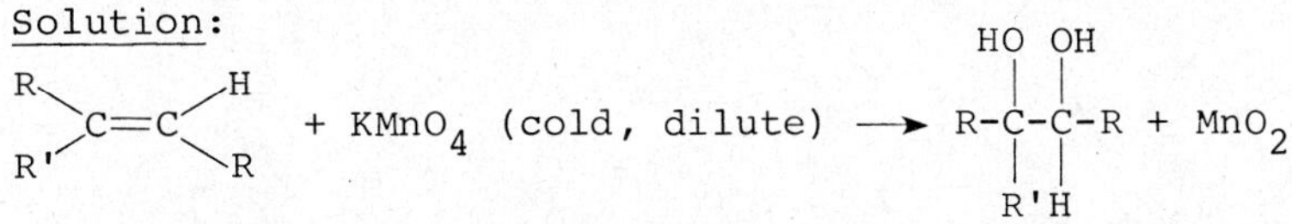

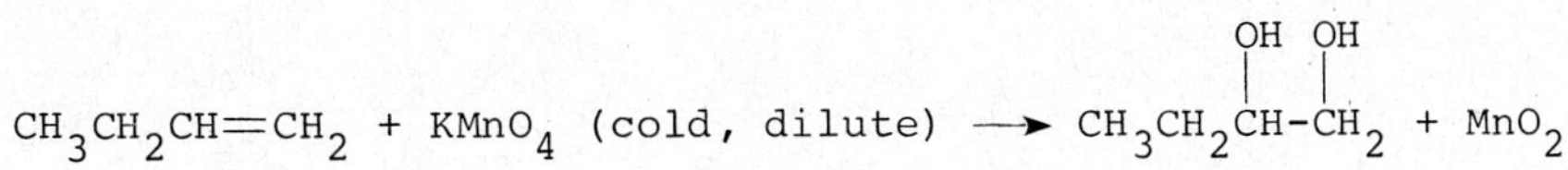

terms: oxidation (Baeyer's test)

---

16. alcohol ⟶ carboxylic acid (use 1-butanol)

Your answer:

Solution:

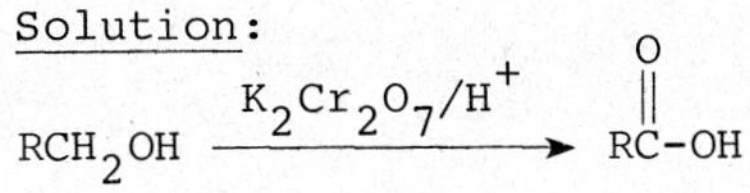

(only 1° alcohols)

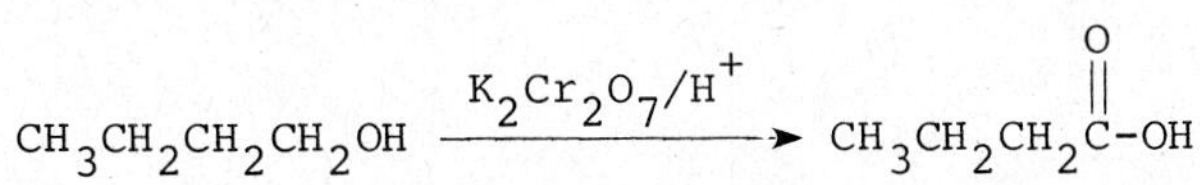

terms: oxidation (dichromate)

---

17. (a) carboxylic acid ⟶ salt (use butanoic acid)

Your answer:

Solution:

$$\mathrm{R{-}\overset{\overset{O}{\|}}{C}{-}OH + NaOH\ (or\ NH_3) \longrightarrow R{-}\overset{\overset{O}{\|}}{C}{-}O^-\ Na^+\ (or\ R{-}\overset{\overset{O}{\|}}{C}{-}O^-\ NH_4{}^+) + H_2O}$$

(or other base)

$$\mathrm{CH_3CH_2CH_2CO_2H + R{-}NH_2 \longrightarrow CH_3CH_2CH_2\overset{\overset{O}{\|}}{C}{-}O^-\ \underset{\underset{R}{|}}{NH_3{}^+}}$$

a substituted ammonium salt

terms: neutralization

(b) salt of carboxylic acid ⟶ carboxylic acid (use sodium butanoate)

Your answer:

Solution:

$$\mathrm{R{-}CO_2{}^-\ Na^+ \xrightarrow[\text{(or other strong acid)}]{HCl} R{-}CO_2H}$$

$$\mathrm{CH_3CH_2CH_2\overset{\overset{O}{\|}}{C}O^-\ Na^+ \xrightarrow{HCl} CH_3CH_2CH_2\overset{\overset{O}{\|}}{C}OH + NaCl}$$

terms: neutralization

---

18. (a) carboxylic acid ⟶ amide

Your answer:

Solution:

$$\mathrm{R\overset{\overset{O}{\|}}{C}{-}OH \xrightarrow{NH_3} R\overset{\overset{O}{\|}}{C}O^-\ NH_4{}^+ \xrightarrow{heat} R\overset{\overset{O}{\|}}{C}{-}NH_2 + H_2O}$$

$$\mathrm{CH_3CH_2CH_2\overset{\overset{O}{\|}}{C}{-}OH \xrightarrow{NH_3} CH_3CH_2CH_2\overset{\overset{O}{\|}}{C}O^-\ NH_4{}^+ \xrightarrow{heat} CH_3CH_2CH_2\overset{\overset{O}{\|}}{C}NH_2 + H_2O}$$

terms: neutralization, followed by dehydration

(b) amide ⟶ acid

Your answer:

Solution:

$$R\text{-}\overset{O}{\overset{\|}{C}}\text{-}NH_2 + H_2O \xrightarrow{H^+} \text{N.R. at room temperature}$$

However, if the amide is heated for several hours in the presence of concentrated HCl, the amide is hydrolyzed to $RCO_2H + NH_4Cl$. Also, hydrolysis of amides is very rapid in the presence of specific enzymes.

---

19. (a) carboxylic acid → acyl halide

Your answer:

Solution:

$$R\text{-}CO_2H + PCl_3 \text{ (or } PCl_5 \text{ or } SOCl_2\text{)} \longrightarrow R\text{-}\overset{O}{\overset{\|}{C}}\text{-}Cl$$

$$CH_3CH_2CH_2\overset{O}{\overset{\|}{C}}\text{-}OH + PCl_3 \longrightarrow CH_3CH_2CH_2\overset{O}{\overset{\|}{C}}\text{-}Cl$$

terms: nucleophilic substitution

(b) acyl halide → carboxylic acid

Your answer:

Solution:

$$R\text{-}\overset{O}{\overset{\|}{C}}\text{-}Cl \xrightarrow{H_2O} R\text{-}\overset{O}{\overset{\|}{C}}\text{-}OH + HCl$$

$$CH_3CH_2CH_2\overset{O}{\overset{\|}{C}}\text{-}Cl \xrightarrow{H_2O} CH_3CH_2CH_2\overset{O}{\overset{\|}{C}}OH + HCl$$

terms: nucleophilic substitution

---

20. (a) carboxylic acid → acid anhydride

Your answer:

Solution:

$$R\overset{\overset{O}{\|}}{C}\text{-}OH + R\text{-}\overset{\overset{O}{\|}}{C}\text{-}Cl \longrightarrow R\text{-}\overset{\overset{O}{\|}}{C}\text{-}O\text{-}\overset{\overset{O}{\|}}{C}\text{-}R + HCl$$

$$CH_3CH_2CH_2\overset{\overset{O}{\|}}{C}OH + CH_3CH_2CH_2\overset{\overset{O}{\|}}{C}\text{-}Cl \longrightarrow CH_3CH_2CH_2\overset{\overset{O}{\|}}{C}\text{-}O\text{-}\overset{\overset{O}{\|}}{C}CH_2CH_2CH_3 + HCl$$

terms: nucleophilic substitution

(b) acid anhydride ⟶ carboxylic acid

Your answer:

Solution:

$$R\text{-}\overset{\overset{O}{\|}}{C}\text{-}O\text{-}\overset{\overset{O}{\|}}{C}\text{-}R + H_2O \xrightarrow{H^+} 2\ R\text{-}\overset{\overset{O}{\|}}{C}\text{-}OH$$

$$CH_3CH_2CH_2\text{-}\overset{\overset{O}{\|}}{C}\text{-}O\text{-}\overset{\overset{O}{\|}}{C}\text{-}CH_2CH_2CH_3 + H_2O \xrightarrow{H^+} 2\ CH_3CH_2CH_2CO_2H$$

terms: acid hydrolysis, catalyzed reaction, nucleophilic substitution

---

21. acyl halide ⟶ amide

Your answer:

Solution:

$$R\text{-}\overset{\overset{O}{\|}}{C}\text{-}Cl + R'NH_2\ \text{(amine)} \longrightarrow R\text{-}\overset{\overset{O}{\|}}{C}\text{-}NHR' + HCl$$

$$CH_3CH_2CH_2\overset{\overset{O}{\|}}{C}\text{-}Cl + CH_3NH_2 \longrightarrow CH_3CH_2CH_2\overset{\overset{O}{\|}}{C}\text{-}NHCH_3 + HCl$$

terms: nucleophilic substitution

---

22. acid anhydride ⟶ amide

Your answer:

Solution:

$$R\text{-}C(=O)\text{-}O\text{-}C(=O)\text{-}R + R'NH_2 \longrightarrow R\text{-}C(=O)\text{-}NHR' + R\text{-}C(=O)\text{-}OH$$

$$CH_3CH_2CH_2C(=O)\text{-}O\text{-}C(=O)CH_2CH_2CH_3 + CH_3NH_2 \longrightarrow CH_3CH_2CH_2C(=O)NHCH_3 + CH_3CH_2CH_2CO_2H$$

terms: nucleophilic substitution

---

23. acyl halide ⟶ ester

Your answer:

Solution:

$$R\text{-}C(=O)\text{-}Cl + R'OH \longrightarrow R\text{-}C(=O)\text{-}OR' + HCl$$

$$CH_3CH_2CH_2C(=O)\text{-}Cl + CH_3OH \longrightarrow CH_3CH_2CH_2C(=O)OCH_3 + HCl$$

terms: nucleophilic substitution

---

24. acid anhydride ⟶ ester

Your answer:

Solution:

$$R\text{-}C(=O)\text{-}O\text{-}C(=O)\text{-}R + R'OH \longrightarrow R\text{-}C(=O)\text{-}OR' + R\text{-}CO_2H$$

$$(CH_3CH_2CH_2C(=O))_2O + CH_3OH \longrightarrow CH_3CH_2CH_2C(=O)OCH_3 + CH_3CH_2CH_2CO_2H$$

terms: nucleophilic substitution

---

25. ester ⟶ carboxylic acid

Your answer:

Solution:

$$R\text{-}\overset{\overset{O}{\|}}{C}\text{-}OR' + NaOH \longrightarrow R\text{-}\overset{\overset{O}{\|}}{C}O^- \; Na^+ + R'OH$$

$$\xrightarrow{HCl} R\text{-}CO_2H + NaCl$$

$$CH_3CH_2CH_2\overset{\overset{O}{\|}}{C}OCH_3 + NaOH \longrightarrow CH_3CH_2CH_2\overset{\overset{O}{\|}}{C}O^- \; Na^+ + CH_3OH$$

$$\xrightarrow{HCl} CH_3CH_2CH_2CO_2H + NaCl$$

terms: basic hydrolysis, saponification (first reaction)

---

26. alcohol + carboxylic acid → ester

Your answer:

Solution:

$$R'OH + RCO_2H \underset{}{\overset{H^+}{\rightleftharpoons}} R\text{-}CO_2R' + H_2O \qquad \text{(equilibrium reaction)}$$

$$CH_3OH + CH_3CH_2CH_2\overset{\overset{O}{\|}}{C}OH \overset{H^+}{\rightleftharpoons} CH_3CH_2CH_2\overset{\overset{O}{\|}}{C}OCH_3 + H_2O$$

terms: catalyzed reaction, esterification

---

27. amine → alcohol (use methylamine, dimethylamine, trimethylamine)

Your answer:

Solution:

1° amine: $R\text{-}NH_2 + HNO_2 \longrightarrow R\text{-}OH + N_2(g) + H_2O$

Alcohols are formed only from primary amines; contrast this reaction to the reactions of secondary and tertiary amines.

2° amine: $\underset{\underset{R}{|}}{R\text{-}NH} + HNO_2 \longrightarrow \underset{\underset{R}{|}}{R\text{-}N\text{-}NO} + H_2O$

a nitrosamine

3° amine: $R_3N + HNO_2 \longrightarrow \underset{\underset{H}{|}}{R_3N^+} \; NO_2^-$

a salt

$CH_3NH_2 + HNO_2 \longrightarrow CH_3OH + N_2(g) + H_2O$

$(CH_3)_2NH + HNO_2 \longrightarrow (CH_3)_2N\text{-}NO + H_2O$

$(CH_3)_3N + HNO_2 \longrightarrow (CH_3)_3\overset{+}{N}(H)\ NO_2^-$

terms: no appropriate terms in the reference list (This is classified as a test for amines.)

---

28. amine $\longrightarrow$ amine salt (use methylamine and acetic acid)

Your answer:

Solution:

RNH_2 (also 2°, 3°) + acid $\longrightarrow$ RNH_3^+ (anion)$^-$

$CH_3NH_2 + CH_3\overset{O}{\overset{\|}{C}}OH \longrightarrow CH_3NH_3^+\ \ ^-O\overset{O}{\overset{\|}{C}}CH_3$

terms: neutralization

II. Biologically Important Compounds

Each of the following structures represents a compound of biological importance. Name the indicated functional groups, and list the reagents with which each can react. (This presumes that you could also write the structure of the product formed.)

1. Progesterone

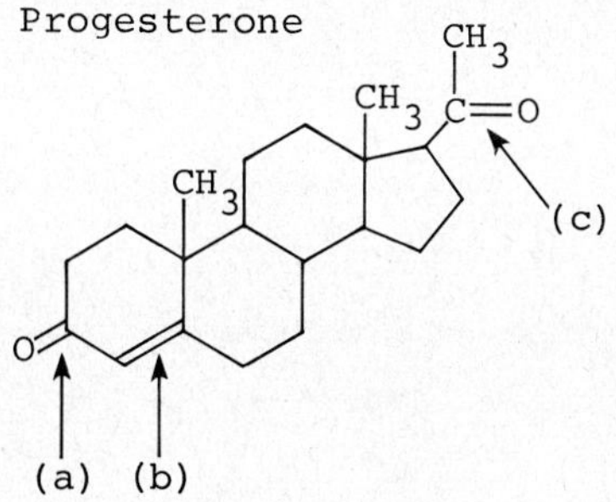

Your answer:

Solution:

Functional groups	Reacts with
(a) ketone	$LiAlH_4$, H_2/Pd, alcohols, amines, DNP*; not with oxidizing agents.
(b) carbon-carbon double bond	Br_2* or Cl_2 (dark), HCl, H_2O/H^+, H_2/Ni*, O_3 with Zn/HCl*.
(c) ketone	Same as for (a); also iodoform* test because $CH_3\overset{\overset{O}{\|}}{C}$- structure is present.

*Tests for identification of the functional group.

---

2. Aspirin

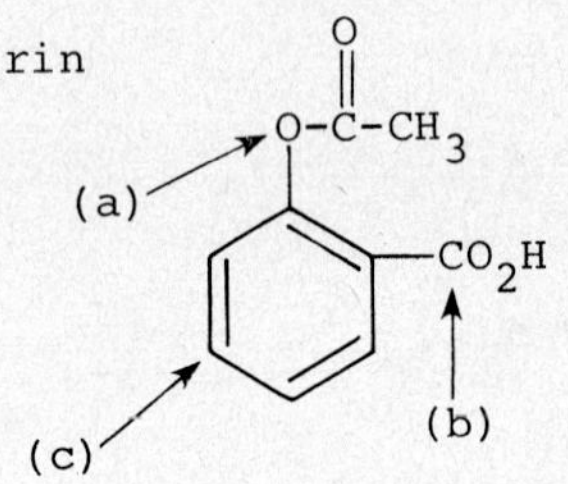

Your answer:

Solution:

Functional groups	Reacts with
(a) ester	H_2O/H^+ (equilibrium mixture), NaOH/heat* (saponification)
(b) carboxylic acid	NaOH*, PCl_3, PBr_5, $SOCl_2$, alcohol, amine/heat
(c) aromatic ring	$Br_2/FeBr_3$*, other test reagents for aromatic rings, such as $R\text{-}Cl/AlCl_3$ and HNO_3/H_2SO_4

*Tests for identification of the functional group.

---

3. DDT (insecticide that has been banned)

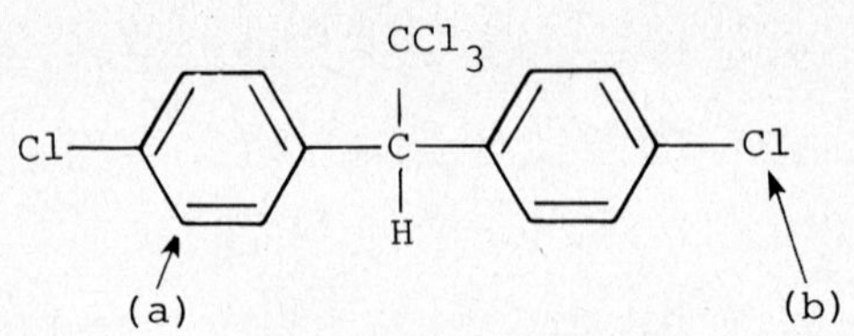

Your answer:

Solution:

Functional groups	Reacts with
(a) aromatic ring	See reagents listed in previous question. No products of these reactions are less toxic than the original compound.
(b) aromatic (aryl) halide	No reactions have been covered in this textbook presentation.

This compound is very stable; it is not easily oxidized or decomposed.

---

4. Tyrosine (an amino acid)

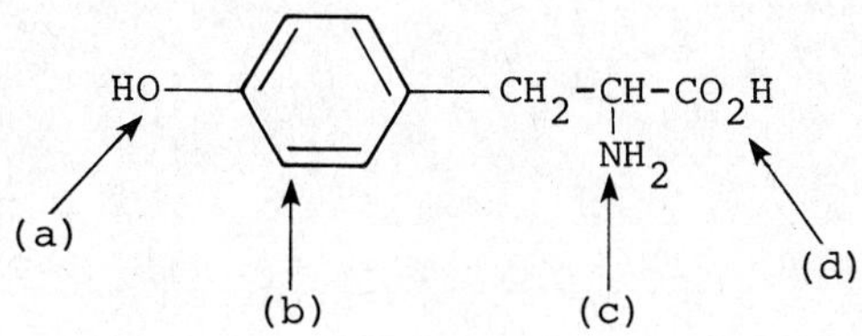

Your answer:

Solution:

Functional groups	Reacts with
(a) phenolic group (aromatic alcohol)	Na*, NaOH, carboxylic acids, acyl chloride, anhydride
(b) aromatic ring	Same as given in 2 (aspirin)
(c) 1° amine	HNO_2*, acids, aldehydes or ketones, carboxylic acids and acid derivatives
(d) carboxylic acid	Same as given in 2 (aspirin).

*Tests for identification of the functional group.

5. α-D-maltose (a carbohydrate)

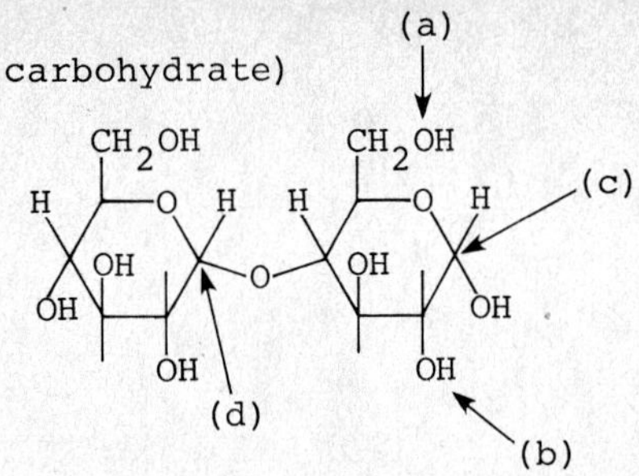

Your answer:

Solution:

Functional groups	Reacts with
(a) 1° alcohol	Na*, $HCl/ZnCl_2$*, PBr_3, $K_2Cr_2O_7/H^+$*, R-CHO, RCOR
(b) 2° alcohol	Same as for 1° alcohol except different results from $HCl/ZnCl_2$ and $K_2Cr_2O_7/H^+$
(c) hemiacetal (-OH and -OR groups on same carbon)	H_2O/H^+ hydrolyzes hemiacetal back to carbonyl and alcohol
(d) acetal (two -OR groups on same C)	H_2O/H^+ hydrolyzes acetal back to carbonyl and alcohol.

*Tests for identification of the functional group.

22 Stereoisomerism

I. Types of Isomerism

A. STRUCTURAL ISOMERISM

Structural isomerism includes variations of how atoms are bonded to each other (branching of the carbon chain, the position of the functional group, and variations of the type of functional group).

1. Draw the structural formula of 2-aminopropanoic acid. Draw structures and write the names of several structural isomers of this acid.

Your answer:

Solution:

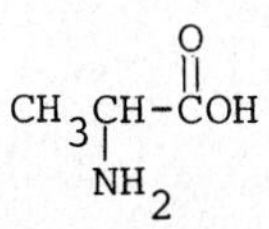

2-aminopropanoic acid

Structural isomers (all possible structures with the molecular formula C_3H_7N) include

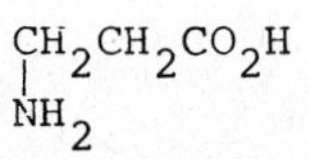

3-aminopropanoic acid

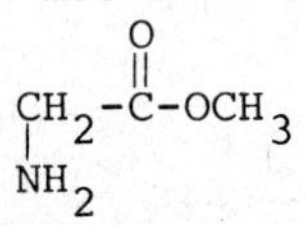

methyl 2-aminoethanoate

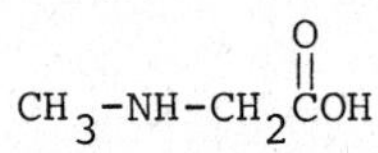

2-(N-methylamino) ethanoic acid

Other structural isomers would include compounds such as aminohydroxyketones or hydroxyamides.

---

2. How would you estimate the differences in the chemical and physical properties of the structures in the previous question?

Your answer:

Solution: Structural isomers are different compounds; they have different physical and chemical properties. The differences in the chemical properties are more pronounced when the functional groups actually differ.

B. STEREOISOMERISM

Stereoisomers have the same sequence of atoms bonded to each other, but the ways in which the atoms are arranged in space differ (i.e., cis-trans or geometric isomers and optical isomers.

1. Construct a model (or visualize one) of 2-aminopropanoic acid (known as alanine, an amino acid). Next construct a model of alanine that is a "mirror image" (as your left hand is a "mirror image" of your right hand) to your first model. Hold the models so that the carboxyl carbon (the most oxidized carbon) is at the top pointing away from you and the methyl group is also pointing away from you. This model represents the conventional position used in the study of stereoisomers. Draw the models that you constructed.

Your answer:

Solution:

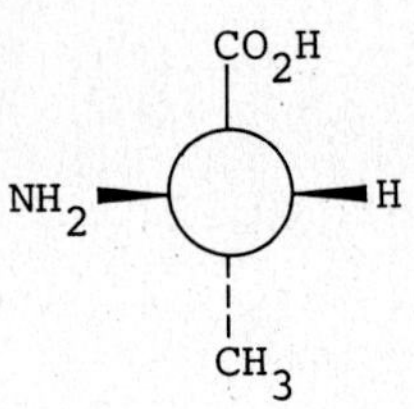

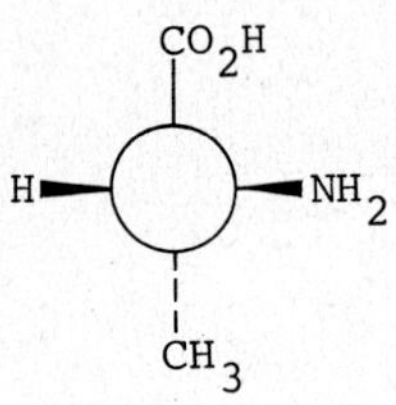

Projection formulas can be drawn that represent each model pushed flat against the paper.

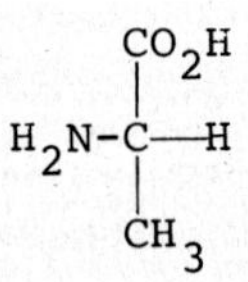

A

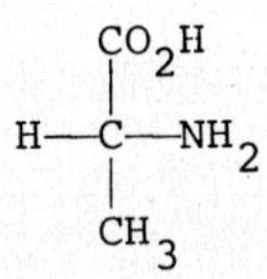

B

---

2. Test your models and drawings for superimposability. If two structures can be superimposed, they represent the same molecule. To test your projection formulas, you can pretend to cut them from the paper and slide the structure (in the same plane) onto the other formula, rotating the structure 180° if necessary. However, do not flip the structure over. Compare structures A and B each with structure C.

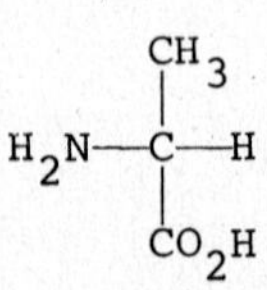

C

Which are superimposable and which are not?

Your answer:

Solution:

A and B are not superimposable (just as right and left hands are not superimposable). B and C are superimposable and therefore represent the same compound.

---

3. What term is used to describe nonsuperimposable mirror images, such as isomers A and B?

Your answer:

Solution: enantiomers

Enantiomers are optical isomers (a type of stereoisomer) in which the same sequence of atoms is bonded to each other (in each molecule), but the molecules differ in the way in which those atoms are arranged in three-dimensional space. Asymmetry in optical isomers is often referred to as a "left-handedness" or a "right-handedness."

---

4. What do enantiomers have in common and how do they differ?

Your answer:

Solution: Enantiomers have the same molecular formula (as do all sets of isomers), they have the same sequence of atoms bonded together, and they have the same physical properties with the exception of one: they differ in their response to plane-polarized light. One member of the enantiomeric pair rotates plane-polarized light to the right (+), and the other rotates plane-polarized light to an equal extent to the left (-). Enantiomers have identical chemical properties unless they are reacted with another asymmetric molecule.

---

5. Geometric isomerism is stereoisomerism occurring because of hindered rotation of bonds (which is due to the presence of double bonds or rings). Identify the isomeric relationships among the three structures shown.

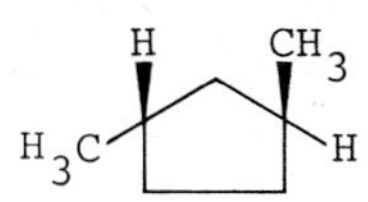

X

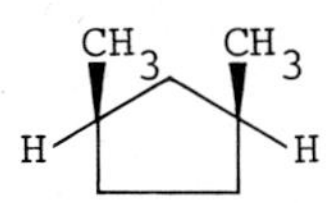

Y

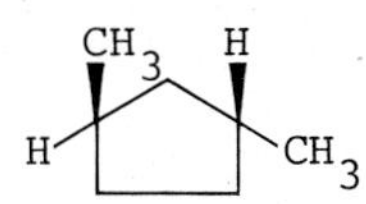

Z

What type of isomerism is exhibited by X and Y? Compare the properties of X and Y. What type of isomerism is exhibited by X and Z? Compare the properties of these two compounds.

Your answer:

Solution: X and Y are, respectively, trans- and cis-1,3-dimethylcyclopentane. They are geometric (or cis-trans) isomers and differ in the spatial arrangement of their atoms. The ring restricts bond rotation and makes these isomers possible. The two compounds have different physical and chemical properties and are called diastereomers. Y and Z (also trans-1,3-dimethylcyclopentane) are another pair of diastereomers. X and Z are enantiomers (optical isomers) because they are nonsuperimposable mirror images. Enantiomers will differ in their response to plane-polarized light; one isomer will rotate the beam of the light to the right, the other isomer will rotate the beam of light to the left. Y does not rotate the plane of polarized light because it is a symmetrical molecule; it is called a meso form.

II. Chiral (Asymmetric) Carbons

1. What structural feature determines whether or not a compound will have an enantiomer?

Your answer:

Solution: Asymmetry (lack of symmetry) at one or more carbons in a molecule provides a structure that can have an enantiomer. A carbon that is bonded to four unique (different) groups is called "chiral" or "asymmetric."

2. For each of the following compounds, draw the structural formula and mark each chiral carbon with an asterisk.

(a) CH_2OH-CHOH-CHOH-CHO

Your answer:

Solution:

You can show either of these structural formulas or the mirror image of either formula.

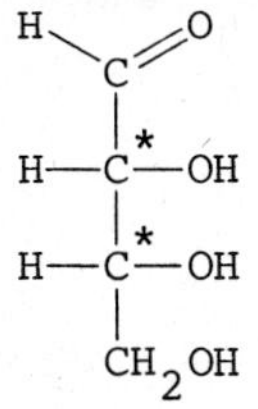

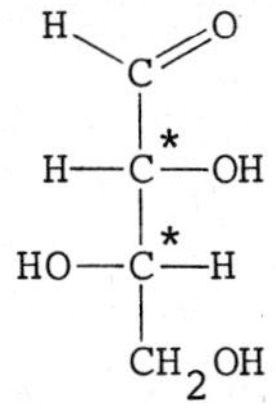

(b) 2-amino-3-hydroxypropanoic acid

Your answer:

Solution:

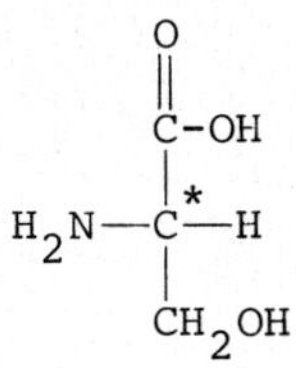

or

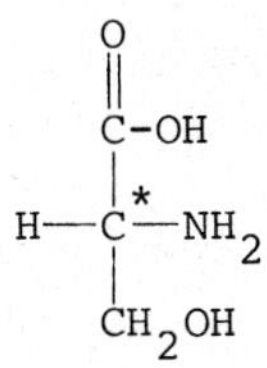

Have you wondered why the substituents on the top and bottom carbons are not written as "mirror" images? You may need to try this on a model to believe it, but the top or bottom carbons can be rotated without changing any of the "handedness" of the chiral carbon.

(c) 2-chloro-3-hydroxybutanal

Your answer:

Solution:

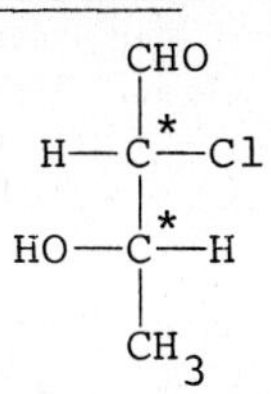

or

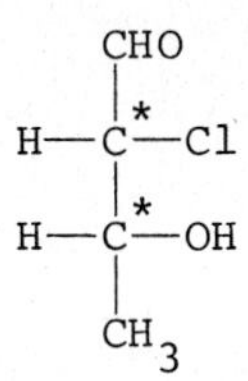

or the mirror images of either of these isomers

(d) trans-1,4-dichlorocyclohexane

Your answer:

Solution:

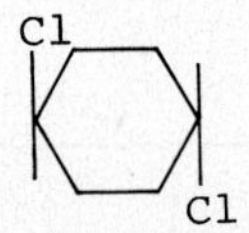

At first appearance, each carbon containing a chlorine might be considered chiral, which implies that the carbon has four unique groups bonded to it. However, if you start at one of these carbons and go around the ring in a clockwise direction, you will "touch" the same atoms in the same order that you would "touch" them if you went around the ring counterclockwise. Thus, there are not actually four unique groups bonded to either of these carbons, and so neither carbon is chiral.

(e) cis-1,3-dichlorocyclopentane

Your answer:

Solution:

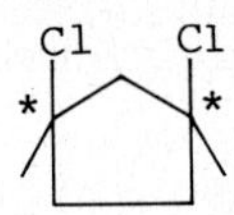

Proceeding around the ring from either marked carbon, in one direction you "meet" $-CH_2-$, then -CHCl-; but proceeding around the ring in the other direction, you "meet" $-CH_2-$, then $-CH_2-$. Thus, there are four unique groups on each marked carbon atom. However, the chiral carbons in this compound are equivalent, in contrast to the "different" chiral carbons in the compound of 2(c). Equivalent chiral carbons will be discussed further in the next section. Another way of approaching this question is to look for symmetry in the molecule. This molecule has a plane of symmetry.

III. Optical Activity

1. A certain compound is said to be "optically active" and have $[\alpha] = -33^{\circ}$ (specific rotation). What do these terms mean?

Your answer:

Solution: When an optically active compound is dissolved in water and the solution is placed in a polarimeter, the molecules rotate the plane of polarized light. A solution containing molecules with a specific rotation of -33° will rotate the plane of polarized light through an angle of 33° in a negative (counterclockwise) direction; such a compound is said to be levorotatory; its rotation is designated as (-) or (l). Its enantiomer is dextrorotatory with a specific rotation of $[\alpha] = +33^{o}$; its rotation is designated as (+) or (d).

---

2. Which of the compounds in problem 2 of Section II is(are) optically active?

Your answer:

Solution: If a structure is optically active, you will be able to answer "yes" to each of these "test" questions.
(1) Is there at least one chiral carbon in the molecule?
(2) If there is more than one chiral carbon, are they "different" (not equivalent)?
Each of the four compounds in problem 2(a) (two structures are shown; each has a mirror image) is optically active. Each compound has two nonequivalent chiral carbons. That there are no more than four optical isomers is verified by the formula 2^n = the number of optical isomers (n = number of nonequivalent chiral carbons).
In problem 2(b) there are two optical isomers ($2^n = 2^1$); each of these structures is optically active.
In problem 2(c) each of the four indicated isomers (2^2) is optically active.

---

3. Why are compounds (d) and (e) in problem II-2 optically inactive?

Your answer:

Solution: A solution of an optically inactive compound will rotate plane-polarized light 0° in a polarimeter.
Compound II-2(d) is optically inactive because it contains no chiral carbons.
Compound II-2(e) is optically inactive because its chiral carbons are equivalent. Observe that there is a plane of symmetry through the molecule and that one half of the molecule is the mirror image of the other half. Such a structure is called a "meso" form. Another test to determine the equivalence of chiral carbons in a symmetrical molecule is the superimposability of the mirror image on the original structure.

A. OPTICAL ACTIVITY OF ENANTIOMERS

1. Draw structures for the two optical isomers of 2-methyl-1-butanol.

Your answer:

Solution:

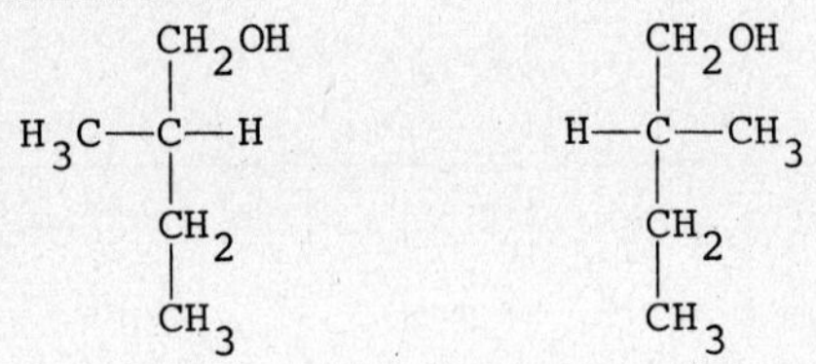

2. What term applies to the relationship between these two compounds?

Your answer:

Solution: One structure is the enantiomer of the other; they are members of an enantiomeric pair.

3. If the specific rotation of one enantiomer is $+5.9^{\circ}$ and the boiling point of that compound is $128^{\circ}C$, what are the specific rotation and the boiling point of the other enantiomer?

Your answer:

Solution: The specific rotation of the other enantiomer is -5.9°, and its boiling point is $128^{\circ}C$.

B. OPTICAL ACTIVITY OF A RACEMIC MIXTURE

1. When butanone reacts with $LiAlH_4$, an equimolar mixture of two optical isomers forms. Draw structures of these isomers. What term applies to this equimolar mixture? What is the specific rotation of the solution containing this mixture?

Your answer:

Solution:

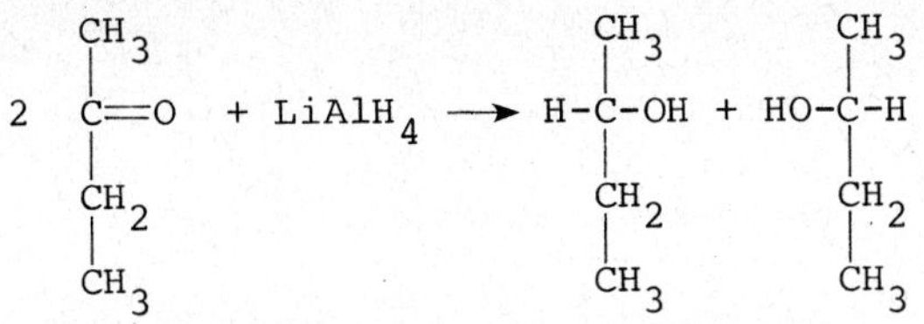

This mixture is called a racemic mixture.

The specific rotation of the solution containing a racemic mixture is 0° because the rotation of one enantiomer cancels the equal but opposite rotation of the other enantiomer.

IV. Prediction of the Number of Optical Isomers

1. How many optical isomers are possible for 2-bromo-3,4-dihydroxybutanal? Draw a projection formula for each isomer. Indicate which structures are enantiomers and which are diastereomers.

Your answer:

Solution:

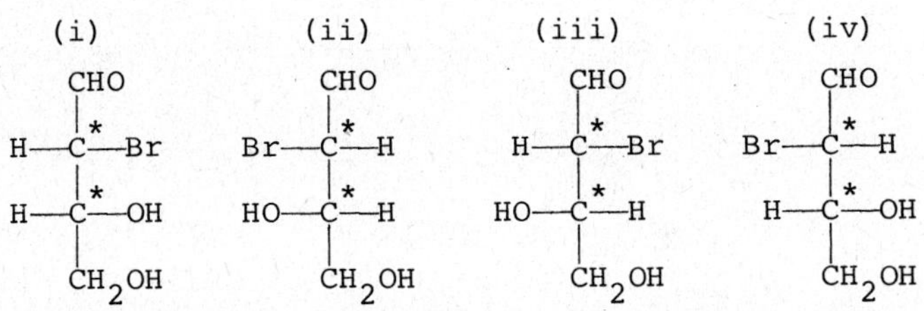

Let n = the number of nonequivalent chiral carbons. Then $2^n = 2^2 = 4$ enantiomers.

(i) and (ii) are enantiomers; (iii) and (iv) are enantiomers; (i) and (iii), (i) and (iv), (ii) and (iii), (ii) and (iv) are diasteriomers.

---

2. When D-erythrose is reduced by $LiAlH_4$, the product is optically inactive. Draw the structure(s) of the product and explain why a solution of the product does not rotate plane-polarized light. The structure of D-erythrose is

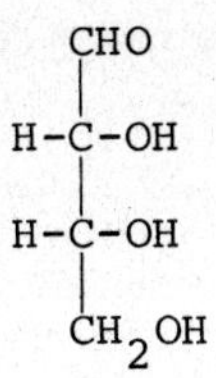

Your answer:

Solution:

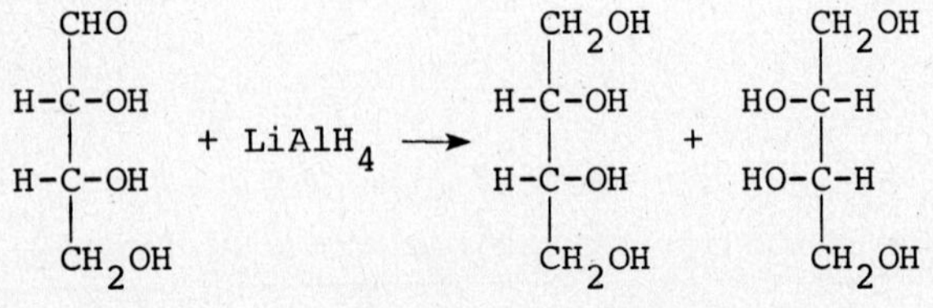

(iv) (v)

Structures (iv) and (v) are superimposable mirror images and therefore represent the same compound. This compound is meso; a mirror plane can be put across the molecule to show that the two halves are mirror images. This structure can also be recognized as a meso structure by the fact that its chiral carbons are equivalent. Whenever chiral carbons are equivalent, the 2^n formula cannot be used to predict the number of enantiomers.

V. Use of D and L to Designate Configuration, (+) and (–) to Designate Specific Rotation

1. D-glucose, D-fructose, D-galactose, and other D-sugars are utilized by our cells; L-sugars are not. L-amino acids are assimilated and are utilized by our bodies; D-amino acids are not. How do you differentiate a D-compound from an L-compound when looking at a projection formula?

Your answer:

Solution: A projection formula is written with the most oxidized carbon at the "top." If the -OH attached to the last chiral carbon is on the right side of the formula, the compound is said to have a D configuration; if the -OH attached to the last chiral carbon is on the left side, the compound is said to have an L configuration.

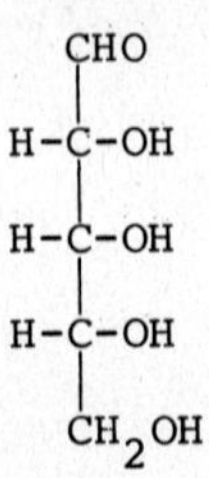

D-ribose

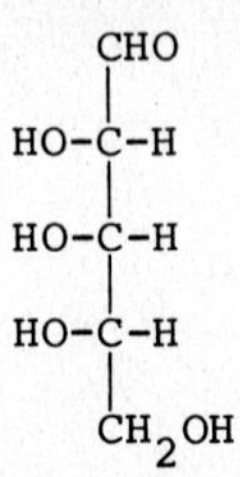

L-ribose

The $-NH_2$ is shown on the left for an L-amino acid and on the right for a D-amino acid.

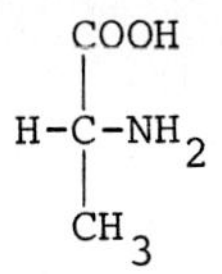

D-alanine

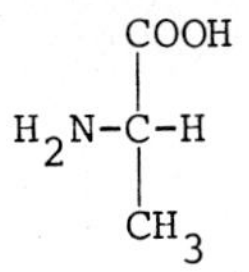

L-alanine

---

2. D-Ribose is levorotatory. Write its complete name (specifying the direction of rotation) and write the complete name for its enantiomer.

Your answer:

Solution: D-(-)-Ribose is levorotatory as indicated by the (-). Its enantiomer, L-(+)-ribose, is dextrorotatory.
It is very important that you realize that the configuration (D or L) is not related to the specific rotation (positive or negative); a D-compound can be either levorotatory or dextrorotatory—the specific rotation can only be determined by the use of a polarimeter.

---

23 Carbohydrates

I. Structure Recognition, Carbohydrate Nomenclature, and Terminology

For each structure shown, write the specific name of the compound and list all the appropriate terms from the following reference list.

monosaccharide	pentose	pyranose ring
disaccharide	hexose	an acetal
polysaccharide	α-anomer	a hemiacetal
phosphate ester	β-anomer	amylose
aldose	reducing sugar	amylopectin
ketose	nonreducing sugar	glycogen
tetrose	furanose ring	starch
		epimer

---

1. 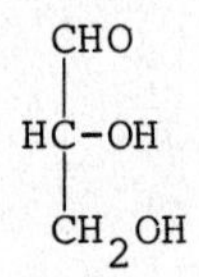

Your answer:

Solution: D-glyceraldehyde

terms: monosaccharide, aldose (contains an aldehyde), reducing sugar (the aldehyde group, which is easily oxidized, reduces Ag^+ to Ag^o in Tollens' test, or Cu^{2+} to Cu_2O in Benedict's or Fehling's test)

---

2. 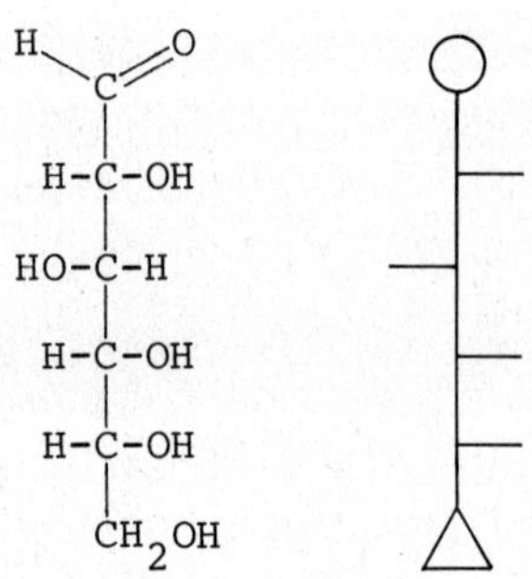

Draw the cyclic (hemiacetal) form, known as a Haworth projection, of the α-anomer for this sugar.

Your answer:

Solution: D-glucose (shown in a Fischer projection and in the abbreviated stick form; also referred to as the free-aldehyde or open-chain form)

terms: monosaccharide, aldose, hexose (a six-carbon sugar), reducing sugar

To draw the cyclic hemiacetal form (a Haworth projection):

The H of the OH on the fifth carbon adds to the carbonyl oxygen and the O of the same OH adds to the carbon of the carbonyl. Carbon 1 is now an asymmetric carbon, called an anomeric carbon. If its OH is down in the cyclic form, the anomer is α; if its OH is up in the cyclic form, the anomer is β.

When the ring is closed, the oxygen is shown in the top, right (northeast) corner of the ring, and the sixth carbon is above the ring for a D-sugar. OH's on the right in the open-chain form are shown down in the cyclic form. (Sometimes the OH's are omitted and only a line is shown to represent the OH.)

α-D-glucopyranose (α-D-glucose)

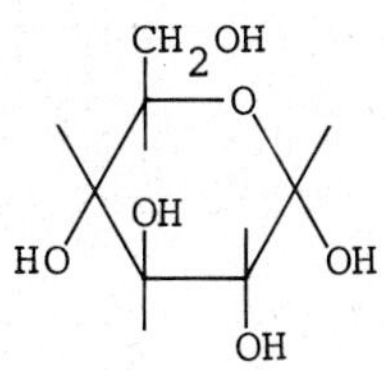

---

3. 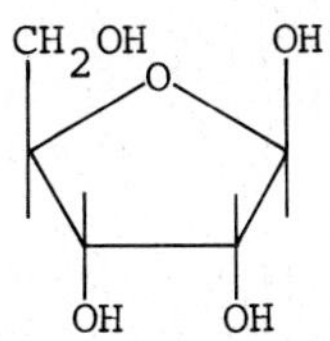

Draw the open-chain form as a Fischer projection and as a stick form.

Your answer:

Solution: β-D-ribofuranose (β-D-ribose) (shown in a Haworth projection)

terms: monosaccharide, aldose, pentose, β-anomer, reducing sugar, furanose ring, hemiacetal

open chain (free-aldehyde) form

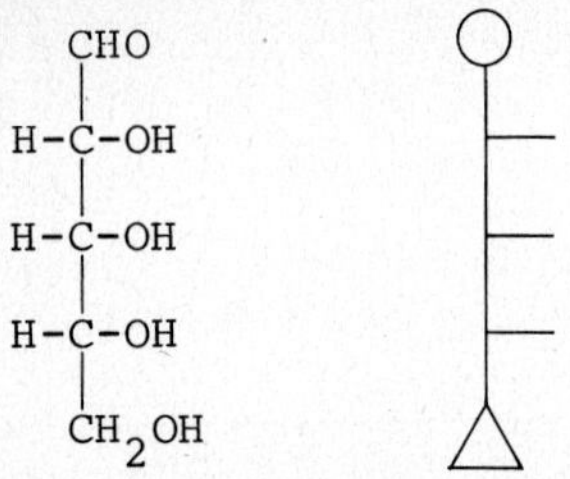

---

4. 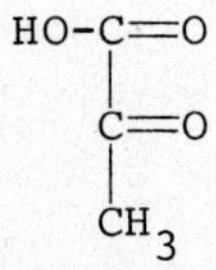

Your answer:

Solution: pyruvic acid, a metabolite (end product of glycolysis)

terms: No terms in the list fit pyruvic acid; it is both a ketone and an acid, but it is not a sugar (so not an "-ose").

---

5. 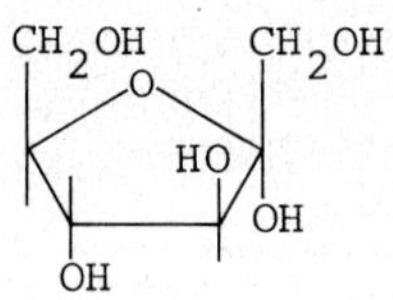

Write the abbreviated open-chain form for this sugar.

Your answer.

Solution: α-D-fructofuranose (α-D-fructose)

terms: monosaccharide, ketose (carbon 2 is anomeric rather than carbon 1 as in an aldose), hexose, α-anomer (OH is down on the anomeric carbon), reducing sugar (a ketone normally will give a negative result for Benedict's, Fehling's, or Tollens' tests because it cannot be oxidized; however, a ketose will yield a positive test as a reducing sugar because it isomerizes to an aldose in the basic solution), furanose ring

abbreviated open-chain structure

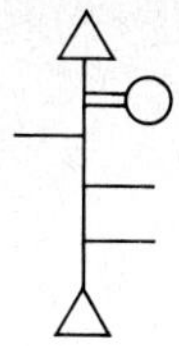

6.

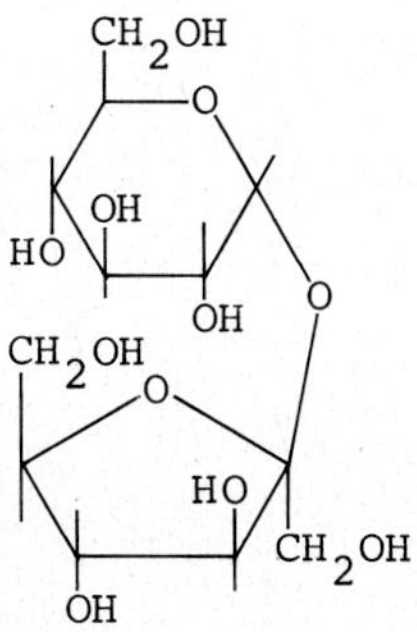

Describe the linkage between the two monosaccharides.

Your answer:

Solution: sucrose (α-D-glucose linked to β-D-fructose through a 1 ⟶ 2 acetal bond)

terms: disaccharide, nonreducing sugar (there is no hemiacetal carbon; consequently, there is no way for the ring to open to yield a free aldehyde structure; you can recognize this because there is no carbon with both OH and OR on it), acetal

The two monosaccharides are bonded by an acetal (glycoside) linkage.

7.

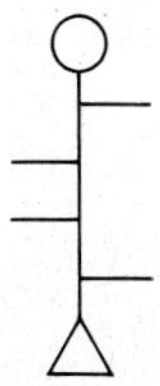

Draw the Haworth projection for the α-anomer in a pyranose ring form.

Your answer:

Solution: D-galactose (in open-chain form)

terms: monosaccharide, aldose, hexose, reducing sugar

α-D-galactose (α-D-galactopyranose) in the Haworth projection is

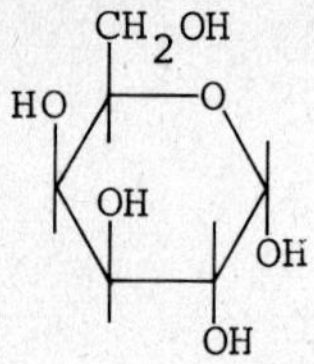

8.

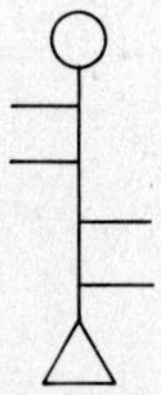

Draw the Haworth projection of the β-anomer in the pyranose form.

Your answer:

Solution: D-mannose (in open-chain or aldo form)

terms: monosaccharide, aldose, hexose, reducing sugar

β-D-mannose (β-D-mannopyranose) in the Haworth projection is

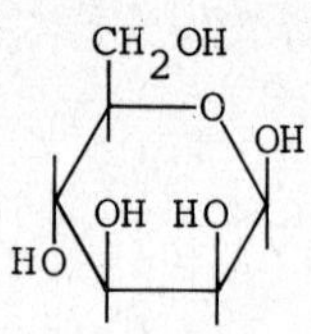

9.

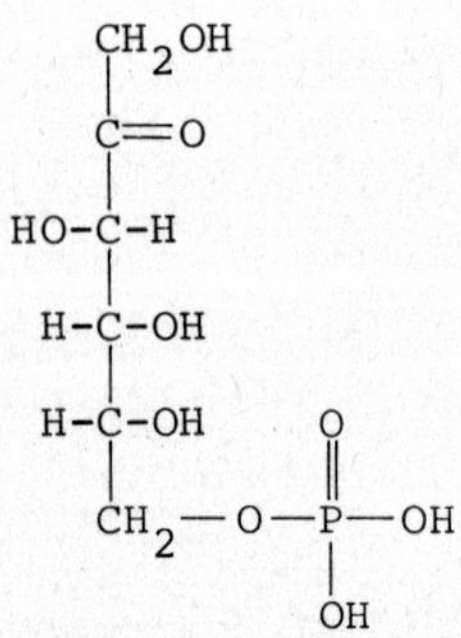

Draw this sugar in cyclic (furanose) form.

Your answer:

Solution: fructose 6-phosphate (F-6-P)

terms: monosaccharide, phosphate ester, ketose, hexose, reducing sugar

Cyclic (furanose) structure is

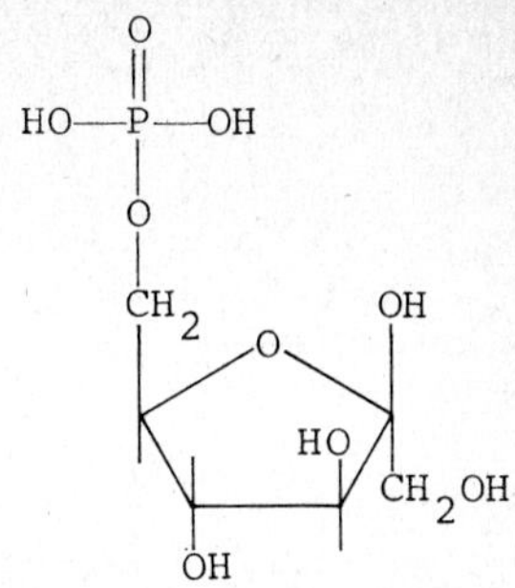

10.

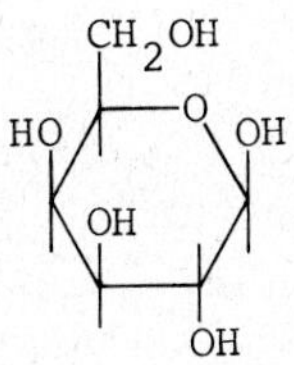

Draw an abbreviated open-chain structure for this sugar.

Your answer:

Solution: β-D-galactose (β-D-galactopyranose)

terms: monosaccharide, aldose, hexose, reducing sugar, β-anomer, pyranose ring, hemiacetal

Abbreviated open-chain form is

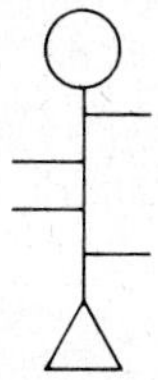

11.

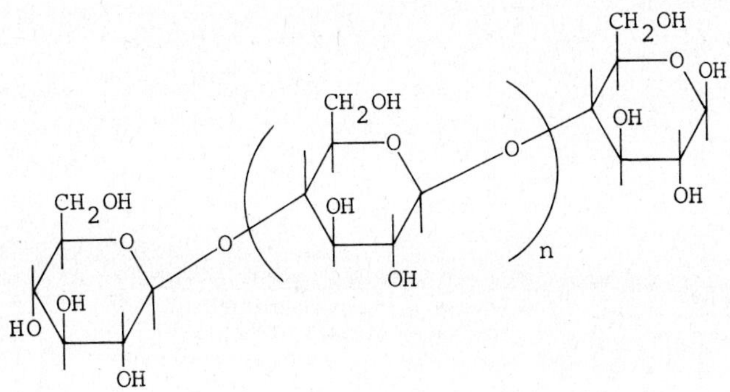

Your answer:

Solution: cellulose (a polymer of β-D-glucose)

terms: polysaccharide, nonreducing sugar (the last unit has a hemiacetal carbon, but "n" is so very large, that cellulose will not yield a positive Benedict's test), acetal (glycoside)

12.

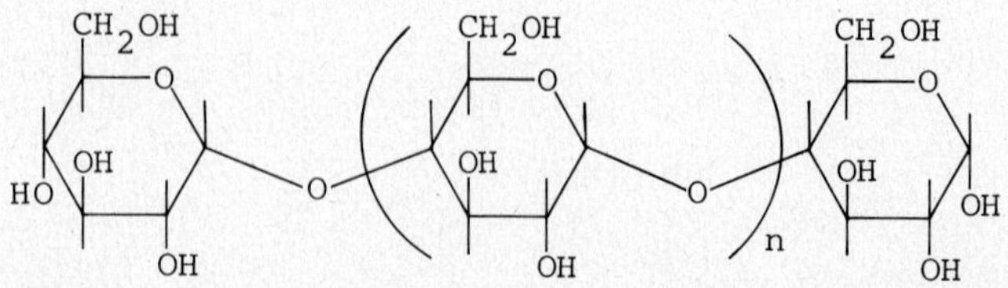

Your answer:

Solution: amylose (a component of starch; a polymer of α-D-glucose)

terms: polysaccharide, nonreducing sugar, acetal

Amylose is one component of starch; the other component is amylopectin, which is also a polymer of α-D-glucoses linked 1 ⟶ 4, but in addition it has α(1 ⟶ 6) branches. Glycogen, the animal storage form of glucose that occurs in the liver, is composed of amylopectin with more frequent branching than occurs in the amylopectin found in starch.

13.

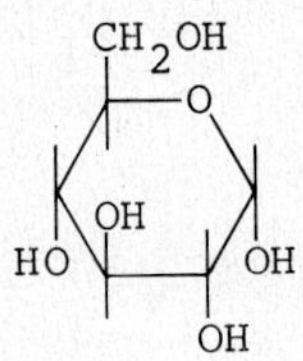

Consider all the monosaccharides of the previous questions. Which ones are epimers of this sugar? Draw the abbreviated stick forms of this sugar and of these epimers.

Your answer:

Solution: α-D-glucopyranose (α-D-glucose)

terms: monosaccharide, aldose, hexose, α-anomer, reducing sugar, pyranose ring, hemiacetal

The three sugars that are epimers (because they differ in the position of OH on just one carbon atom are

D-galactose D-glucose D-mannose

Notice the β-D-mannose is presented in problem 8; its anomer, α-D-mannose, is an epimer of α-D-glucose.

---

14. 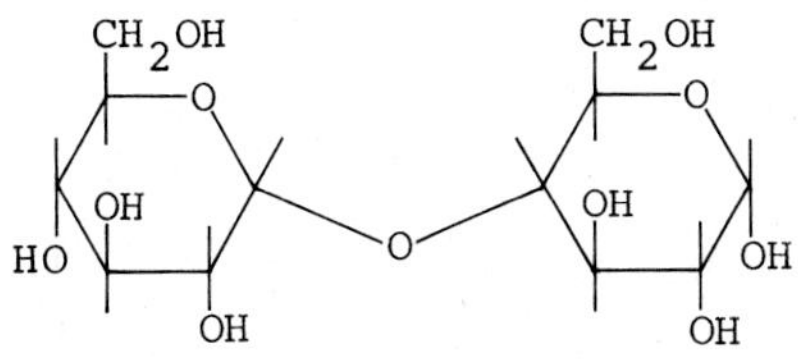

Your answer:

Solution: α-maltose (consists of α-D-glucose units bonded α(1 ⟶ 4)
terms: disaccharide, reducing sugar, acetal (glycoside)

---

15. 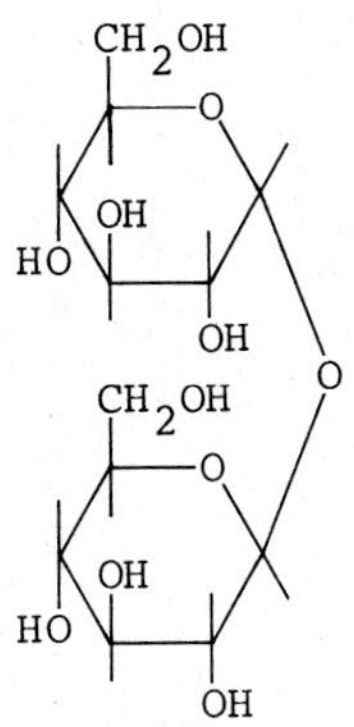

Your answer:

Solution:
terms: disaccharide (of α-D-glucose bonded 1 ⟶ 1 with β-D-glucose) nonreducing sugar, acetal (glycoside)

---

16. 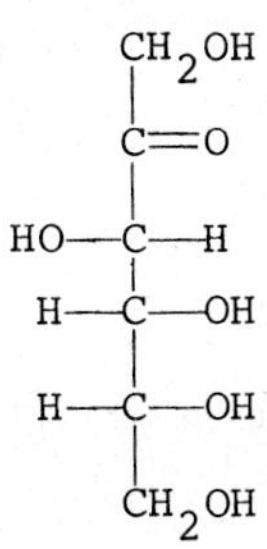

Your answer:

Solution D-fructose
terms: monosaccharide, ketose, hexose, reducing sugar

II. Stereoisomerism and Mutarotation in Carbohydrates

1. Draw the Fischer projections of this sugar and of its enantiomer.

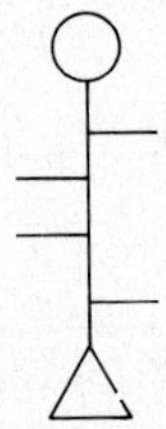

The compound shown is dextrorotatory and melts at 167°C. Name each of the enantiomers and compare their optical rotations and melting points. Can both compounds be utilized by our bodies?

Your answer:

Solution:

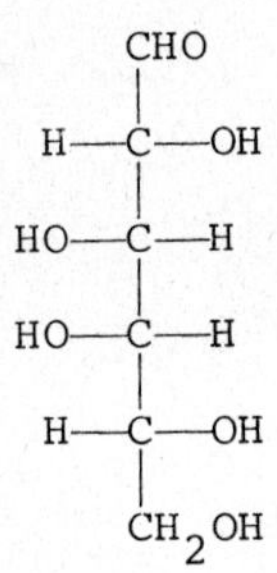

D (+) galactose
m.p. = 167°C

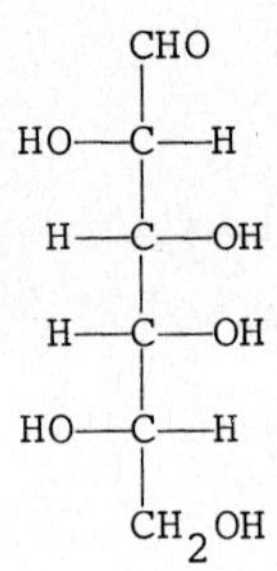

L (-) galactose
m.p. = 167°C

Only D-sugars occur in nature and can be utilized by our bodies; if ingested, L-sugars pass through our digestive tracts and are eliminated.

---

2. How many stereoisomers are possible for galactose?

Your answer:

Solution: 32

Galactose (as written in the aldose form) has four chiral (asymmetric) carbons that permit 2^4 or 16 optical isomers (8 pairs of enantiomers). However, the aldehyde form is very unstable, and only a small percentage of the sugar actually occurs in this form in solution. In the more stable cyclic form, there are five chiral centers, so each of the 16 isomers (already mentioned) has an α and a β isomer (called anomers). Thus there are a total of 32 (2^5) optical isomers of galactose.

---

3. Write an equation showing the equilibrium among the three forms of glucose.

Your answer:

Solution:

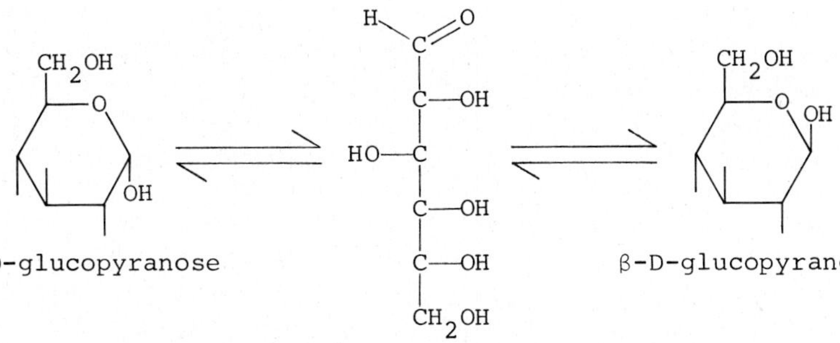

α-D-glucopyranose — aldose form of D-glucose — β-D-glucopyranose

4. If a sample of α-D-glucose is dissolved in water and is allowed to equilibrate, it is found that the specific rotation changes from 113° to 52.7^{o}. If a sample of β-D-glucose is dissolved in solution, the specific rotation changes from 19° to 52.7^{o}. Name and define the process that occurs. Include the equilibrium equation of problem 3 in your description.

Your answer:

Solution: The process described is mutarotation—a gradual change in rotation (of polarized light) to an equilibrium point. Aldohexoses and aldopentoses undergo mutarotation.

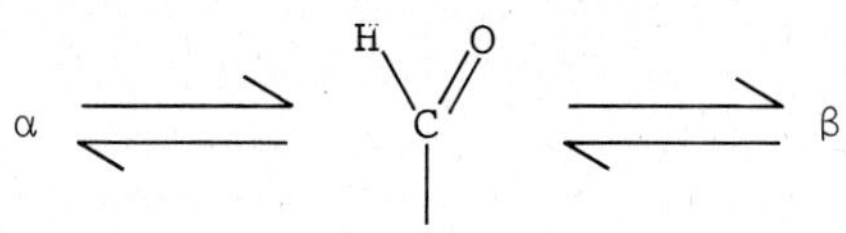

When α is dissolved in solution, it gradually changes into the aldose form and on into the β form. Reaction to the right is initially favored, but eventually some β changes back to α through the aldose form. When the rate of forming α equals the rate of forming β, an equilibrium is established. In an equilibrated solution of glucose 37% of the mixture is α and 63% is β. (The β anomer is more stable because the OH is away from the ring in the equatorial position and interferes with none of the other OH's.)

5. Which pairs are epimers, anomers, diastereomers, or enantiomers?
   (a) α-D-galactose and β-D-galactose
   (b) α-D-mannose and α-L-mannose
   (c) β-D-glucose and β-D-fructose

Your answer:

epimers:

anomers:

diastereomers:

enantiomers:

Solution: epimers: (a); anomers: (a); diastereomers: (a) and (c); enantiomers: (b)

III. Reactions of Carbohydrates

A. OXIDATION

Write an equation for each reaction. Show each reactant sugar as an α-anomer.

1. D-glucose + Tollens' reagent ⟶

Your answer:

Solution:

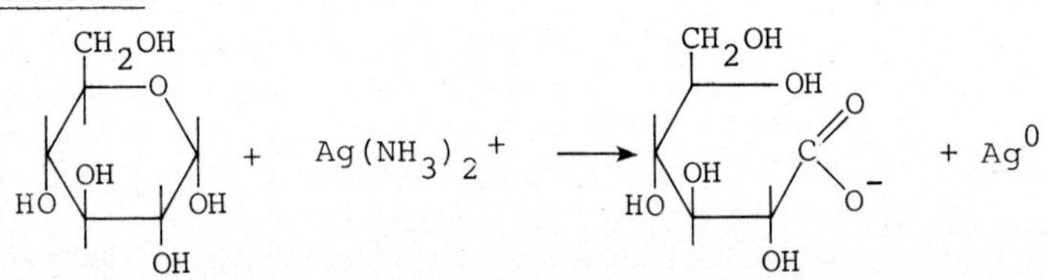

The ring opens to the aldose form to enable the reaction to occur.

---

2. acetone + Benedict's reagent ⟶

Your answer:

Solution: $CH_3\underset{\underset{O}{\|}}{C}CH_3 + Cu^{2+}/OH^- \longrightarrow$ N.R.

---

3. D-fructose + Benedict's reagent ⟶

Your answer:

Solution:

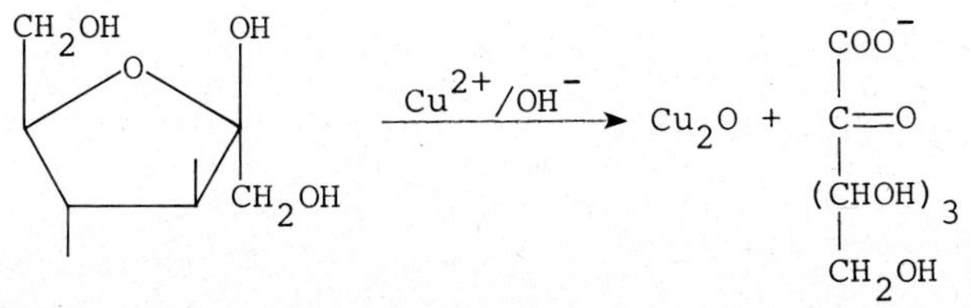

A ketose in basic solution isomerizes to an aldose and is therefore a reducing sugar (gives a positive Benedict's test).

---

4. maltose + Benedict's solution ⟶

Your answer:

Solution:

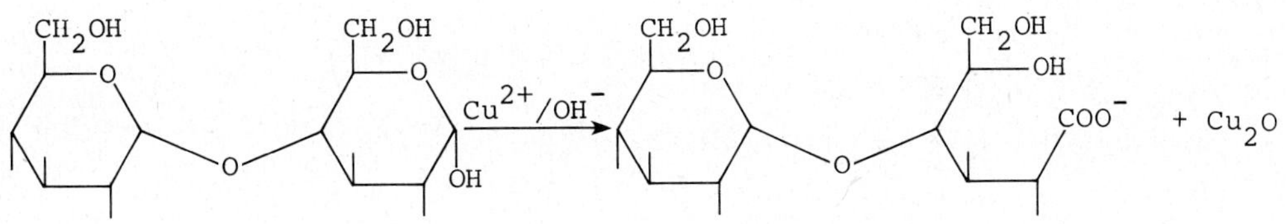

---

5. sucrose + Benedict's solution ⟶

Your answer:

Solution:

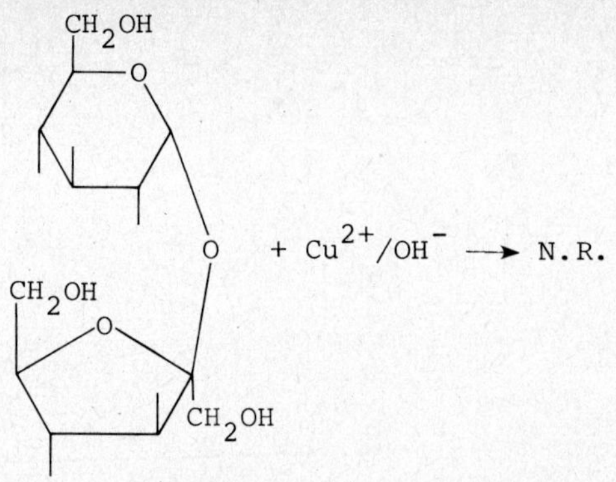

No hemiacetal carbon is present so the ring cannot open and, therefore, cannot be oxidized.

---

6. hydrolyzed sucrose + Benedict's solution ⟶

Your answer:

Solution: Hydrolyzed sucrose is an equimolar mixture of glucose and fructose. Both glucose and fructose are reducing sugars (problems 2 and 5 in Section I) and so will yield positive results when reacted with Benedict's solution.

B. ENZYMATIC REACTIONS

Name the products of each reaction. Where does each reaction occur in the digestive tract? If a reaction does not occur, explain why it doesn't.

1. amylose + amylase ⟶

Your answer:

Solution: amylose + amylase ⟶ maltose (units of maltose)
Amylase (ptalin) is an enzymatic component of the saliva secreted into the mouth.

---

2. cellulose + amylase ⟶

Your answer:

Solution: cellulose + amylase ⟶ N.R.
Humans have no specific enzyme to hydrolyze the β-linkage of cellulose; therefore, all cellulose ingested becomes the "bulk" component of the diet.

---

3. maltose + maltase ⟶

Your answer:

Solution: maltose + maltase $\longrightarrow$ 2 glucose units
Maltase is found in the small intestine.

---

4. sucrose + sucrase $\longrightarrow$

Your answer:

Solution: sucrose + sucrase $\longrightarrow$ glucose + fructose
Sucrase is found in the small intestine.

---

24 Lipids

I. Formula Recognition and Classification

For each compound in this section, answer these questions

(a) Which biological terms (from list one) describe this molecule? (In the solutions, the most specific biological term is underlined.)

(b) What chemical functional groups (from list 2) are present in the molecule?

(c) What is the name of the compound? (Or what are the names of its components?)

1. Biological category	2. Functional group
fat (saturated or unsaturated)	acetal
fatty acid (saturated or unsaturated)	acid
glycerol	acyl halide (acid halide)
lecithin	alcohol (1°, 2°, 3°)
lipid	aldehyde
not a lipid	alkene (or diene, triene)
oil	alkyl halide (haloalkane)
phospholipid	amide
phosphatide	amine
phosphatidic acid	anhydride
prostaglandin	aromatic
soap	hemiacetal
sphingolipid	hemiketal
sphingomyelin	ether
sphingosine	ester
steroid	ketal
triglyceride	ketone
wax	phenol
	salt

1. 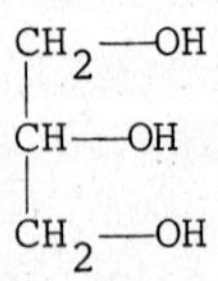

Your answers:

(a)

(b)

(c)

Solution:

(a) not a lipid (because it is water soluble)

(b) 1° and 2° alcohol (a triol)

(c) glycerol

---

2. $CH_3(CH_2)_{14}CO_2H$

Your answers:

(a)

(b)

(c)

Solution:

(a) fatty acid (saturated), lipid (because it is soluble in nonpolar solvents, not in the polar solvent, water)

(b) acid

(c) palmitic acid

---

3. $CH_3(CH_2)_{16}CO_2H$

Your answers:

(a)

(b)

(c)

Solution:

(a) fatty acid (saturated), lipid

(b) acid

(c) stearic acid

---

4. 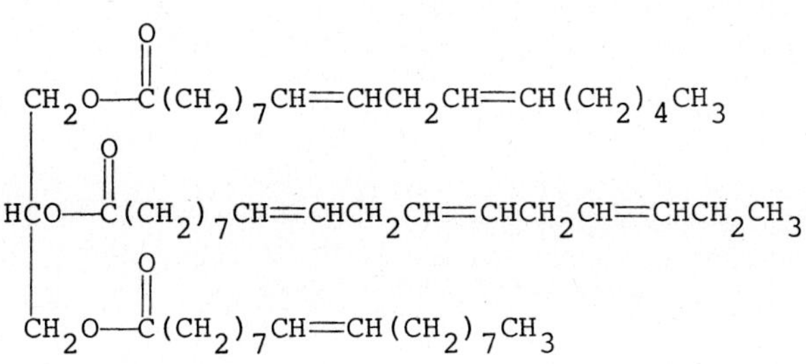

Your answers:

(a)

(b)

(c) Names of components:

Solution:
(a) lipid, oil, triglyceride (unsaturated)
(b) alkene (—hexaene), ester
(c) glycerol, linoleic acid, linolenic acid, oleic acid

---

5. $HO{-}CH_2CH_2{-}N^+(CH_3)_3$

Your answers:

(a)

(b)

(c)

Solution:
(a) not a lipid
(b) 1° alcohol
(c) choline

6.

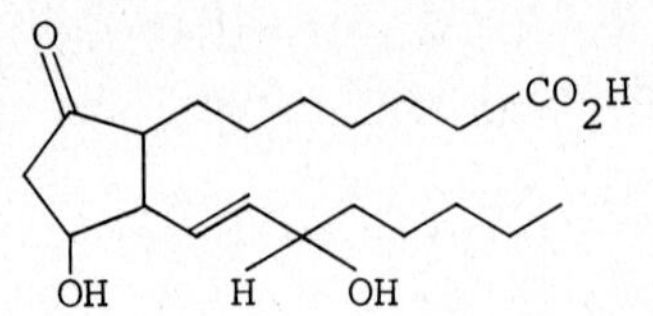

Your answers:

(a)

(b)

(c) Omit

Solution:
(a) lipid, prostaglandin
(b) acid, alkene, 2° alcohol, ketone

---

7. $CH_3(CH_2)_{12}{-}\overset{O}{\overset{\|}{C}}{-}OCH_2(CH_2)_{24}CH_3$

Your answers:

(a)

(b)

(c)

Solution:

(a) lipid, wax

(b) ester

(c) long-chain carboxylic acid esterified to long-chain alcohol

---

8. 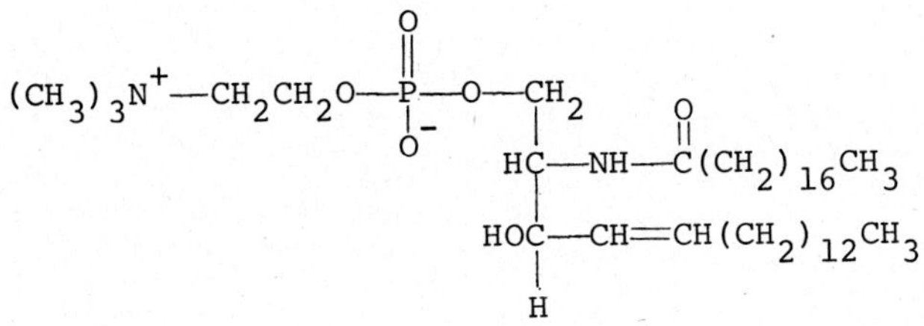

Your answers:

(a)

(b)

(c) Names of components:

Solution:

(a) lipid, phospholipid, sphingomyelin, sphingolipid

(b) acid (on phosphorus), 2° alcohol, alkene, amide, ammonium salt, ester

(c) sphingosine, phosphoric acid, choline, long-chain fatty acid

---

9. 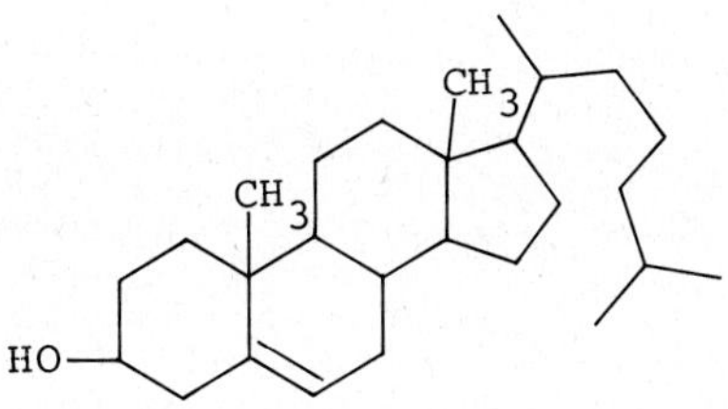

Your answers:

(a)

(b)

(c)

Solution:

(a) lipid, steroid

(b) 2° alcohol, alkene

(c) cholesterol

10.

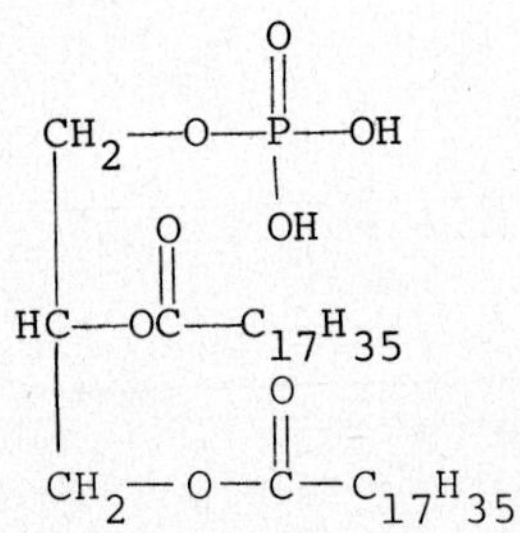

Your answers:

(a)

(b)

(c) Kinds of components:

Solution:

(a) lipid, phospholipid, phosphatidic acid

(b) acid (phosphoric), ester (of phosphoric acid and carboxylic acid)

(c) glycerol, phosphoric acid, stearic acid

11.

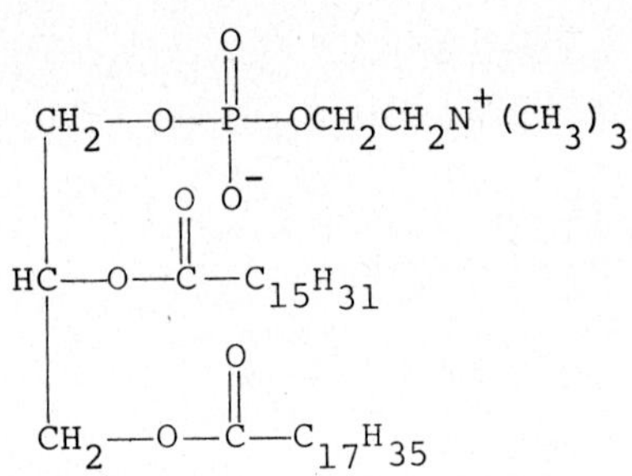

Your answers:

(a)

(b)

(c) Names of components:

Solution:

(a) lipid, lecithin (also called a phosphatidyl choline), phospholipid, phosphatide

(b) acid (phosphoric), ester (of phosphoric acid and carboxylic acids), ammonium salt

(c) glycerol, choline, phosphoric acid, palmitic acid, stearic acid

---

12. 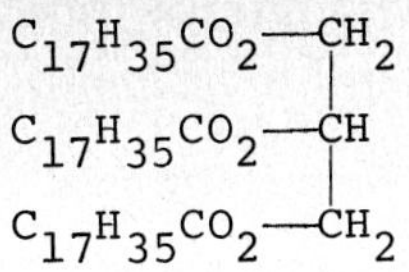

Your answers:

(a)

(b)

(c) Names of components:

Solution:

(a) fat (saturated), lipid, triglyceride

(b) ester (carboxylic acid esterified to glycerol)

(c) glycerol, stearic acid

---

13. $HO—CH_2CH_2—NH_2$

Your answers:

(a)

(b)

(c)

Solution:

(a) not a lipid

(b) 1° alcohol, 1° amine

(c) ethanolamine

---

14. 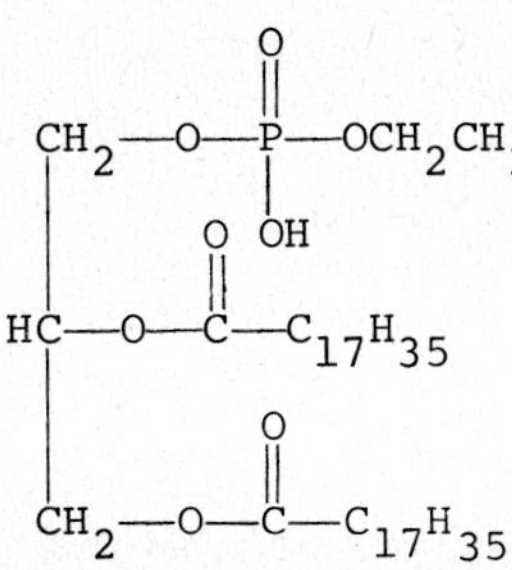

Your answers:

(a)

(b)

(c) Names of components:

Solution:

(a) phosphatide (known as phosphatidyl ethanolamine), lipid, phospholipid

(b) acid (phosphoric), amine, ester (carboxylic and phosphoric acids esterified to glycerol)

(c) glycerol, phosphoric acid, ethanolamine, stearic acid

---

15. 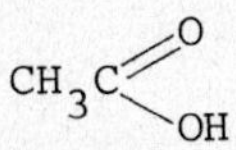

Your answers:

(a)

(b)

(c)

Solution:

(a) not a lipid

(b) acid (carboxylic)

(c) acetic acid

---

16. 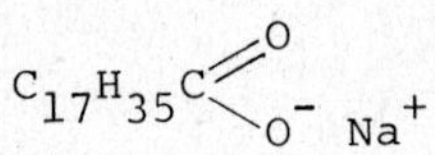

Your answers:

(a)

(b)

(c)

Solution:

(a) a soap

(b) salt (sodium salt of stearic acid)

(c) sodium stearate

II. Terminology Related to Polar and Nonpolar Aspects of Lipid Molecules

1. Illustrate the following with labeled sketches. Indicate where each occurs and its importance. Let 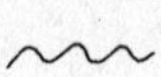 represent the hydrophobic (water-hating or nonpolar) parts of molecules and  represent the hydrophilic (water-loving or polar) parts of molecules.

(a) Micelle

Your answer:

Solution: 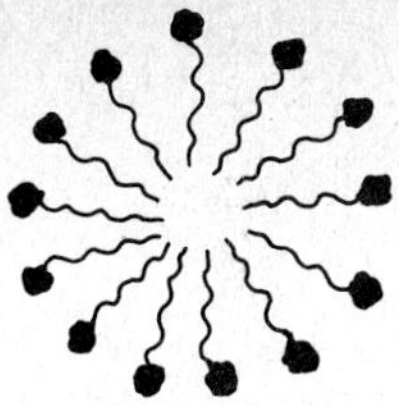

Other nonpolar material may associate with the nonpolar tails in the middle of the spherical micelle (for example, oily dirt in a micelle of soap or detergent or triglycerides in the center of micelles of phospholipids in our bodies).

(b) bilayer

Your answer:

Solution:

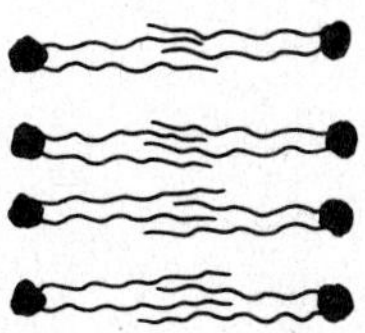

The phospholipids are arranged in the bilayer so that the polar phosphate "heads" face the polar aqueous environment and shield the nonpolar lipid "tails" from this polar environment. Protein molecules are also found in the phospholipid bilayers that comprise the membranes of cells.

---

2. Consider each molecule in the groups of molecules shown. Label the hydrophobic and hydrophilic portions of each molecule, and indicate whether the molecule is soluble in water, acids, bases, or nonpolar solvents.

(a) $C_{17}H_{35}CO_2H$, $C_{17}H_{35}CO_2^- \ Na^+$,

$$\begin{array}{l} C_{17}H_{35}CO_2CH_2 \\ \quad\quad\quad\quad\quad | \\ C_{17}H_{35}CO_2CH \\ \quad\quad\quad\quad\quad | \\ C_{17}H_{35}CO_2CH_2 \end{array}$$

Your answer:

Solution:

$C_{17}H_{35}$—CO_2H

hydrophobic hydrophilic

The polarity and H bonding of the carboxyl group will enable only a carbon chain of five (or fewer) carbons to be soluble in water. Therefore, this compound is soluble in nonpolar solvents or in NaOH solution. (It will form a salt in basic solution.)

$C_{17}H_{35}$—$CO_2^-\ Na^+$

hydrophobic hydrophilic

Sodium stearate is a soap; its ionic bond is the most extreme case of "polarity" and will enable even a long carbon chain to have water solubility.

$C_{17}H_{35}$—CO_2CH_2
$C_{17}H_{35}$—CO_2CH
$C_{17}H_{35}$—CO_2CH_2

hydrophobic polar, but less polar than carboxyl group

A triglyceride (fat) is only soluble in nonpolar solvents.

(b) $CH_3(CH_2)_{12}CO_2(CH_2)_{25}CH_3$, HO—$CH_2CH_2NH_2$

Your answer:

Solution:

$CH_3(CH_2)_{12}CO_2(CH_2)_{25}CH_3$

In this wax there is very low polarity in the ester bond that is sandwiched between two long (hydrophobic) chains; therefore, this compound is soluble only in nonpolar solvents. (Wax on duck feathers prevents the ducks from getting soggy!)

HO—CH_2CH_2—NH_2

Ethanolamine contains a very short carbon chain with two H-bonding functional groups, so is very soluble in water but not in nonpolar solvents.

(c) 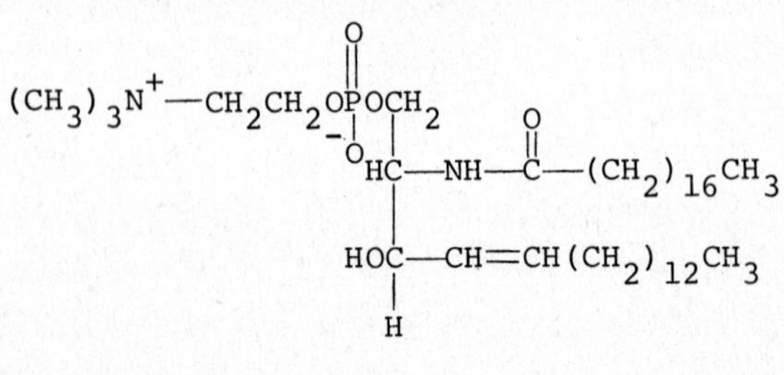

Your answer:

Solution: This compound is sphingomyelin; it is found in the myelin sheath that insulates our nerves. The molecule can be represented as  because of the presence of the two long hydrophobic chains that are attached to the polar "head." Because of the diverse characteristics of the molecule, its solubility is not easily predicted.

III. Reactions

Write the products for each reaction shown. Label any compounds and describe reactions when possible.

1. $$3\ CH_3(CH_2)_{16}\overset{\overset{O}{\|}}{C}{-}OH + PCl_3 \longrightarrow ? \xrightarrow{\text{glycerol}} ?$$

Your answer:

Solution: lab preparation of a triglyceride

$$3CH_3(CH_2)_{16}CO_2H \xrightarrow{PCl_3} 3CH_3(CH_2)_{16}\overset{\overset{O}{\|}}{C}Cl \xrightarrow{\text{glycerol}} \begin{array}{l} CH_3(CH_2)_{16}CO_2CH_2 \\ CH_3(CH_2)_{16}CO_2CH \\ CH_3(CH_2)_{16}CO_2CH_2 \end{array}$$

---

2. $$CH_3(CH_2)_{14}\overset{\overset{O}{\|}}{C}{-}OH + H_2 \xrightarrow{Ni}$$

Your answer:

Solution: $CH_3(CH_2)_{14}\overset{O}{\overset{\|}{C}}\text{—}OH + H_2 \xrightarrow{Ni}$ N.R.

The carbon chain in palmitic acid is already saturated; the carbonyl of the acid is not reduced by H_2/Ni. Polyunsaturated oils (vegetable oils) like the compound in the next exercise are hydrogenated commercially to produce margarines.

---

3. 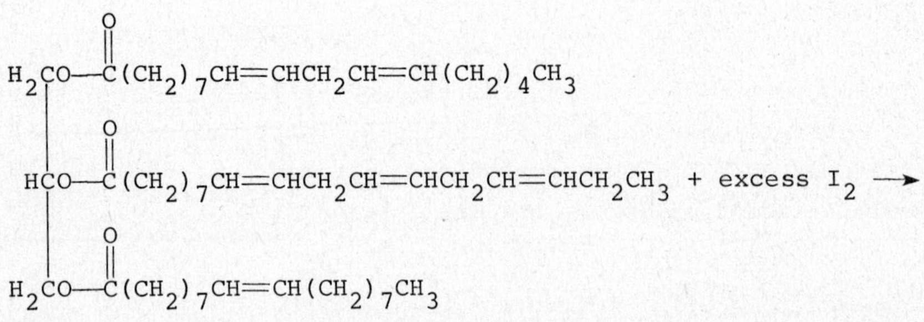

Your answer:

Solution: triglyceride + excess I_2 $\longrightarrow$ 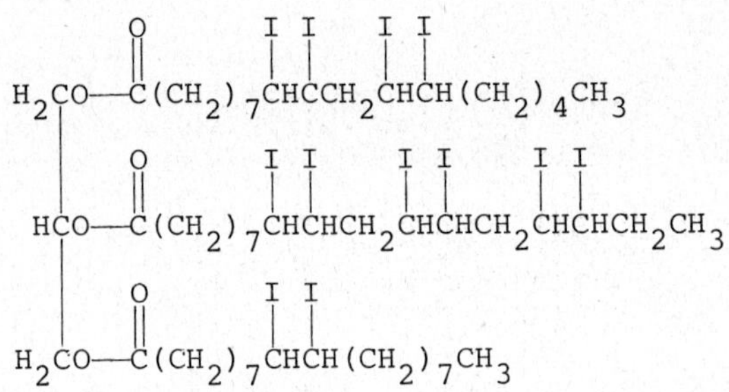

The iodine saturates the double bonds. This reaction is often used to quantitatively measure the number of double bonds per mole of triglyceride (six double bonds require 6 moles of I_2).

---

4. Reactant in problem 3 $\xrightarrow{\text{lipase}}$ (name of each product)

Your answer:

Solution: linoleic acid + linolenic acid + oleic acid + glycerol

Triglycerides are enzymatically hydrolyzed by lipase in the small intestine.

---

5. $$\begin{array}{l} C_{17}H_{35}CO_2—CH_2 \\ C_{17}H_{35}CO_2—CH \quad + \text{NaOH/heat} \longrightarrow \\ C_{17}H_{35}CO_2—CH_2 \end{array}$$

Your answer:

Solution: triglyceride + NaOH/heat $\longrightarrow$ 3 $C_{17}H_{35}\ CO_2^-\ Na^+$ + glycerol

sodium stearate

(a soap)

The hydrolysis of an ester in a basic solution is called saponification ("soap making").

---

6. $CH_3(CH_2)_{12}\overset{O}{\overset{\|}{C}}—OCH_2(CH_2)_{24}CH_3$ + lipase $\longrightarrow$

Your answer:

Solution: $CH_3(CH_2)_{12}CO_2CH_2(CH_2)_{24}CH_3$ + lipase $\longrightarrow$ N.R.

Lipase will not hydrolyze waxes; hence wax is not "digestible."

IV. Biological Significance

For each compound in Section I, briefly describe the biological function, occurrence, and/or importance.

1. Glycerol

Your answer:

Solution: Glycerol acts as the "backbone" for the triglycerides and phosphatides; the component fatty acids (or phosphoric acid) are each esterified to its alcohol groups.

---

2. & 3. palmitic and stearic acids

Your answer:

Solution: Fatty acids are used by cells for energy; but if no energy is required during a given time period, the fatty acids are stored as triglycerides in the adipose (fat) tissue.

---

4. Triglyceride

Your answer:

Solution: Polyunsaturated triglycerides occur in vegetable oils; saturated triglycerides occur predominately in animal fats. Triglycerides can be hydrolyzed by lipase to fatty acids (which can be used to provide energy to the cells) and glycerol.

---

5. Choline

Your answer:

Solution: Choline is a component of the lecithins (phosphatidyl choline) and sphingomyelin—very important in our bodies.

---

6. Prostaglandins

Your answer:

Solution: Prostaglandins are found in many organs and tissues of the body; they stimulate the contraction of smooth muscles (in labor) and have other varied functions.

---

7. Wax

Your answer:

Solution: Waxes are very hydrophobic and so protect birds (ducks), leaves, our kitchen floors, and our cars from water.

---

8. Sphingomyelin

Your answer:

Solution: Sphingomyelin forms a protective (insulating) covering, which is known as the myelin sheath, around our nerves. (Stability due partly to amide bond.)

---

9. Cholesterol

Your answer:

Solution: Cholesterol is the precursor for the important steroids required by our bodies, including the sex hormones and the bile acids. It has been suggested that cholesterol is responsible (at least in part) for arterial disease.

---

10. A phosphatidic acid

Your answer:

Solution: Different phosphatides are formed by esterifying various alcohols to the phosphate on phosphatidic acid.

---

11. A lecithin

Your answer:

Solution: Lecithins are found in cell membranes. Also, because of their dual solubility (in fatty materials and in polar solvents), lecithins are useful in transporting fatty materials in our blood.

---

12. A saturated triglyceride

Your answer:

Solution: A saturated triglyceride (fat) is found in animal fat and is considered to be less "healthy" in the diet than are polyunsaturated fats (vegetable oils).

---

13. Ethanolamine

Your answer:

Solution: Ethanolamine is a component of phosphatidyl ethanolamine.

---

14. Phosphatidyl ethanolamine

Your answer:

Solution: Phosphatidyl ethanolamines are found in membranes. They are a significant constituent of nerve and brain tissue.

---

15. Acetic acid

Your answer:

Solution: Acetic acid is present in vinegar.

---

16. A soap

Your answer:

Solution: A soap can effectively "cleanse" skin, laundry, and so on due to its dipolar nature; the hydrophobic portion of the molecule "solubilizes" the oily grime, and the hydrophilic portion permits the molecule to "mix" with water and to "wash away" the grime.

---

25 Amino Acids and Proteins

Before you answer these questions, you should memorize the names and structures of the following amino acids: glycine, alanine, valine, serine, cysteine, proline, phenylalanine, tyrosine, aspartic acid, glutamic acid, and lysine.

I. Names, Structural Formulas, and Reactions of Amino Acids and Peptides

1. Draw the structure of the tetrapeptide Ala-Lys-Pro-Glu. Which amino acid is the "N-terminal" amino acid and which is the "C-terminal" amino acid?

Your answer:

Solution:

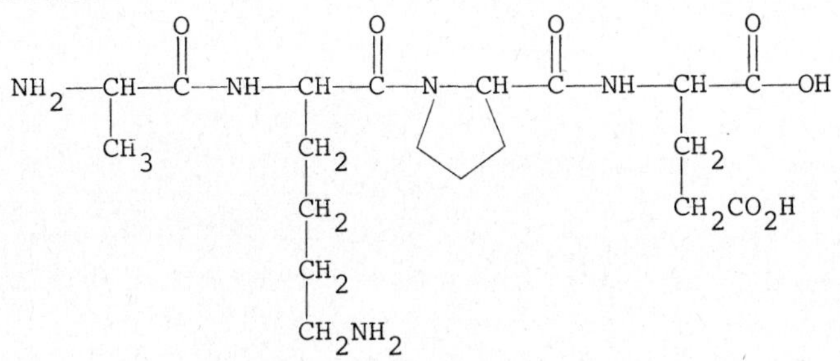

Ala - Lys - Pro - Glu (alanyllysylprolinylglutamic acid)

The peptide bond is the amide bond formed by the removal of H_2O between each pair of amino acids (-OH from the carboxyl of one amino acid, and -H from the α-amino of another amino acid). In peptide formulas, the free amino end is written on the left, and the free carboxyl end is written on the right. Thus alanine is the "N-terminal" and glutamic acid is the "C-terminal" amino acid.

---

2. Write the structural formulas of the products formed when the tetrapeptide of the previous problem reacts with each of the following reagents. Start with a new sample of the tetrapeptide for each reaction.

(a) Nitrous acid, HNO_2 (Specify the number of moles of products that form per mole of tetrapeptide.)

Your answer:

Solution:

Ala-Lys-Pro-Glu + 2 HNO_2 $\longrightarrow$ 2 $N_2(g)$ +

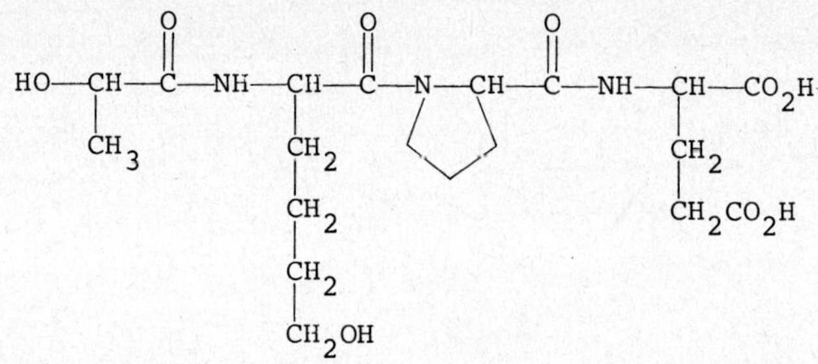

Two moles of N_2 form, one from the 1° amine at the N-terminal amino acid (Ala) and one from the 1° amine in lysine, a basic amino acid.

(b) Dilute NaOH solution

Your answer:

Solution:

Ala-Lys-Pro-Glu + 2 NaOH $\longrightarrow$

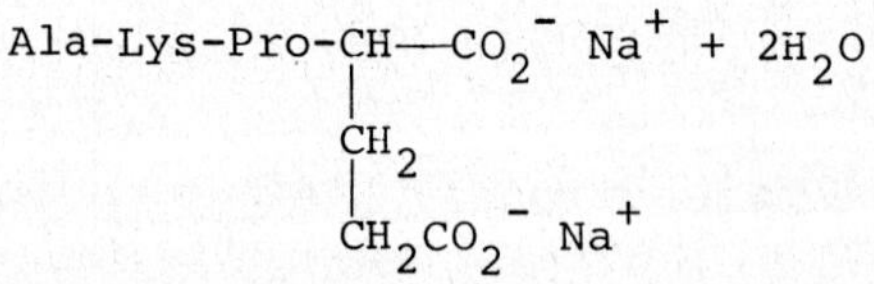

(c) 6 M HCl and prolonged heating (complete hydrolysis)

Your answer:

Solution:

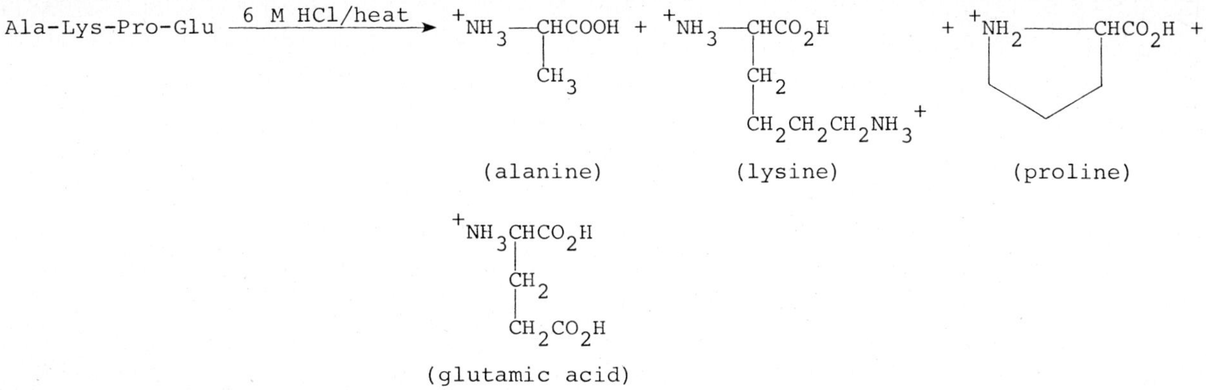

Remember that there are approximately 55 moles of H_2O per liter of solution—plenty of water (under acid conditions) to hydrolyze each peptide bond, as long as sufficient time and heat are permitted.

Notice that the hydrolyzed amino acids each have protonated amino groups; this is due to the presence of the acid.

(d) 2,4-Dinitrofluorobenzene followed by complete hydrolysis

Your answer:

Solution:

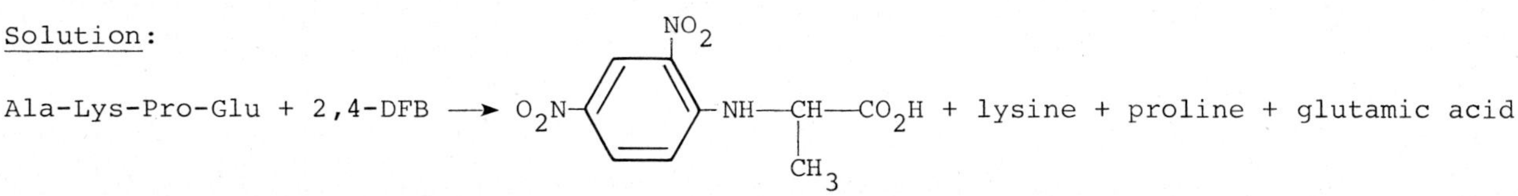

The first product shown can be extracted by a nonpolar solvent, and the amino acid attached to the ring can be identified as alanine, the N-terminal amino acid; hence, this reaction is known as N-terminal analysis.

(e) Hydrazine

Your answer:

Solution:

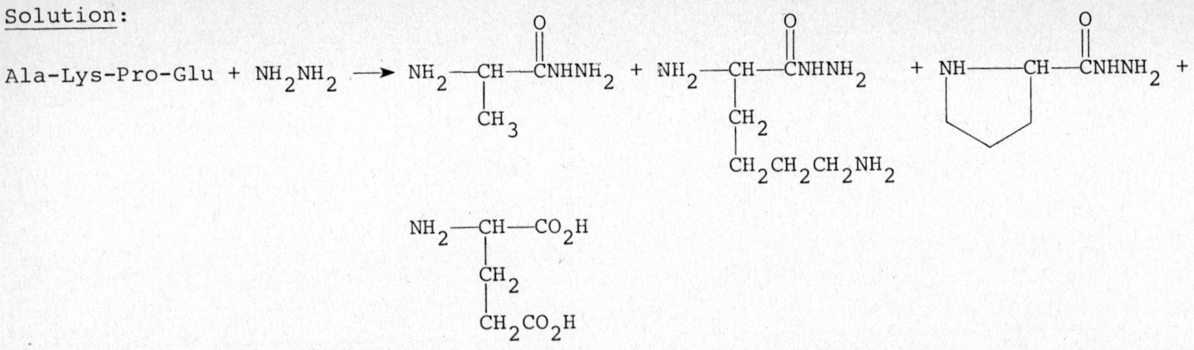

Notice that hydrazine bonds to the carbonyl carbon at each amide bond. However, the C-terminal amino acid is unaffected and thereby can be identified.

II. Determination of Amino Acid Sequence in a Decapeptide

1. A decapeptide is analyzed for amino acid sequence. After analyzing the following clues, write the amino acid sequence for this decapeptide.
   (a) After complete hydrolysis, it is found that lysine occurs twice, glycine twice, and each of the other amino acids occurs once in the decapeptide.
   (b) Reaction of the peptide with 2,4-dinitrofluorobenzene, followed by hydrolysis, yields eight amino acids and

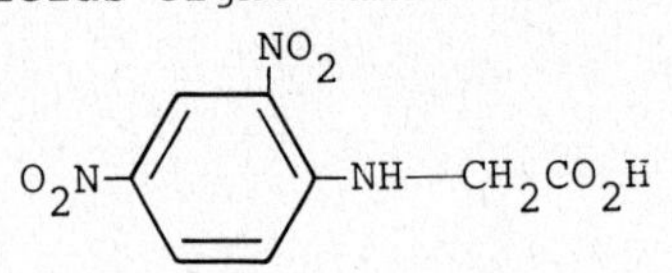

   (c) Reaction of the decapeptide with hydrazine yields hydrazides of all amino acids except phenylalanine (which occurs uncombined in the product).
   (d) Partial hydrolysis yields these peptide fragments: Ser-Glu-Ala, Gly, Pro-Lys-Phe, Gly-Lys-Cys, Glu-Ala, Ala-Gly-Pro, Glu, and Cys-Ser-Glu. (Assume that for each fragment written, the left amino acid is the N-terminal for that fragment.)

Your answer: (Respond to these clues)

Solution:

(a) This information eliminates confusion when piecing together the fragments from partial hydrolysis.

(b) Glycine is bonded to the dinitrobenzene and therefore is the N-terminal amino acid in the decapeptide.

(c) When a peptide reacts with hydrazine, each amino acid except the C-terminal one becomes bonded to the hydrazine. Therefore, phenylalanine is the C-terminal amino acid in the decapeptide.

(d) Overlap the fragments and utilize the information in (a)-(c). Glycine is the N-terminal, and so one of the two glycines is shown on the far left; phenylalanine is the C-terminal, and so it must appear on the far right.

```
Gly-Lys-Cys
        Cys-Ser-Glu
            Ser-Glu-Ala
                Glu-Ala
                    Ala-Gly-Pro
                            Pro-Lys-Phe
-------------------------------------------
Gly-Lys-Cys-Ser-Glu-Ala-Gly-Pro-Lys-Phe    (the decapeptide)
```

2. Some enzymes hydrolyze specific peptide bonds within a peptide chain and so are called endopeptidases; other enzymes hydrolyze any amino acid at a specific end of a peptide chain and so are called exopeptidases. In the following list identify whether an enzyme is an endopeptidase or an exopeptidase. Indicate the specific peptide bonds attacked by each enzyme.
   (a) pepsin (localized in the stomach)
   (b) trypsin (localized in the small intestine)
   (c) chymotrypsin (localized in the small intestine)
   (d) aminopeptidase (found in the small intestine)
   (e) carboxypeptidase (found in the small intestine)

Your answer:

Solution:

Endopeptidases:

(a) pepsin—the amino (nitrogen) end of the amino acids containing aromatic rings (phenylalanine, tyrosine, tryptophan)

(b) trypsin—the carboxy (carbon) end of basic amino acids (lysine, arginine)

(c) chymotrypsin—the carboxy end of aromatic amino acids (phenylalanine, tyrosine, tryptophan)

Exopeptidases:

(d) aminopeptidase—the carboxy end of the N-terminal amino acid

(e) carboxypeptidase—the amino end of the C-terminal amino acid

---

3. Write the amino acid sequence of the decapeptide in problem 1. On it mark the bonds that could be hydrolyzed by the enzymes in problem 2.

Your answer:

Solution:

```
              Gly-Lys-Cys-Ser-Glu-Ala-Gly-Pro-Lys-Phe
                 ↑   ↑                          ↑
aminopeptidase┘   └trypsin                      ├─pepsin
                                                │
                                                ├─trypsin
                                                │
                                         carboxypeptidase
```

Even though pepsin, trypsin, and carboxypeptidase are each capable of hydrolyzing the bond between lysine and phenylalanine, the decapeptide would first encounter pepsin (in the stomach) if it were undergoing digestion.

---

4. How many different decapeptides are possible that contain the same amino acids as the decapeptide in problem 1?

Your answer:

Solution: 907,200 possible decapeptides

$$\frac{10!}{2! \times 2!} = \frac{10 \times 9 \times 8 \times 7 \times 6 \times 5 \times 4 \times 3 \times 2 \times 1}{(2 \times 1)\ (2 \times 1)} = 907{,}200$$ (Aren't you glad you don't have to draw them all!)

In general, the possible number of different peptide chains can be calculated by:

$$\frac{n!}{(x!)\ (y!)\cdots}$$

where ! = factorial

n = the number of amino acids in the peptide

x = the number of times a given amino acid is repeated

y = the number of times another amino acid is repeated

---

5. How many different decapeptides are possible containing four alanines, three glycines, and three other amino acids (each different)?

Your answer:

Solution: 25,200 possible decapeptides

$$\frac{10!}{(4!)(3!)} = \frac{10 \times 9 \times 8 \times 7 \times 6 \times 5 \times 4 \times 3 \times 2 \times 1}{(4 \times 3 \times 2 \times 1)(3 \times 2 \times 1)} = 25{,}200$$

III. Primary, Secondary, Tertiary, and Quaternary Structures of Proteins

For these questions assume that the decapeptide of problem 1 in Section II is repeated many times to form a protein.

1. What is the primary structure of this protein?

Your answer:

Solution: For this protein (and for any other protein), the primary structure is the sequence of amino acids (your answer to II-1 repeated many times). An important facet of primary structure is the cross-linking of peptide chains due to the presence of the amino acid cysteine, which forms a disulfide bond with another cysteine.

2. Use generalized structural formulas to illustrate at least three possible hydrogen bonds involved in the sheetlike secondary structure. Is this hydrogen bonding intermolecular or intramolecular?

Your answer:

Solution:

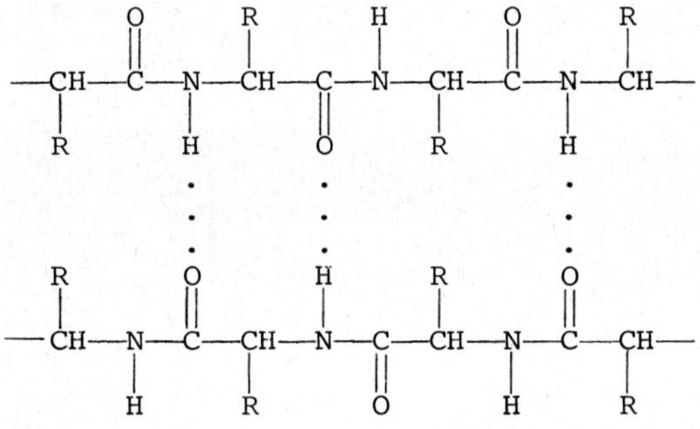

The hydrogen bonds involved in sheet formation are intermolecular (between molecules).

3. Another type of secondary structure is the alpha helix. Refer to a picture in your text of the alpha helix and observe the hydrogen bonds. Are these hydrogen bonds intermolecular or intramolecular? How does proline affect the secondary structure of a protein?

Your answer:

Solution: The hydrogen bonds are intramolecular. The presence of proline in a sequence destabilizes the alpha helix. The ring structure of proline prevents free rotation of the bonds and forces the nitrogen to have different bond angles than it normally does. Also, because proline is a secondary amine, no hydrogen is present on its amide nitrogen, and thus the hydrogen bonding normally found in the secondary structure of the peptide is not possible.

---

4. Describe the four types of interactions that determine the tertiary structure of proteins.

Your answer:

Solution:

(1) hydrogen bonding between side-chain groups of the amino acids (for example, the -OH of serine with the -OH of another serine or the -OH of tyrosine)
(2) ionic interactions between side-chain groups (for example, the ammonium ion of lysine attracted to the carboxylate of glutamic acid or aspartic acid)
(3) disulfide bridges—two cysteines can form a disulfide bond
(4) hydrophobic interactions in which the peptide chain will surround polar groups (such as those of the amino acids phenylalanine, tyrosine, and tryptophan) to protect them from the polar environment of the cell

The extent to which any of these interactions occurs depends on the proximity of the interacting groups, which ultimately depends on the sequence of the amino acids (the primary structure).

---

5. Can the decapeptide of problem II-1 have a quaternary structure? Why or why not?

Your answer:

Solution: no

Quaternary structure cannot form from a single peptide chain; it forms from several chains (referred to as subunits) that are "held" together by the same types of interactions that can occur in tertiary structure.

IV. Denaturation and Digestion

1. How does digestion of proteins differ from denaturation of proteins? What levels of structure are affected by each process?

Your answer:

Solution: Digestion refers to the hydrolysis of the peptide bonds and can be accomplished either enzymatically or by heating in the presence of 6 M HCl. Hydrolysis of the peptide bonds implies the destruction of the primary structure of the protein and, naturally, also the destruction of the secondary, tertiary, and quaternary (when it occurs) structures since they are dependent on the primary structure. Denaturation destroys the secondary, tertiary, and quaternary structures, but leaves the primary structure intact.

2. List several ways in which proteins can be denatured.

Your answer:

Solution: Denaturation, which destroys the biological functions of proteins, can be accomplished by any of the following ways.

(1) change in pH of the solution (addition of acid or base)
(2) heating or freezing
(3) shaking or vigorously stirring a solution of protein
(4) addition of heavy metals (Pb^{2+}, Hg^{2+}, etc.) to a solution of protein
(5) addition of formaldehyde to a protein solution
(6) high pressure
(7) high-energy radiation

26 Nucleic Acids and Review of Carbohydrates, Lipids, and Proteins

I. Structure Recognition and Terminology

For each of the following compounds,

(a) Name the compound unless otherwise instructed.

(b) Identify the bond that is marked with an arrow. (Choose from the list of chemical bonds.)

(c) Choose all the appropriate biological terms from the second list.

Chemical bonds	Biological terms
ester bond	protein
acetal bond	tripeptide, amino acid
disulfide bond	carbohydrate
peptide (amide) bond	monosaccharide, disaccharide, polysaccharide, reducing sugar, nonreducing sugar
N-glycosidic bond	lipid
ionic bond	fatty acid, triglyceride, phospholipid, sphingolipid, prostaglandin, steroid
hydrogen bond	nucleic acid
	DNA (part of DNA), RNA (part of RNA), nucleotide, purine base (adenine, guanine), pyrimidine base (cytosine, thymine, uracil)
	optically active
	polymer

1. 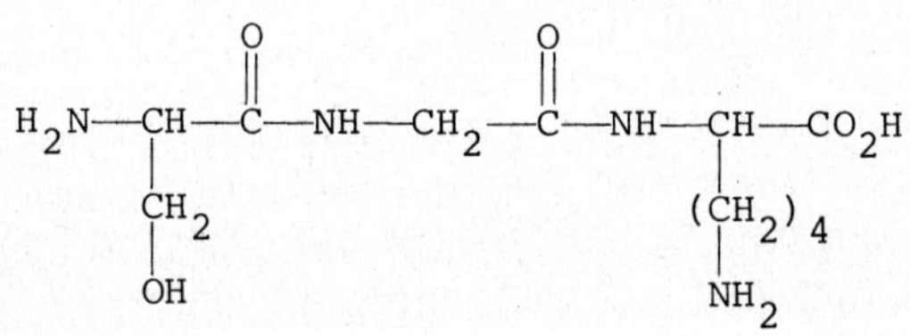

Your answers:

(a)

(b)

(c)

Solution:

(a) Ser-Gly-Lys

(b) peptide (amide) bond

(c) protein (if it were to contain more amino acids), tripeptide, optically active

---

2. 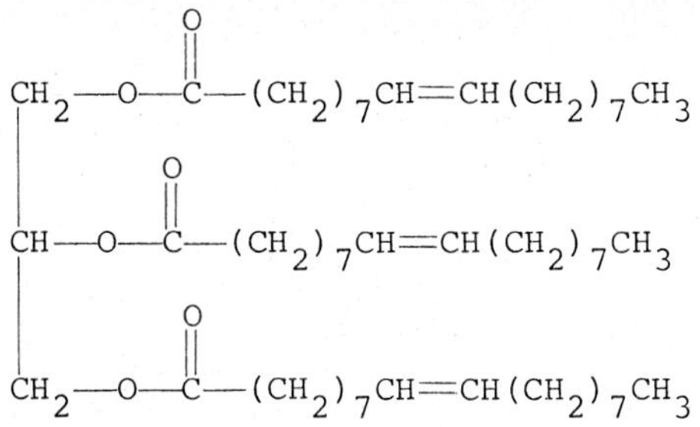

Your answers:

(a) Names of components:

(b)

(c)

Solution:

(a) glycerol, three oleic acids

(b) ester bond

(c) lipid–triglyceride, not optically active

---

3. 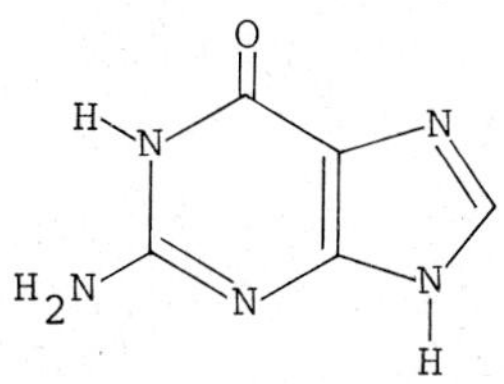

Your answers:

(a)

(b)

(c)

Solution:

(a) guanine

(b) no bond marked

(c) a purine base, found in both DNA and RNA, not optically active

---

4. 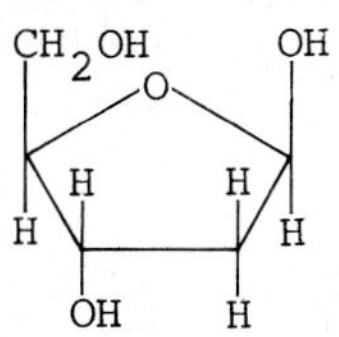

Your answers:

(a)

(b)

(c)

Solution:

(a) β-D-2-deoxyribose (usually called deoxyribose)

(b) no bond marked

(c) carbohydrate–monosaccharide, reducing sugar, found in DNA, optically active

---

5.

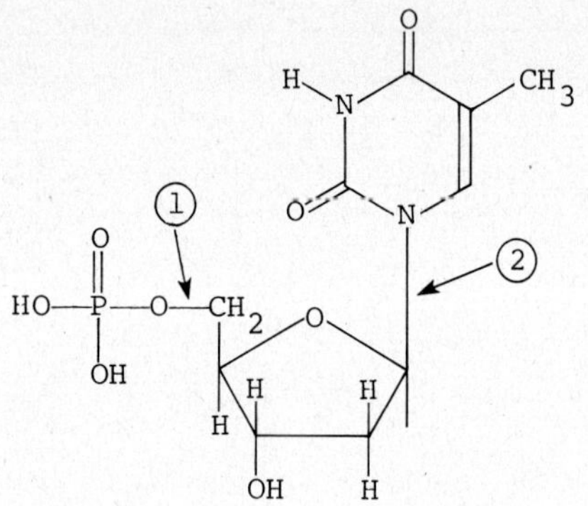

Your answers:

(a) Names of components:

(b)

(c)

Solution:

(a) thymine, 2-deoxyribose, phosphoric acid

(b) (1) ester (phosphate ester bond to deoxyribose)

(2) N-glycosidic bond from thymine to deoxyribose

(c) portion of DNA, nucleotide (a monomer of nucleic acid), optically active

---

6.

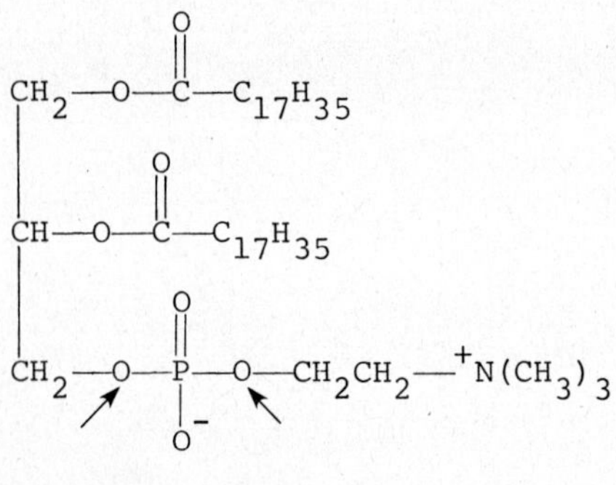

Your answers:

(a) Names of components:

(b)

(c)

Solution:
(a) a lecithin composed of glycerol, two stearic acids, phosphoric acid, and choline
(b) two phosphate ester bonds
(c) lipid—a phospholipid, optically active

7.

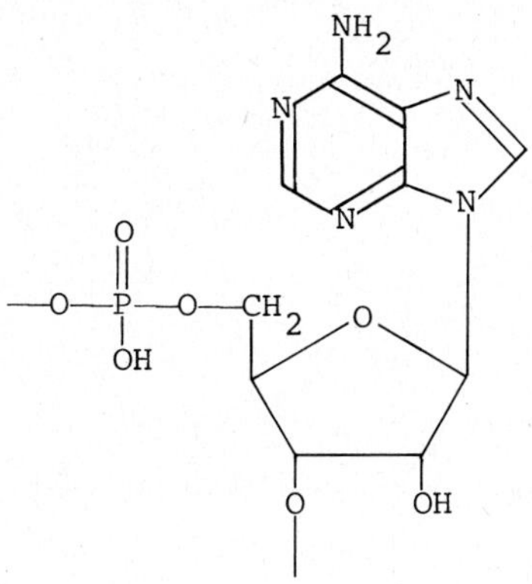

Your answers:

(a) Names of components:

(b)

(c)

Solution:
(a) a nucleotide composed of adenine, ribose, phosphoric acid
(b) N-glycosidic bond
(c) portion of RNA, nucleotide (a monomer of nucleic acid), optically active

---

8.

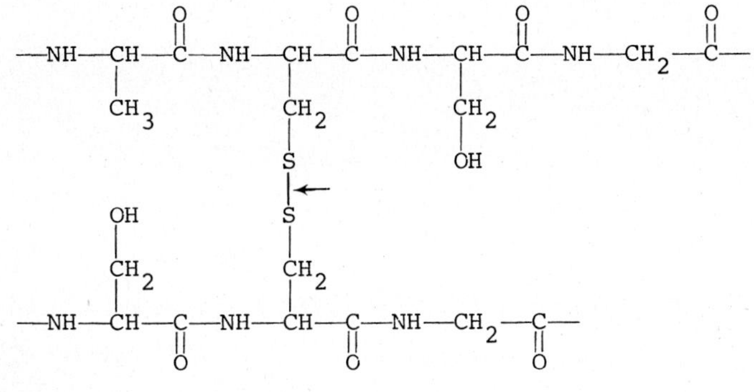

Your answers:

(a) omit

(b)

(c)

Solution:

(b) disulfide bond (shown as intermolecular, can be intramolecular also)

(c) protein, polymer, optically active

9.

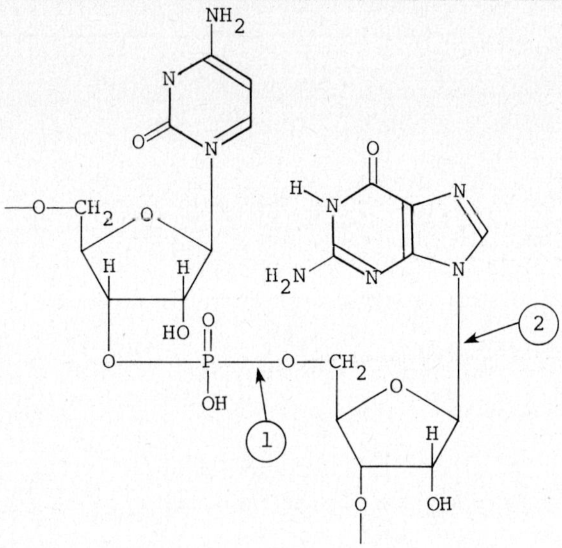

Your answers:

(a) Names of components

(b)

(c)

Solution:

(a) cytosine, guanine, ribose, phosphoric acid

(b) (1) ester bond; (2) N-glycosidic bond

(c) portion of RNA, two nucleotides sharing a phosphate group, optically active, polymer

---

10.

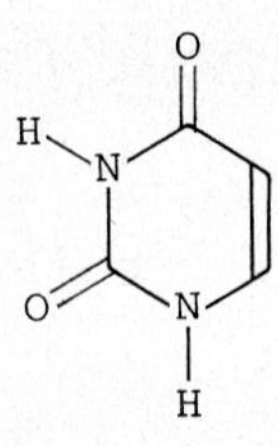

Your answers:

(a)

(b)

(c)

Solution:
(a) uracil
(b) no bond marked
(c) pyrimidine base (uracil) found only in RNA, not optically active

---

11.

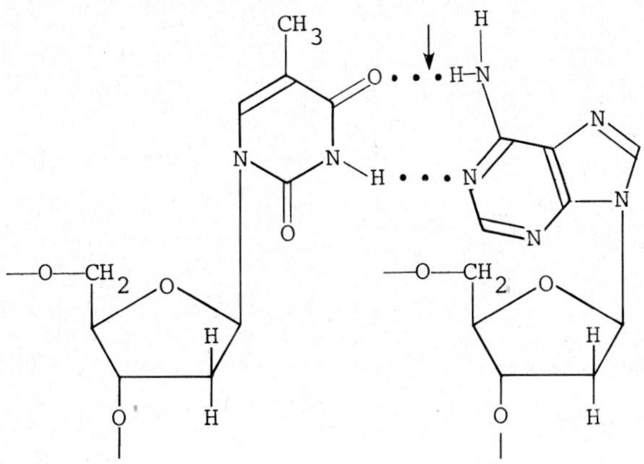

Your answers:

(a) Names of components

(b)

(c)

Solution:
(a) deoxyribose, thymine, adenine
(b) hydrogen bond between complementary pairs of bases in two DNA strands
(c) portion of DNA, two nucleosides in complementary strands of DNA, optically active, polymer

---

II. Nucleic Acids

1. Which of the following nucleic acids

DNA mRNA tRNA rRNA

(a) is responsible for the transfer of genetic information from one generation to the next?
(b) contains the instructions for the synthesis of all proteins?
(c) is (each) responsible for directing the construction of one specific peptide chain?

(d) enables mRNA and tRNA to form a complex so that protein chains may be synthesized?
(e) carries the three base sequences called codons that each specifies an amino acid?
(f) each brings one specific amino acid to the site of protein synthesis; its anticodon base-pairs with the codon of mRNA to position the amino acids it carries on the ribosome?
(g) carries the anticodons that "code" for all three types of RNA?

Your answers:

(a)

(b)

(c)

(d)

(e)

(f)

(g)

Solution: (a) DNA; (b) DNA; (c) mRNA; (d) rRNA; (e) mRNA; (f) tRNA; (g) DNA

---

2. A base sequence on a messenger RNA is AUC CAG UGC
(a) are AUC, CAG, and UGC called codons or anticodons? What do these sequences tell you?

Your answer:

Solution: Each "triplet" on the mRNA is called a codon. A given codon specifies an amino acid that is brought to the ribosome by a tRNA.

(b) This base sequence on mRNA is complementary to the base sequences on the ends of three tRNA's. What are the base sequences on those three tRNA's? Are these codons or anticodons? How do you "figure out" which amino acid is to be brought by a given tRNA?

Your answer:

Solution: UAG GUC ACG, anticodons
To determine the amino acid that is specified, look up the amino acid in a codon table in your text or in a reference book. The codon triplet AUC specifies isoleucine, CAG specifies glutamine, and UGC specifies cysteine.

(c) What is the base sequence on the DNA strand from which the mRNA (of the previous questions) is replicated? Are these sequences codons or anticodons?

Your answer:

Solution: TAG GTC ACG, anticodons (only the triplets on mRNA are called codons)

III. Biopolymers – Proteins, Carbohydrates, Lipids, and Nucleic Acids

1. What biopolymer(s) is(are) used as sources of energy?

Your answer:

Solution: lipids and carbohydrates

---

2. Which biopolymer(s) constitute the major components of muscles, skin, bones, and enzymes?

Your answer:

Solution: proteins

---

3. Which kind(s) of biopolymer(s) are the building materials for cell membranes?

Your answer:

Solution: proteins and lipids

---

4. Which biopolymer contains the material of heredity?

Your answer:

Solution: nucleic acids

This workbook is at an end. We hope that it has been helpful in your chemistry courses and perhaps in the development of your study methods.

The Authors